QCD at 200 TeV

QCD at 200 TeV

Edited by

Luisa Cifarelli

University of Naples
Naples, Italy
and CERN
Geneva, Switzerland

and

Yuri Dokshitzer

University of Lund
Lund, Sweden
and LNPI, Gatchina
St. Petersburg, Russia

Springer Science+Business Media, LLC

Library of Congress Cataloging-in-Publication Data

QCD at 200 TeV / edited by Luisa Cifarelli and Yuri Dokshitzer.
      p.   cm. -- (Ettore Majorana international science series.
  Physical sciences ; v. 60)
    Includes bibliographical references and index.
    ISBN 978-1-4613-6522-8   ISBN 978-1-4615-3440-2 (eBook)
    DOI 10.1007/978-1-4615-3440-2
    1. Quantum chromodynamics--Congresses. 2. Particles (Nuclear
  physics)--Congresses.   I. Cifarelli, L. (Luisa) II. Dokshitzer,
  Yuri. III. Series.
  QC793.3.Q35Q23 1992
  539.7'.548--dc20   92-14547
                                                                 CIP

Proceedings of the Seventeenth Workshop of the INFN Eloisatron Project
on QCD at 200 TeV, held June 11–17, 1991, in Erice, Sicily, Italy

ISBN 978-1-4613-6522-8

© 1992 Springer Science+Business Media New York
Originally published by Plenum Press, New York in 1992
Softcover reprint of the hardcover 1st edition 1992

PREFACE

This volume contains the Proceedings of the 17th Workshop of the INFN ELOISATRON Project on "QCD at 200 TeV", held at the "Ettore Majorana" Centre for Scientific Culture, EMCSC, Erice, Trapani, Italy, in the period 11-17 June 1991.

The new multi-TeV frontiers of Subnuclear Physics are no more beyond our imagination. A conceptual design of the highest energy (100+100 TeV) proton-proton collider - the ELOISATRON - already exists. Intensive R&D studies are on the go to develop the most promising and innovative detector technologies for the highest energy and luminosity. QCD (Quantum Chromo-Dynamics) will be the theory to describe the expected Physics scenario of future Supercolliders.

The purpose of the Workshop was therefore to review the recent status of QCD in High Energy interactions and to discuss the novel aspects of Perturbative and Non-Perturbative QCD with special emphasis on future experimental studies at Super-High Energy Colliders, up to the 200 TeV limit.

The topics were:

- Classical QCD: particle multiplication, multiplicities and spectra, jet profiles, coherence effects, etc.
- Hadron interaction cross-sections and structure functions at Super-High Energies, small-x behaviour, QCD Pomeron, "hot spots".
- QCD fragmentation models, present and future.
- Artificial neural networks in High Energy Physics.
- New theoretical aspects of QCD at Super-High Energies (instanton-induced large cross-sections, baryon number violation and peculiar multi-quark production events, etc.).

Emphasis was also given to the crucial problem of disentangling new Physics from the overwhelming standard QCD processes, both from the theoretical and experimental point of view.

The Workshop was sponsored by the Italian National Institute for Nuclear Physics (INFN), the Italian Ministry of University and Scientific Research, and the Sicilian Regional Government. It was held in the framework of the three-year Galilean Celebrations (1991-1993), promoted by the Galileo Galilei Foundation (GGF), the World Federation of Scientists (WFS), the EMCSC and the International Centre for Theoretical Physics (ICTP), in collaboration with the World Laboratory, on the occasion of the 400th anniversary of the birth of Modern Science. We are thankful to the staff of the EMCSC for their - as always - kind and efficient support.

L. Cifarelli
University of Naples, Italy and
CERN, Geneva, Switzerland

Yu. Dokshitzer
University of Lund, Sweden and
LNPI, Gatchina, St. Petersburg, USSR

CONTENTS

MULTIPARTICLE PRODUCTION IN HADRONIC INTERACTIONS AT SUPERHIGH ENERGIES

A.B.Kaidalov

ITEP, Moscow

Abstract

Theoretical approach, based on 1/N-expansion in QCD and string model is presented. A comparison with experimental data on multiparticle production in hadronic interactions is carried out and predictions for future supercolliders are given.

1. Introduction

Consequences of QCD for hard processes, where the perturbative calculations are applicable, are now well verified experimentally. However many important problems of the theory including the famous problem of color confinement, are not yet solved. Investigation of "soft" hadronic interactions at high energies can give an information on the properties of QCD at large distances. An interesting problem of the theory is a relation between "soft" and "semihard" dynamics at very low x. The problem of low x-region becomes especially important at superhigh energies. This is connected also to a practical reason, - there are many hard processes, including heavy-quark production, which at the energies $\sqrt{s} \sim 100$ TeV are governed by the low x behavior of structure functions. In the central rapidity region $x \sim M / \sqrt{s}$ and for energies of Eloisatron $\sqrt{s} = 2 \cdot 10^5$ GeV, $x \leq 10^{-5}$ for $c\bar{c}$ and $x \leq 5 \cdot 10^{-5}$ for $b\bar{b}$-states. Even for $t\bar{t}$-states (with $M \sim 200$ GeV) the values of $x \leq 10^{-3}$. However our knowledge of structure functions $f_{q(g)}(x, Q^2)$ in this region is very limited. There are many talks at this workshop devoted to this problem. On the other hand for light quarks (u, d, s)

QCD at 200 TeV, Edited by L. Cifarelli
and Y. Dokshitzer, Plenum Press, New York, 1992

-states the small x-region is important already at present energies $\sqrt{s} \leq 10^3$ GeV and their study can give an insight to this problem. In particular I think that it is possible to find a reasonable initial condition at $Q^2 \leq 1$ GeV^2 for QCD evolution of structure functions in the low x-region.

A nonperturbative approach should be used for investigation of the "soft" processes at large distances $r \leq 1/\Lambda_{QCD}$, where the perturbation theory is no longer valid. I shall discuss the approach, based on the 1/N-expansion in QCD[1,2] and the color-tube (string) model[3]. Powerful method of the reggeon calculus[4,5,6] in conjunction with the QCD-based ideas will be applied to multiparticle production at very high energies. This approach, known as the dual parton model[7,8,9] or the quark-gluon strings model[10], gives a good quantitative description to a large amount of experimental information on high energy hadronic interactions. In this talk after a short description of the model the comparison of its predictions with available experimental data on the main characteristics of the multiparticle production will be presented. The predictions for future colliders, including Eloisatron, will be given.

II. 1/N-expansion and the color-tube model

Now I will shortly remind some ideas of the approach, based on the 1/N- expansion n QCD[1,2] and the string model.

Method of the 1/N-expansion leads to a useful classification of diagrams in a theory with the symmetry group SU(N) for large values of N. In QCD the value of N can correspond to either number of colors[1] ($N=N_c$) or number of flavors[2] ($N=N_f \sim N_c$). All amplitudes are presented as series in a small parameter 1/N. Each term of the series corresponds to an infinite sum of Feynman diagrams, which are characterized by a certain topology. The first terms of this expansion is connected to the planar diagrams, shown in fig.1. Wavy lines in this figure are gluons, full lines are quarks. It is assumed that at high energies the diagrams for amplitudes of binary reactions (fig.1a) correspond to the secondary Regge-exchanges R (ρ, f,...) in the t-channel. The contribution to the total cross section of the planar

diagrams for multiparticle production, shown in fig .1b)
decreases with energy as $1/s^{(1-\alpha_R(0))}$. The space-time picture
of interaction in this case[9] correspond to the process of
annihilation of valence quarks of two colliding hadrons,
formation of a color-tube intermediate state and its
subsequent fission into two or more white pieces (hadrons).

The Pomeranchuk singularity is related in QCD to gluonic
exchanges in the t-channel and in the frame of 1/N-expansion
approach is associated to the cylinder type graphs (fig. 2a).
Cutting of the cylinder diagram of fig. 2a) leads to
production of two color-tubes, - chains of hadrons (fig. 2b).
It is usually assumed, that the bare Pomeron, corresponding
to the cylinder diagrams, is a simple Regge-pole with $\alpha_P(0)$
> 1 - supercritical Pomeron. This conclusion follows, in
particular, from an analysis of experimental data on high
energy hadronic interactions.

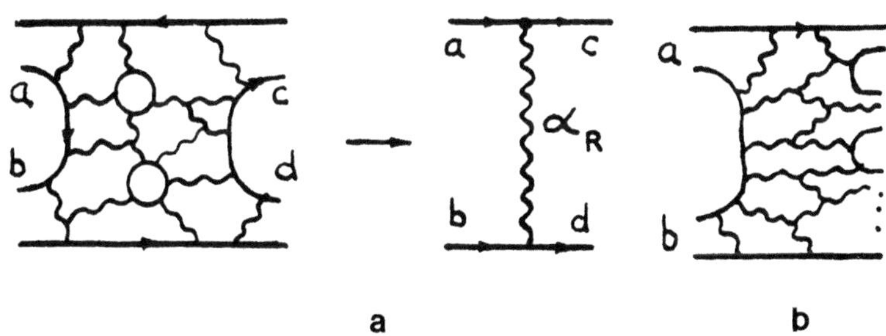

Fig.1. Planar diagram for binary reaction - a) and for
    multiparticle production - b).

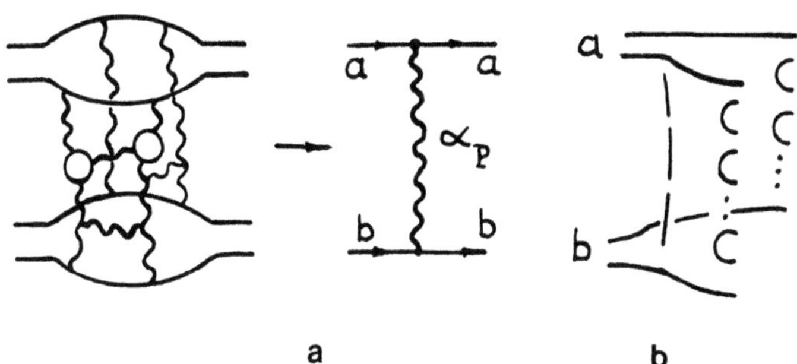

Fig.2. Cylinder type graph for elastic scattering - a)
    and its cutting in the s- channel - b).

In QCD the Pomeron was studied in the framework of perturbation theory[11,12] and the rightmost singularity has an intercept above one: $\Delta \equiv 1 - \alpha_P = \frac{\alpha_s}{\pi} 12 \ln 2$. An analysis [13], based on the nonperturbative method, also confirms a supercritical nature of Pomeron., In the following the value of Δ will be considered as a free parameter and will be determined from the fit to an energy behavior of total interaction cross sections.

More complicated processes, related to exchange in the t-channel of several Pomerons (fig.3) should be taken into account in a study of high energy scattering. They are especially important for supercritical case, as their contribution to the elastic scattering amplitude increases with energy as $\left[\frac{s}{s}\right]^{n\Delta}$ (n is the number of exchanged Pomerons). The multipomeron graphs correspond to higher terms of the topological expansion ($T_n \sim 1/(N^2)^n$) - fig. 3b). The s-channel discontinuities of these diagrams (fig.4) are related to processes of production of 2k (k $\leq$ n) - strings (chains of particles). Let us note that it is possible only if fast colliding hadrons contain besides valence quarks also additional color objects, - $q\bar{q}$-pairs and gluons.

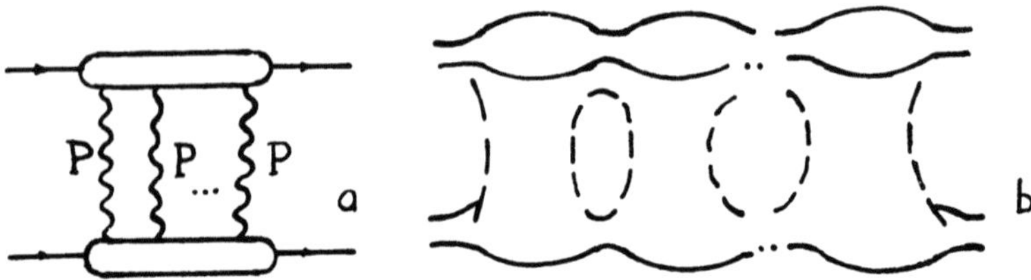

Fig.3. Multipomeron-exchange diagrams.

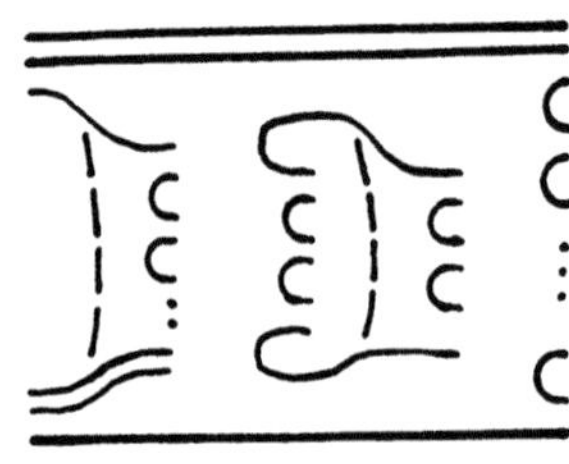

Fig.4. Multichain production diagram.

The AGK-cutting rules[5] give a possibility to determine the cross sections for 2k-strings production (with arbitrary number of uncut Pomerons) if the contributions of all rescatterings to the forward elastic scattering amplitude are known. In the following we will calculate them using the reggeon diagram technique[4].

III. The model of quark-gluon strings

The approach, based on 1/N-expansion and string model described above, has been used to formulate the quark-gluon strings model (QGSM)[10] for high energy hadronic interactions.

Inclusive cross sections and multiplicity distributions can be obtained in this model by summation over hadronic production in processes of 2k-chains (strings) formation

$$\frac{d\sigma^h}{dy} = \sum_{k=0}^{\infty} \sigma_k(\xi)\, \rho_k^h(\xi, y) \tag{1}$$

$$\sigma_N(\xi) = \sum_{k=0}^{\infty} \sigma_k(\xi)\, W_N^k(\bar{N}_k(\xi)) \tag{2}$$

where $\rho_k(\xi, y) = \dfrac{1}{\sigma_k}\dfrac{d\sigma_k^h}{dy}$ and W_N^k are rapidity and multiplicity distributions for 2k-chains production ($\xi = \ln\frac{s}{s_o}$) The term with k=0 corresponds to the process of diffraction dissociation $\sigma^{(DD)}$.

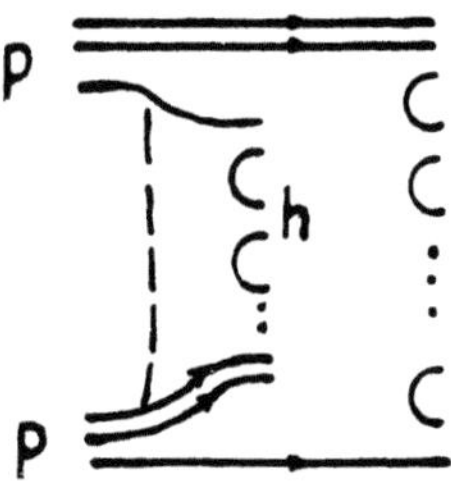

Fig.5. Cylinder diagram contribution to the inclusive cross section of hadron h production.

Consider as an example the inclusive cross section of the reaction pp ⇒ hX. A contribution of the two-chain diagram of fig.5 to this cross section, in the fragmentation region can be written as a convolution of the quark (diquark) distribution functions in the colliding hadron with fragmentation functions, which describe a transformation of a string into hadrons

$$\frac{X}{\sigma_2} \frac{d\sigma_2^h}{dX} \approx \int_X^1 dx_1\, f_p^{q(2)}(x_1)\, D_q^h\left(\frac{X}{X_1}\right)\frac{X}{X_1} \quad + \tag{3}$$

$$+ \text{Contrib. of the second chain}$$

where $\quad X = \dfrac{2p_\parallel^h}{\sqrt{S}}$.

The function $f_p^{q(2)}(x_1)$ determines how the energy of the initial proton is divided between q and qq-chains and $f_p^{qq(2)}(x_1) = f_p^{q(2)}(1-x_1)$.

For arbitrary configuration, which contains both valence and "sea"-chains, functions $\rho_k^h(\xi, x)$ can be written in the form

$$\rho_k^h(\xi,x) = a^h \left\{ F_{qq}^{h(k)}(X_+)\, F_{qq}^{h(k)}(X_-) + F_{qq}^{h(k)}(X_+)\, F_q^{h(k)}(X_-) + \right.$$

$$\left. + 2(k-1)\, F_{q_{sea}}^{h(k)}(X_+)\, F_{q_{sea}}^{h(k)}(X_-) \right\} \tag{4}$$

where $x_\pm = \dfrac{1}{2}\left[\sqrt{x_\perp^2 + x^2} \pm x\right]$,

$$X_\perp = \frac{2\,\bar{m}_\perp^h}{\sqrt{S}} \ , \quad X_+ = \exp(y - y_{max}^h) .$$

$$F_i^{h(k)}(x) = \int_X^1 dx_1\, f_p^{i(k)}(x_1)\, D_i^h\left(\frac{X}{X_1}\right)\frac{X}{X_1} \tag{5}$$

$$i = q\ ,\ qq,\ q_{sea}$$

The first two terms in eq. (4) correspond to chains, connected to valence quarks and diquarks and generalize eq. (3) to all values of rapidity. The last term is due to extra chains, connected to sea quarks.

An analysis of planar diagrams allows one to determine the behavior of the quark distribution functions at $x \Rightarrow 0$ and $x \Rightarrow 1$ [10]. They are expressed in terms of intercepts of known

$$
f_p^{q(1)}(x) = \begin{cases} c_1 \, x^{-\alpha_R(0)} & , \quad x \Rightarrow 0 \\[2mm] c_2 \, (1-x)^{\alpha_R(0)-2\alpha_N(0)} & , \quad x \Rightarrow 1 \end{cases} \tag{6}
$$

where $\alpha_R(0) = 0.5$ and $\alpha_N(0) \approx -0.5$ are the bosonic and fermionic Regge-intercepts.

The formulas, which interpolate the behavior given by eq. (6) are used in the calculations[10]. For example

$$
f_p^{u_v(k)}(x_1) = C_k^{u_v} \, x_1^{-\alpha_R(0)} (1-x_1)^{\alpha_R(0)-2\alpha_N(0)+(k-1)} \tag{7}
$$

The coefficients C_k are determined from the normalization condition.

All variety of inclusive spectra of different hadrons is determined by the fragmentation functions $D_i^h(z)$. The behavior of these functions in the limits $z \Rightarrow 0$ and $z \Rightarrow 1$ can be derived from analysis of planar diagrams[10]. The general rules for construction of fragmentation functions for quarks and diquarks are given in ref.[14]. For example

$$
z \, D_u^{\pi^+}(z) = \begin{cases} a^\pi & , \quad z \Rightarrow 0 \\[2mm] c^\pi \, (1-z)^{-\alpha_R + \lambda} & , \quad z \Rightarrow 1 \end{cases} \tag{8}
$$

where $\lambda = 2 \, \alpha_R' \cdot \overline{p_{\perp\pi}^2} \approx 0.5$.

The functions $\tilde{D}_i^h(z)$, which enter eq. (5), differ from functions $D_i^h(z)$ only by the constants a^h, which determine the density of hadrons of a given type produced in the central rapidity region in a single chain. Using interpolation formulas for $D_i^h(z)$ and sum rules, which follow from conservation of momentum, charge, strangeness e.t.c., it is possible to determine the fragmentation functions for many types of hadrons in a fill kinematical region[14].

In this approach a suppression of strange-quark production can be determined theoretically and turns out to be not a constant but a function of z [14]. Thus, contrary to other models, where fragmentation functions are determined from experimental data, in QGSM practically all parameters are fixed theoretically. This makes the model very predictive. The inclusive spectra in the model automatically have the correct triple Regge limits for $x \Rightarrow 1$, double-Regge limit for $x \Rightarrow 0$ and satisfy to all conservation laws.

In order to determine the cross sections σ_k of 2k-chains formation the AGK-cutting rules[5] are used and the contributions of diagrams with exchange of many Pomerons to elastic amplitudes are calculated with the help of reggeon diagram technique[4] . The diagrams of "nonenhanced" type, which do not take into account interaction between Pomerons, leads to the formulae of the eikonal type. In the "quasieikonal" model, which accounts for both elastic and inelastic diffractive rescattering, the cross sections of 2k-chains production have the form[15]

$$\sigma_k \ (\xi) \ = \ \frac{\sigma_P}{k \cdot z} \left[1 \ - \ \exp \ (-z) \cdot \sum_{i=0}^{k-1} \frac{z^i}{i \ !} \right] \ , \ i \geq 1 \qquad (9)$$

where $\sigma_P = 8 \ \pi \gamma_P \exp \ (\xi \Delta)$, $z = \dfrac{2C \gamma_P}{R^2 + \alpha' \cdot \xi} \exp \ (\xi \Delta)$, $\xi = \ln \dfrac{s}{s_0}$.

Parameters γ_P , R^2 characterize the values of the Pomeron residue at t=0 and its t-dependence. The quantity $C = 1 + \dfrac{\sigma^{(DD)}}{\sigma^{(el)}}$ ($\sigma^{(DD)}$ is the cross section of diffraction dissociation) takes into account deviations from eikonal approximation. The parameters of the model were determined from the fit to experimental data on $\sigma^{(tot)}$ and $\dfrac{d\sigma^{(el)}}{dt}$ at high energies. The best value for the main parameter of the model $\Delta = 0.12 \pm 0.02$.

The total interaction cross section in this model has the form

$$\sigma^{(tot)} = \sum_{k=0}^{\infty} \sigma_k (\ \xi \ ) = \sigma_P f (\frac{z}{2}) \ ; \ f(\frac{z}{2}) = \sum_{n=1}^{\infty} \frac{(-z)^{n-1}}{n \cdot n \ !} \qquad (10)$$

where $\sigma_0 (\xi) \ = \ \sigma^{(el)} \ + \ \sigma^{(DD)}$ is the cross section of diffractive processes.

$$\sigma_0(\xi) = \sigma_P \left[f\left(\frac{z}{2}\right) - f(z) \right] \tag{11}$$

For superhigh energies, when $\xi \gg 1$ $\sigma^{(tot)}(\xi)$ and $\sigma_0(\xi)$ have a Froissart type behavior

$$\sigma^{(tot)}(\xi) \approx \frac{8\pi\alpha'_P \Delta}{C} \xi^2 \tag{12}$$

Thus interaction of hadrons correspond asymptotically to scattering on a black disk with logarithmically increasing radius.

IV. Comparison with experiment

All the parameters of the model, which determine the contribution of the Pomeranchuk-pole were determined from the data on elastic scattering of hadrons at high energies. The description of experimental data on $\sigma^{(tot)}(s)$ for pp and $\bar{p}p$ -interactions[16] is shown in fig.6 and the energy dependence of the diffraction cone slope B in pp-elastic scattering is given in fig.7. The model predicts for $\sigma^{(tot)}_{pp}$ at Eloisatron energy the value around 150 mb and for the slope $B \approx 30$ GeV^2. Let us note that asymptotically $B \sim \xi^2$, but at present energies the slope B is approximately linear in ξ (fig.7). The deviations from this linear dependence will be essential only for energies of future colliders $\sqrt{s} > 10$ Tev. It will be very interesting to check this prediction of the theory.

The model does not contain any "odderon"-type[17] singularities and at energies $\sqrt{s} \geq 10^2$ GeV all characteristics of pp and $\bar{p}p$- interactions practically coincide. So in the following we will often compare data on $\bar{p}p$- interactions at energies of the Sp$\bar{p}$S - collider and Tevatron with data on pp-interactions at lower energies.

Rapidity (and pseudorapidity) distributions of charged particles in pp($\bar{p}p$)-interactions at different energies are shown in fig.8. In the supercritical Pomeron theory with account of "nonehanced" diagrams inclusive cross sections $\frac{d\sigma^h}{dy}$ at very high energies and $y \approx 0$ increases with energy as $\left(\frac{s}{s_0}\right)^\Delta$. At energies $\sqrt{s} \sim 10^5$ GeV density of charged hadrons in the central region reaches the values $7 \div 8$.

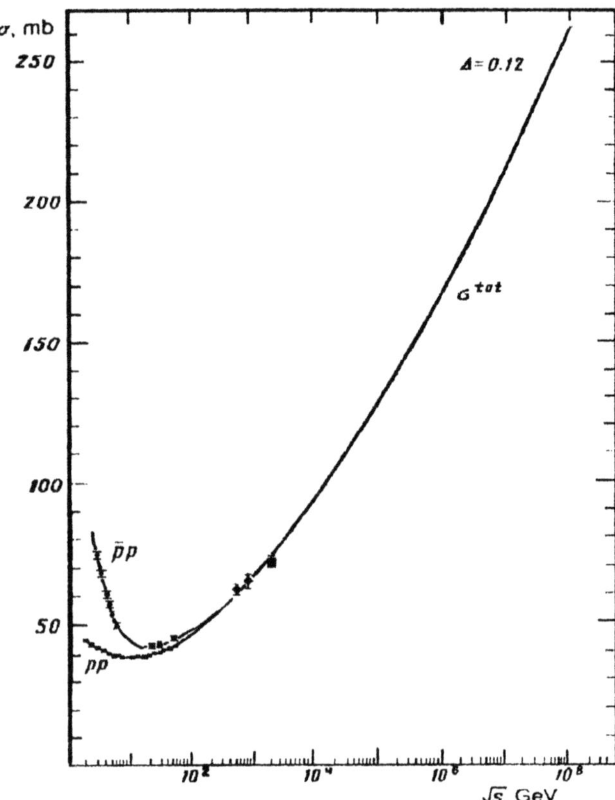

Fig.6. $\sigma^{(tot)}$ for pp($\bar{p}$p)-interactions.

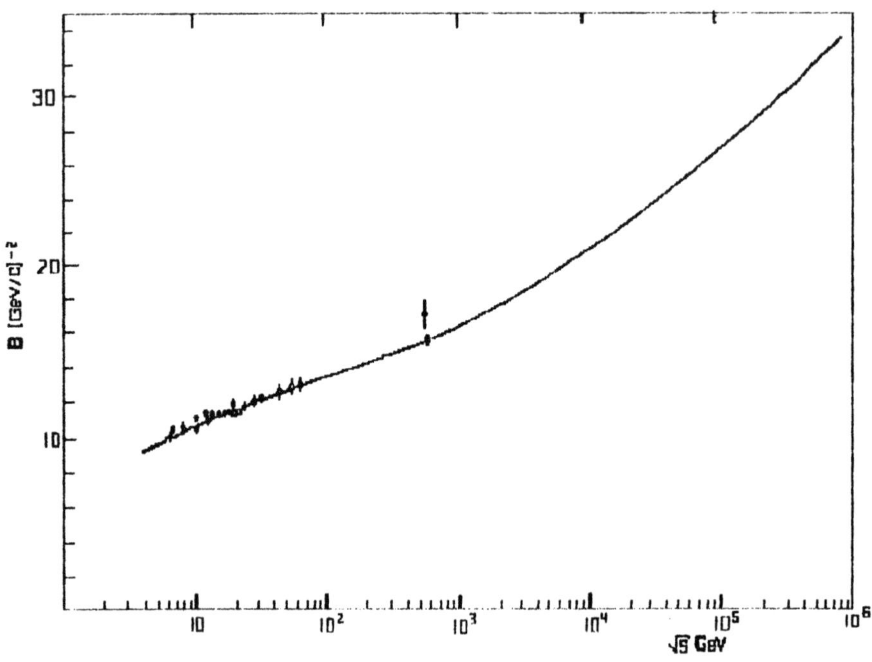

Fig.7. Energy dependence of the slope in pp-elastic scattering

10

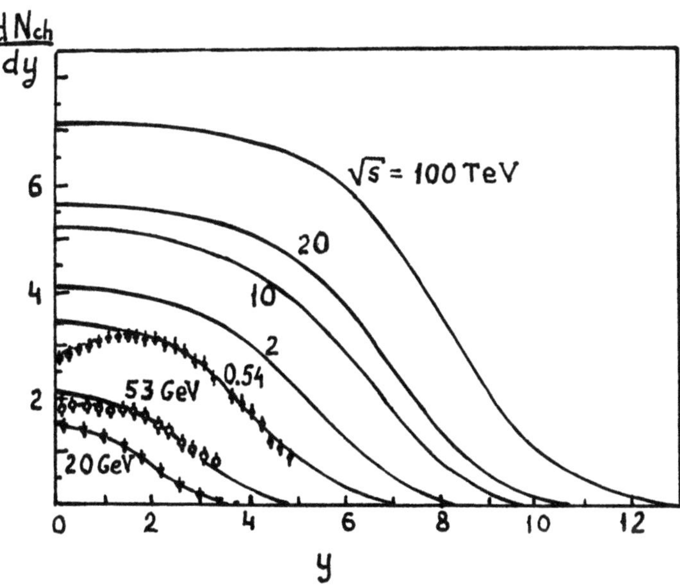

Fig.8. Rapidity distributions for charged hadrons in
pp(p̄p)-interactions at different energies.

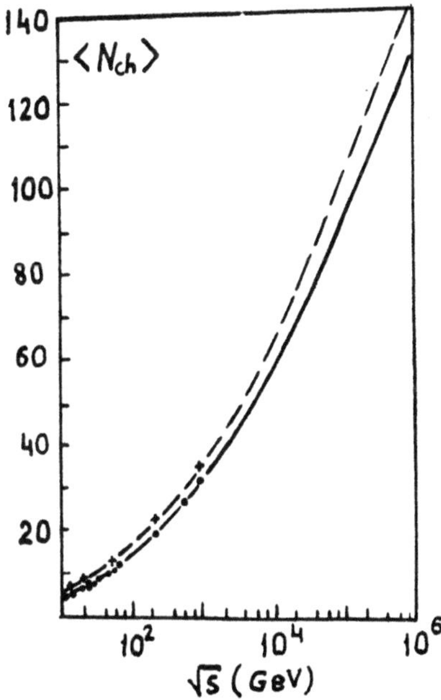

Fig.9. Energy dependence of $\langle N_{ch} \rangle$ in pp-interactions.
Full curve for all inelastic and the dashed one
for nondiffractive events.

An integral over the rapidity distributions gives an average multiplicity of charged hadrons $< N_h >$. Energy dependencies of $< N_{ch} >$ for both all inelastic collisions and the nondiffractive ones are shown in fig.9. The model well describes the rapidity distributions and $< N_{ch} >$ at accessible energies. $< N_{ch} >$ increases with energy much faster than $\ln \frac{s}{s_0}$ and reaches the value about 120 at the Eloisatron energy.

The multiplicity distributions in the model are given, according to eq.(2),by a sum of contributions, connected to a different number of cut-Pomerons. Each of these contributions have a Poisson-like form (only short range correlations inside chains), but their sum has a peculiar dependence on energy. At energies $\sqrt{s} \leq 10^2$ GeV different contributions overlap strongly and lead to an approximate KNO-scaling. The model reproduces well[10] experimental data at these energies. However the mean number of produced chains increases with energy and the model predicts[10] the violation of KNO-scaling. This prediction was confirmed by the data of UA5 and UA1 groups at Sp$\bar{\text{p}}$S-collider. The multiplicity distributions in pp($\bar{\text{p}}$p)-interactions at energies $\sqrt{s}$ = 540 GeV and $\sqrt{s}$ = 10^5 GeV are shown in fig. 10. As energy increases the maximum of a distribution in a variable $z = N_{ch}/< N_{ch}>$ moves to the left and the distribution increases in the region of large z. It is interesting that at superhigh energies $\sqrt{s} \sim 10^5$ GeV different terms in the sum of eq.(2) start to be separated and the distribution has the corresponding maxima and minima. At present energies $\sqrt{s} \geq 10^3$ GeV only the first maximum and a "shoulder" start to appear.

The model also well reproduces the semiinclusive rapidity distributions and the dependence of multiplicity distributions on rapidity[10] . Thus the study of multiplicity distributions confirms the multicomponent structure of the model.

Another consequence of this structure is an existence of the long range rapidity correlations. It leads in particular to strong forward-backward correlations. The dependence of a mean multiplicity of charged hadrons, produced in the backward rapidity region (-4 to -1), on a number of charged

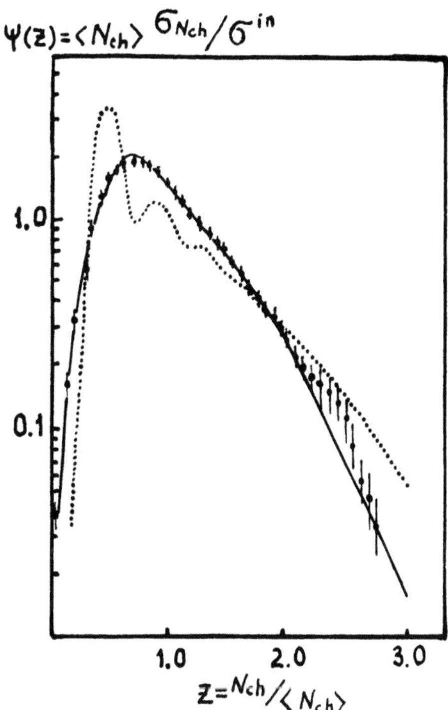

Fig.10. Multiplicity distributions in pp($\bar{\text{p}}$)-interactions. Full curve corresponds to $\sqrt{s}$ = 540 GeV the dotted curve is for $\sqrt{s}$ = 10^5 GeV. The data at $\sqrt{s}$ = 540 GeV are from UA5

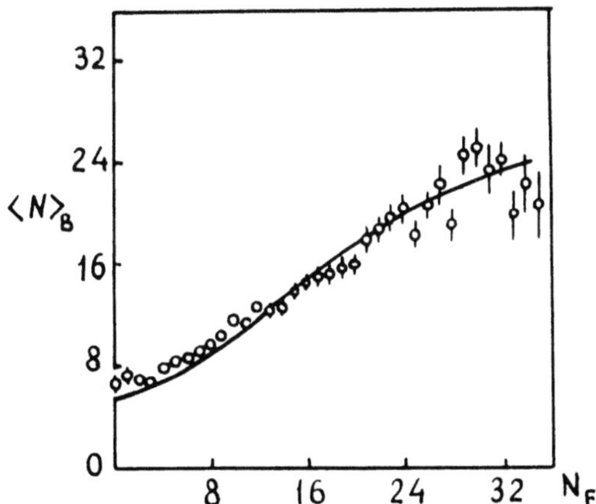

Fig.11. Forward-backward correlations at $\sqrt{s}$ = 540 GeV and the prediction of QGSM.

particles in the forward rapidity interval (1 to 4) is shown in fig.11. The model well describes a strong correlation, observed experimentally.

Let us consider now inclusive spectra of different hadrons at high energies. MQGS gives a possibility to calculate spectra for all values of variable x. Spectra of π-mesons in pp-interactions are shown in fig.12. Note that the Feynman scaling is strongly violated in the region of small x. On the other hand in the fragmentation region for $x \geq 0.1$ a violation of scaling is rather weak. This has important implications for experiments in cosmic rays. Predictions of the model for inclusive cross sections of $K^{\pm}$ -mesons[18] are given in fig.13. The strange-quark suppression is predicted theoretically[14] and is confirmed by experimental data. An energy dependence of mean multiplicities of kaons is shown in fig.14 and its ratio to the mean multiplicity of pions is given in fig.15. The increase of the last ratio with energy in the model is mainly due to the mass difference of kaons and pions and the ratio tends to a constant value ≈ 0.12 asymptotically.

The spectra of protons and antiprotons and the corresponding theoretical curves[18] are shown in fig.16. The spectra of protons have a clear "leading" behavior due to fragmentation of diquarks which have a distribution, concentrated at large values of x_{qq}. On the contrary the antiprotons are mainly determined by the "central" fragmentation of the valence chains and also arise from sea-chains. Asymptotically in the central region the spectra of protons and antiprotons should be equal (same is true for all particles and antiparticles).

Another example of "leading" behavior is the spectrum of Λ-hyperons, shown in fig.17. As it was emphasized above the strange-quark suppression in the fragmentation functions of the model is a function of z. This leads to a shift of the maximum of the spectrum for Λ to smaller values of x than for nonstrange baryons (like n or Δ) in a good agreement with experimental data. This difference in fragmentation functions is connected to a difference for Regge-trajectories made of strange (α_{ϕ}) and nonstrange (α_{ρ})-quarks.

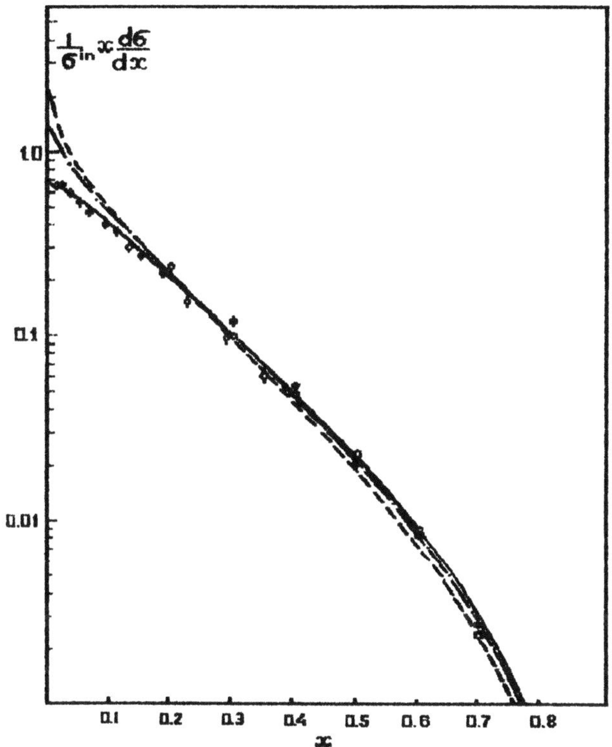

Fig.12. Inclusive cross sections for π^- - mesons at various energies.The data at $\sqrt{s} \simeq 20$ GeV are given. The full curve is for $\sqrt{s} = 20$ GeV, the dashed-dot one for $\sqrt{s} = 540$ GeV and the dashed one for $\sqrt{s} = 10^4$ GeV.

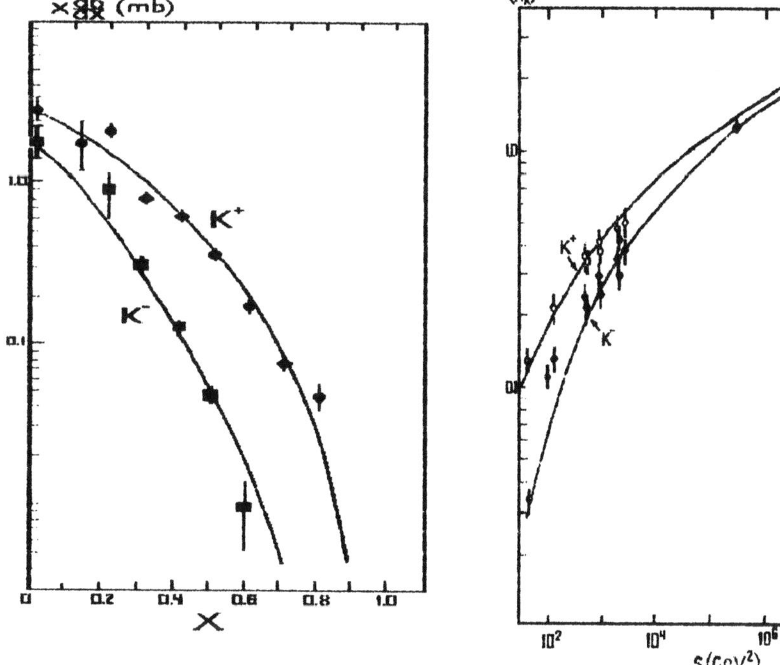

Fig.13. K^+ and K^- spectra in pp interactions at $p_L = 175$ GeV.

Fig.14. Energy dependence of mean multiplicities for charged kaons.

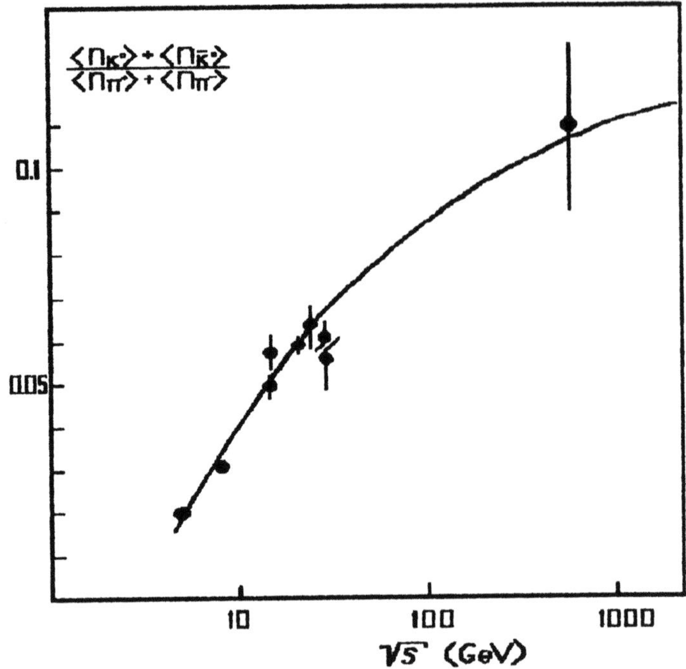

Fig.15. Ratio of multiplicities of K and π -mesons as a function of energy.

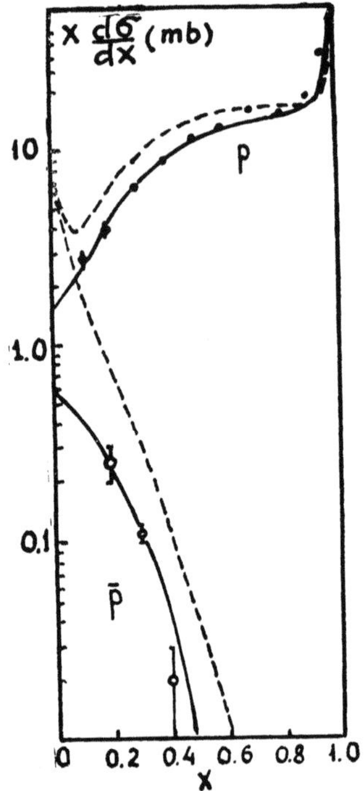

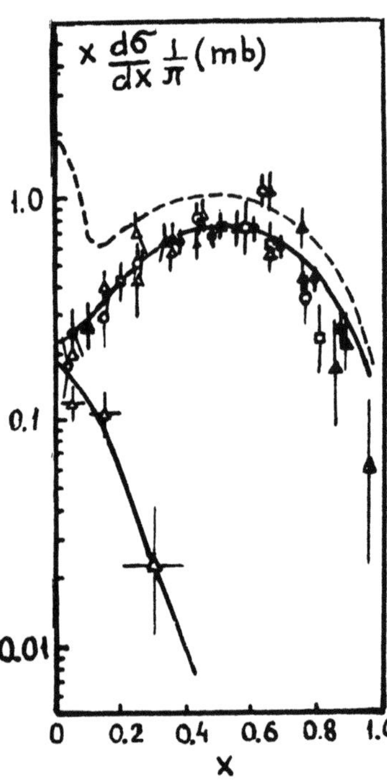

Fig.16. p and p̄ -spectra. Experimental points for p = 175 GeV (full curve). The dashed curve for √s = 540 GeV.

Fig.17. Spectra of Λ and Λ̄ hyperons. Theoretical curves for √s = 27.5 GeV (full line) and √s = 540 GeV (dashed one)

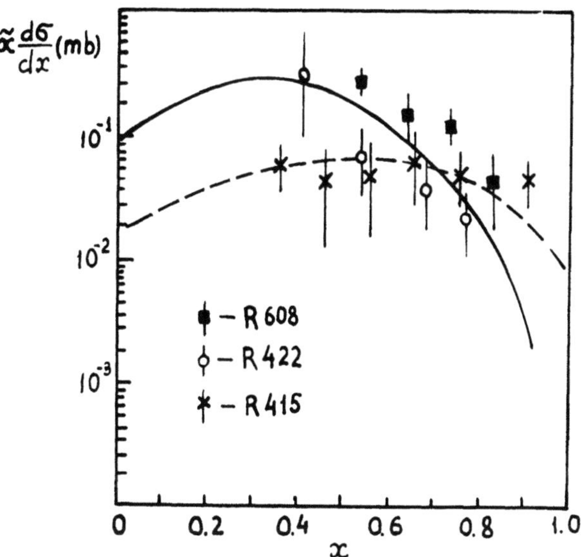

Fig.18. Spectra of Λ_c-baryons in pp-collisions.

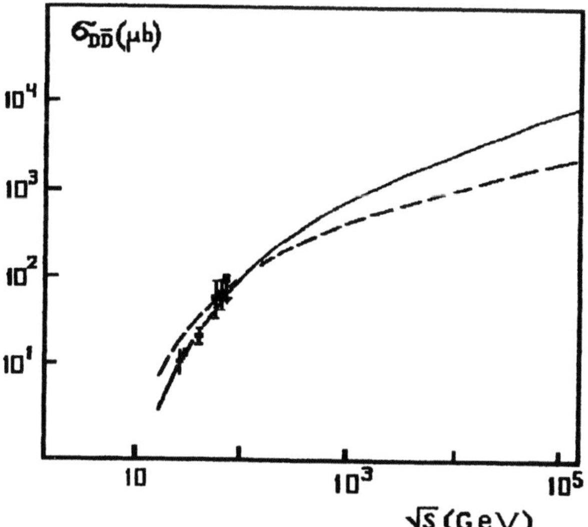

Fig.19. Total cross section for $D\bar{D}$ -production in pp-interactions.

The model has been also used to predict inclusive cross sections for heavy-quark (c,b) production[19].The form of the x-behavior depends on the intercepts of the Regge-trajectories made of c and b-quarks correspondingly. There are no experimental information on these trajectories in the region $t \leq 0$ .Two values of $\alpha_{c\bar{c}}(0)$ have been used in ref.[19] to calculate spectra of charmed particles,- $\alpha_{c\bar{c}}(0)=-2$,which follows from the mass spectrum of $c\bar{c}$-states under the assumption of linear trajectories, and $\alpha_{c\bar{c}}(0)=0$,which was taken from perturbative QCD. The theoretical calculations for these two values of $\alpha_{c\bar{c}}(0)$ are shown in figs.18,19 as full and dashed curves correspondingly. The nonperturbative value of $\alpha_{c\bar{c}}(0)$ is preferred by recent data on Λ_c production. The model can rather reliably predict energy behavior of inclusive cross sections for heavy particles at superhigh energies, where the perturbative calculations have a big uncertainty due to a small x region. Note that it corresponds to the quarks and gluons structure functions, which behave for $x \Rightarrow 0$ as $1/x^{1+\Lambda}$.

The predictions of the QGSM for Eloisatron energy are summarized in Table 1.

The model was generalized in ref.[20] in order to include interactions between Pomerons. It was shown that the more complete theoretical scheme leads to practically same results for most characteristics of hadronic interactions at energies up to 10^6GeV as the simple model, discussed above.

In conclusion, the approach based on 1/N-expansion in QCD, string model and Regge theory gives an adequate description of soft hadronic interactions at high energies. It gives an explanation for many properties of multiparticle production and can be used to obtain predictions for future colliders.

Table 1

$\sigma^{(tot)}$ (mb)	$\sigma^{(in)}$ (mb)	B (GeV^{-2})	$\frac{dN_{ch}}{dy}\big|_{y=0}$	$\langle N_{ch}\rangle$	$\langle N_{\pi^+}\rangle$	$\langle N_{K^+}\rangle$	$\langle N_p\rangle$	$\langle N_{\bar{p}}\rangle$
150	100	30	7.6	120	48	5.6	7.2	6

$\langle N_\Lambda\rangle$	$\langle N_{\bar{\Lambda}}\rangle$	$\sigma_{D\bar{D}}$ (mb)	$\sigma_{B\bar{B}}$ (mb)
2.3	2	3 – 9	0.3– 1

R E F E R E N C E S

1. t'Hooft. Nucl.Phys., B72, 461 (1974)
2. G.Veneziano, Phys.Lett. 52B, 220 (1974); Nucl.Phys. B117, 519 (1976)
3. A.Casper,J.Kogut,L.Susskind. Phys.Rev.Lett. 31, 792 (1973)
   X.Artru, G.Mennessier. Nucl.Phys. B70, 93 (1974)
   A.Casher,H.Neuberger,S.Nussinov. Phys.Rev. D20, 179 (1979)
   B.Andersson, G.Gustafson, C.Petersson. Nucl.Phys. B135, 273 (1978)
4. V.N.Gribov. JETP 53, 654 (1967)
5. V.A.Abromovski, O.V.Kancheli, V.N.Gribov. Yad.Fiz. 18, 595 (1973)
6. M.Baker, K.A.Ter-Martirosyan. Phys.Rep. 28C, 1 (1976)
7. A.Capella, U.Sukhatme, J.Tran-Thanh-Van et al. Zeit. fur Phys., C3, 329 (1980);
   A.Capella, J.Tran-Thanh-Van, Phys.Lett. 114B, 450 (1982);
   Zeit. fur Phys., C18, 85 (1983); ibid C23, 165 (1984)
8. Cohen-Tannoudji, F.Hayot, R.Peschanski. Phys.Rev. 17, 2390 (1978); G.Cohen Tannoudji et al ibid D21, 2699 (1980)
9. P.Auranche, F.W.Bopp. Phys.Lett. 114B, 363 (1982)
   P.Auranche, F.W.Bopp, J.Ranft. Zeit. fur Phys., C23, 85 (1983); ibid C23, 165 (1984)
10. A.B.Kaidalov. JETP Lett. 32, 494 (1080): Yad.Fiz. 33, 1369, (1981) ; A.B.Kaidalov.Phys. Lett. 116B, 459 (1982)
    A.B.Kaidalov,K.A.Ter-Martirosyan. Phys. Lett. 117B, 247 (1982)
    Yad. Fiz. 39, 1545 (1984); ibid 40, 211 (1984)
    A.B.Kaidalov,K.A.Ter-Martirosyan,Yu.M.Shabelski. Sov. Journ. Nucl. Phys. 43, 822 (1986)
11. E.A.Kuraev,L.N.Lipatov,V.S.Fadin. Sov.Phys.JETP 44, 433 (1976) 45, 199 (1977)
12. L.N.Lipatov. Sov. Phys.JETP 63, 904 (1986)
13. Yu.A.Simonov. Nucl.Phys.23B, 293 (1991)
14. A.B.Kaidalov.Sov. Jorn. Nucl Phys. 45, 902 (1987)
15. K.A.Ter-Martirosyan. Phys. Lett. 44B, 377 (1973)
16. K.A.Ter-Martirosyan. Sov. Journ. Nucl. Phys. 44, 817 (1986)
17. K.Kang, B.Nicolescu. Phys. ReV. D11, 2461 (1975)
    D.Joynson,E.Leader,C.Lopez,B.Nicolescu, Nuovo Cim. 30A, 345 (1975)
18. A.B.Kaidalov,O.I.Piskunova. Yad.Fiz. 41, 1278 (1985); Zeit. fur Phys. C30, 145 (1986)
19. A.B.Kaidalov,O.I.Piskunova. Yad. Fiz. 43, 1545 (1986);
    O.I.Piskunova. Preprint FIAN 118, Moscow, (1990)
20. A.B.Kaidalov, L.A.Ponomarev,K.A.Ter-Martirosyan. Sov. Journ. Nucl. Phys. 44, 468 (1986)

JET TOPOLOGY AND NEW JET COUNTING ALGORITHMS

S. Catani

Cavendish Laboratory, University of Cambridge, Cambridge, UK
and INFN, Sezione di Firenze, Firenze, Italy

Abstract

A QCD theoretical analysis of jet cross sections is presented. We discuss in detail the JADE (invariant mass-type) and $k_\perp$-algorithms for multijet cross sections in e^+e^--annihilation, with emphasis on hadronization effects and large perturbative corrections at small values of the jet resolution parameter y_{cut}. For the $k_\perp$-algorithm we present results of calculations which include resummation of leading and next-to-leading logarithms of y_{cut} to all orders in QCD perturbation theory. We also discuss the recently-proposed Geneva algorithm and show that it suffers from some of the same difficulties as the JADE algorithm.

1. Introduction

A jet can be defined as a large amount of hadronic energy in a small angular region. According to this *qualitative* definition, the first *Evidence for Jet Structure in Hadron Production by e^+e^- Annihilation* was reported in 1975 [1]. Since then, hadronic jet production in lepton [2,3] and hadron [4,5] colliders has become a main tool in investigating strong interaction physics and testing Quantum Chromodynamics (QCD).

We are nowadays in a position to use jet cross section data both for

- precise quantitative tests of QCD (measurement of the QCD coupling α_S and scale $\Lambda_{\overline{\mathrm{MS}}}$ [3], study of QCD coherence [6]);

- looking for breakdown of the standard model and new physics [5].

In order to do that, the qualitative definition of a jet given above is no longer sufficient and must be replaced by a precise *quantitative* definition. We need a jet algorithm able to specify unambiguosly a jet configuration starting from hadrons detected in the final state. A *jet algorithm* is defined giving

j_1) a test variable (energy-angle resolution) y_{kl};

j_2) a recombination procedure.

QCD at 200 TeV, Edited by L. Cifarelli
and Y. Dokshitzer, Plenum Press, New York, 1992

The test variable y_{kl} is needed in order to specify whether or not two hadrons h_k, h_l belong to the same jet, whilst the recombination procedure tells us how jet properties are related to the ones of hadrons belonging to it.

The jet defining conditions j_1) and j_2) have to fulfill the requirements of being

i) infrared (IR) and collinear safe

ii) simple to implement in the experimental analysis

iii) simple to implement in the theoretical calculation

iv) subject to small hadronization corrections.

Requirements ii) and iii) are self-evident.

The other requirements follow from the motivation of comparing data with theory, namely QCD. So far, we are able to perform QCD calculations, using perturbation theory, only for the region of small distances in which high energy collisions produce partons (quarks, gluons). At larger distances the produced quarks and gluons are confined by the colour force field and are forced to dress themselves up into colourless hadrons. Although it is not possible to describe the hadronization process by perturbation theory, the "preconfinement" property [7] (or "local parton hadron duality" [8]) of QCD implies that hadron jets should mantain the kinematic features of the underlying quarks and gluons. That is because, according to preconfinement, the hadronic flow in the final state follows the partonic flow quite closely, with transfers of momentum and other quantum numbers that are local in phase space.

The way to enhance preconfinement effects (or, equivalently, to reduce hadronization corrections) is to define jet cross sections which at parton level are not very sensitive to large distance physics, namely any small value of parton masses. Therefore jet algorithms must satisfy the property i) above, i.e. jet cross sections at parton level must be finite order by order in perturbation theory in the limiting case of final state massless partons [9,10]. Obviously, hadronization corrections still affect jet algorithm satisfying the property i). Therefore one should try to minimize hadronization effects for instance using Monte Carlo event generators to compare parton level and hadron level results for different jet algorithms.

Still within the class of jet algorithms fulfilling the requirements i) $- iv$), one has ample freedom of choosing *test variables* and *recombination procedures*. The aim of this paper is not to provide a comprehensive review of existing jet algorithms [11] but to present a QCD inspired critical analysis of them. Our goal is to advocate a QCD-based jet algorithm, i.e. a definition of a jet whose dynamics follows QCD theory as closely as possible.

We start our analysis in Sec. 2, discussing the jet algorithm mostly used at exclusive level, namely the "JADE algorithm" for e^+e^--annihilation [12]. The description of good and bad features of the JADE algorithm performed in Secs. 2 and 3 will provide us with the physical motivations to introduce the QCD inspired jet algorithm which we call $k_\perp$-algorithm [13]. The $k_\perp$-algorithm is defined in Sec. 4 where we present our main theoretical results. In Sec. 5 we draw some conclusion. Some results on another jet algorithm, the recently proposed "Geneva algorithm" [14,3], are presented in Appendix A.

2. The JADE algorithm

2.1 Definition

The JADE algorithm is a full exclusive algorithm to define jet cross sections in e^+e^--annihilation. Let us consider a n-hadron final state

$$e^+(p) + e^-(\bar{p}) \to h_1(p_1) + \cdots + h_n(p_n) \ , \qquad p + \bar{p} \equiv Q \ . \tag{1}$$

The JADE algorithm is defined according to the following iterative procedure.

1) Define a resolution parameter y_{cut}.

2) For every pair of hadrons h_k, h_l compute the corresponding "invariant mass" M_{kl}^2 and define

$$y_{kl} = M_{kl}^2/Q^2 \ . \tag{2}$$

3) If y_{ij} is the smallest value of y_{kl} computed in 2) and $y_{kl} < y_{\text{cut}}$, combine (p_i, p_j) in a single jet ('pseudoparticle') p_{ij} according to a recombination prescription.

4) Repeat this procedure from step 2) until all pairs of objects (particles and/or pseudoparticles) have $y_{kl} > y_{\text{cut}}$. Whatever objects remain at this stage are called jets.

According to the classification given by j_1), j_2) in Sec. 1, the JADE algorithm uses the scaled invariant mass (2) as test variable. In the experimental as well as theoretical analyses several different recombination procedures may be used [12,15]. They differ among themselves in the way in which the "invariant mass" and the 4-momentum p_{ij} of a pseudoparticle are defined in terms of the momenta (p_i, p_j) of the recombined particles and/or pseudoparticles. There are essentially two classes of recombination schemes (Table 1) whose main difference is in the definition of the "invariant mass". In the first class (E-scheme) the invariant mass M_{ij}^2 is the true invariant mass, whilst in the massless recombination schemes (JADE, E0, P, P0) M_{ij}^2 is computed defining the pseudoparticle momenta in such a way to have massless pseudoparticles. The various schemes in this latter class correspond to various possibilities of rescaling energy and 3-momentum in order to define a massless pseudoparticle.

The JADE algorithm obviously fulfils the requirements (i) and (ii) in Sec. 1. In the following we discuss in detail the points (iii) and (iv).

2.2 Hadronization corrections

Hadronization effects for the JADE algorithm can be investigated using Monte Carlo event generators [16-18]. These Monte Carlo generators include a parton shower, describing parton production according to perturbative QCD, plus hadronization models. Comparing jet cross sections obtained by Monte Carlo simulations both from partons at the end of the QCD shower and from particles after hadronization, one is able to estimate the size of hadronization corrections.

Fig. 1 shows the results of the analysis carried out by the OPAL collaboration [19] at LEP using the JETSET QCD shower model. Similar results have been obtained from other Collaborations at LEP and lower energies as well as using different Monte Carlo generators.

It can be seen from Fig. 1 that hadronization corrections are large for the E-recombination scheme, moderate for the P- and P0-scheme and small for the E0-scheme (the results for the JADE-scheme are similar to the ones for the E0-scheme). The fact

Table 1. Recombination procedures for the JADE algorithm: massive (E) and massless (JADE, E0, P, P0) recombination schemes.

scheme	M_{ij}^2/Q^2	recombination
E	$(p_i + p_j)^2/Q^2$	$p_{ij} = p_i + p_j$
JADE	$2\epsilon_i\epsilon_j(1 - \cos\theta_{ij})$	$\epsilon_i = E_i/Q$ $p_{ij} = p_i + p_j$
E0		$\epsilon_i = E_i/Q$ $E_{ij} = E_i + E_j$ $\mathbf{p}_{ij} = \frac{E_{ij}}{|\mathbf{p}_i+\mathbf{p}_j|}(\mathbf{p}_i + \mathbf{p}_j)$
P		$\epsilon_i = E_i/Q$ $\mathbf{p}_{ij} = \mathbf{p}_i + \mathbf{p}_j$ $E_{ij} = |\mathbf{p}_{ij}|$
P0		$\epsilon_i = E_i/\sum_k E_k$ $\mathbf{p}_{ij} = \mathbf{p}_i + \mathbf{p}_j$ $E_{ij} = |\mathbf{p}_{ij}|$

that hadronization effects are relevant for most of the recombination schemes may signal that something is going wrong with the JADE algorithm and call for a better theoretical understanding of jet definition.

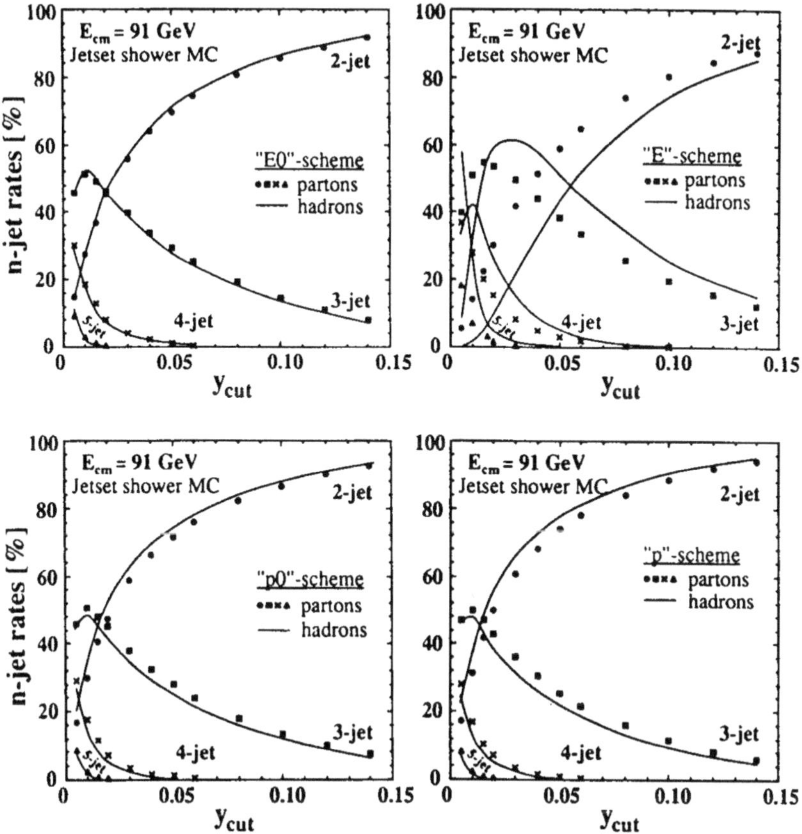

Fig. 1. OPAL results on relative production rates of n-jet events, determined from model calculations before and after the hadronization process. Jets are defined by the JADE algorithm and four different jet recombination schemes are used.

2.3 Theoretical calculations and renormalization scale problem

Let R_n represent the n-jet fraction

$$R_n = \frac{\sigma_{n\text{-jet}}}{\sigma_{\text{TOT}}} \ . \tag{3}$$

Since the JADE algorithm is IR and collinear safe, R_n can be computed in QCD perturbation theory. So far the n-jet fractions ($n = 2, 3, 4$) have been evaluated up to the second order in α_S and the result can be written as follows

$$R_n(\alpha_S(\mu), Q^2/\mu^2; y_{\text{cut}}) = \delta_{n2} + \alpha_S(\mu)C_n^{(1)}(y_{\text{cut}}) \tag{4}$$
$$+ \ \alpha_S^2(\mu)\left[C_n^{(2)}(y_{\text{cut}}) - C_n^{(1)}(y_{\text{cut}})\beta_0 \ln Q^2/\mu^2\right] + O(\alpha_S^3(\mu), \alpha_S^3(\mu)\ln^2 Q^2/\mu^2) \ ,$$

where $12\pi\beta_0 = 11C_A - 2N_f$, N_f is the number of flavours, $C_A = N_c = 3$ is the number of colours and the coupling $\alpha_S(\mu)$ can be expressed as a function af the QCD scale $\Lambda_{\overline{\text{MS}}}$ according to the two loop expression

$$\alpha_S(\mu) = \frac{1}{\beta_0 \ln(\mu^2/\Lambda_{\overline{\text{MS}}}^2)}\left[1 - \frac{6(153 - 19N_f)}{(33 - 2N_f)^2}\frac{\ln[\ln(\mu^2/\Lambda_{\overline{\text{MS}}}^2)]}{\ln(\mu^2/\Lambda_{\overline{\text{MS}}}^2)}\right] \ . \tag{5}$$

In one loop order $C_4^{(1)} = 0$, and $C_2^{(1)}, C_3^{(1)}$ are given by [†]

$$
\begin{aligned}
C_2^{(1)}(y_{\mathrm{cut}}) \;=\; -C_3^{(1)}(y_{\mathrm{cut}}) = \frac{C_F}{2\pi}\Bigg[&-2\ln^2 y_{\mathrm{cut}} - 3\ln y_{\mathrm{cut}} + \frac{\pi^2}{3} - \frac{5}{2} \\
+\;\; & 6y_{\mathrm{cut}}(1 + \ln y_{\mathrm{cut}}) + 4\ln y_{\mathrm{cut}}\ln(1 - y_{\mathrm{cut}}) - 2\ln^2(1 - y_{\mathrm{cut}}) \\
+\;\; & 3(1 - 2y_{\mathrm{cut}})\ln(1 - 2y_{\mathrm{cut}}) + \frac{9}{2}y_{\mathrm{cut}}^2 - 4\mathrm{Li}_2\Big(\frac{y_{\mathrm{cut}}}{1 - y_{\mathrm{cut}}}\Big)\Bigg] \,,
\end{aligned}
\tag{6}
$$

where

$$
C_F = \frac{N_c^2 - 1}{2N_c} \,, \qquad \mathrm{Li}_2(z) = \sum_{n=1}^{\infty} \frac{z^n}{n^2} \,.
\tag{7}
$$

The two loop functions $C_n^{(2)}(y_{\mathrm{cut}})$ (note that $C_2^{(2)}$ and $C_3^{(2)}$ are recombination scheme dependent) have been computed numerically [15,20] using the four-parton matrix elements in Ref. [21].

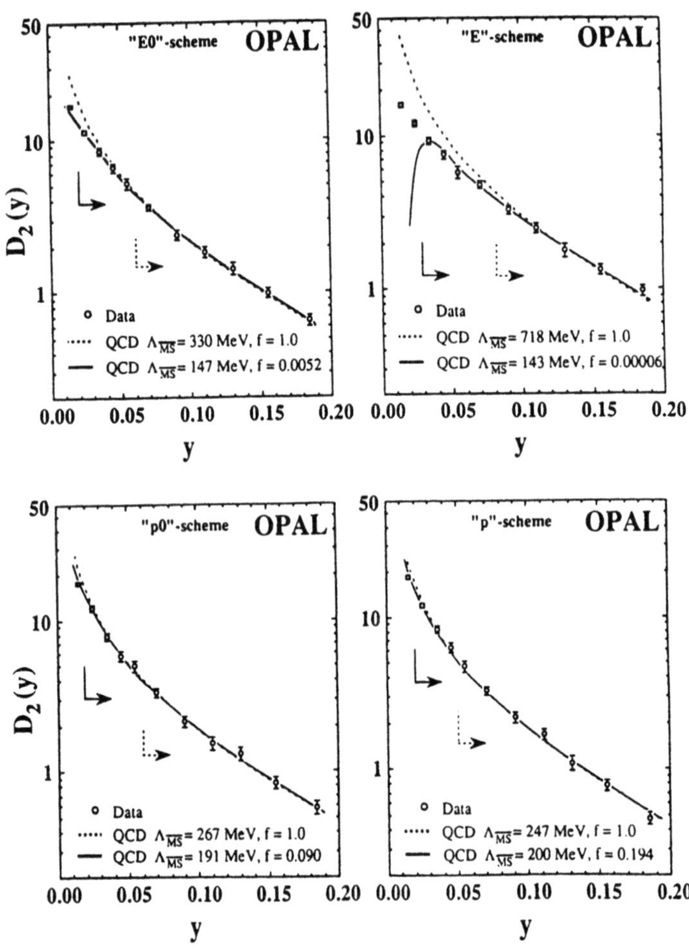

Fig. 2. Measured distributions of $D_2(y)$ for the JADE algorithm, corrected for detector acceptance and hadronization effects, compared to the corresponding analytic $O(\alpha_S^2)$ QCD calculations. The QCD parameters are taken from the fit results of $\Lambda_{\overline{MS}}$ with $f = 1$ ($f \equiv \mu^2/Q^2$) and of $\Lambda_{\overline{MS}}$ and μ^2 in the region of y indicated by the arrows.

[†] Our result (6) differs from that in Ref. [20] in the coefficient of the dilogarithm Li_2.

The two loop expression in eq. (4) depends on the renormalization scale μ^2 at which the running coupling $\alpha_S(\mu)$ is evaluated. This dependence is an artifact of the fixed order perturbative expansion in α_S because $R_n(\alpha_S(\mu), Q^2/\mu^2; y_{\mathrm{cut}})$ is a renormalization group invariant quantity if computed to all orders in α_S. We have taken into account this property writing explicitly a correction term of order $\alpha_S^3(\mu)\ln^2 Q^2/\mu^2$ on the r.h.s. of eq. (4).

In comparing eq. (4) with data, one has to fix the value of μ^2. Since the only physical scale in R_n is the centre of mass energy Q^2, the "natural" value to be used for μ^2 is $\mu^2 \simeq Q^2$.

The two loop QCD prediction for R_n has been compared with data (after correction for detector acceptance and hadronization effects) at Z^0 energies [19,22] and below [12,23,24]. As an illustrative example we show in Fig. 2 the result for $D_2(y_{\mathrm{cut}}) = \mathrm{d}R_2(y_{\mathrm{cut}})/\mathrm{d}y_{\mathrm{cut}}$ of the OPAL Collaboration [19].

The main conclusion is that eq. (4) with $\mu^2 = Q^2$ agrees with data in the large y_{cut} region ($y_{\mathrm{cut}} > 0.05$), whilst a good description of the data in the small y_{cut} region ($y_{\mathrm{cut}} < 0.05$) requires the use of very small renormalization scales $\mu^2 \ll Q^2$ [19]. The need of such "unnatural" renormalization scale to fit experimental data, already noticed for other quantities [25,19],.requires some physical explanation.

A first tentative explanation is that for $y_{\mathrm{cut}} \ll 1$, jet cross section definition involves two very different scales Q^2 and $y_{\mathrm{cut}}Q^2$. Therefore one might argue that $\mu^2 \sim y_{\mathrm{cut}}Q^2$. However that implies that R_n are functions of $\alpha_S(\sqrt{y_{\mathrm{cut}}}Q)$ and hence for $y_{\mathrm{cut}} \ll 1$ they depend on large distance physics and, eventually, become IR sensitive. It follows that $\mu^2 \sim y_{\mathrm{cut}}Q^2$ has to be excluded on theoretical grounds.

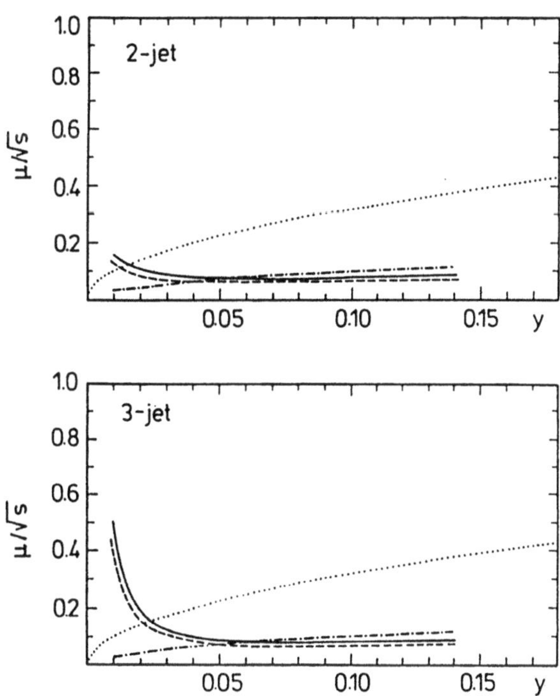

Fig. 3. The scale $\mu/\sqrt{s}$ ($s = Q^2$) computed in Ref. [29] according to the $\sqrt{y}$ procedure (dotted) and to the optimization procedures in [26] (dashed), [27] (full) and [28] (dashed-dotted). The results are obtained for $Q = 91$ GeV and $\Lambda_{\overline{\mathrm{MS}}} = 0.1$ GeV.

A second tentative explanation might use the recourse to some optimization procedure. Several prescriptions to fix the value of the renormalization scale μ^2 have been proposed [26-28] and are nowadays widely used. Their application to jet cross sections in the JADE algorithm has been investigated in [29]. It turns out that the optimized renormalization scale μ_{op}^2 is much smaller than Q^2 for $y_{\rm cut} > 0.05$ but one can get $\mu_{op}^2 \gg Q^2$ reducing $y_{\rm cut}$ to very small values (Fig. 3).

Although neither $\mu^2 \sim y_{\rm cut} Q^2$ nor $\mu^2 \sim \mu_{op}^2$ have theoretical justifications, one can try to compare the corresponding predictions with data. From the analysis of Ref. [29], we conclude (Fig. 4) that both prescriptions are not able to improve the agreement with data in the relevant region of small $y_{\rm cut} < 0.05$.

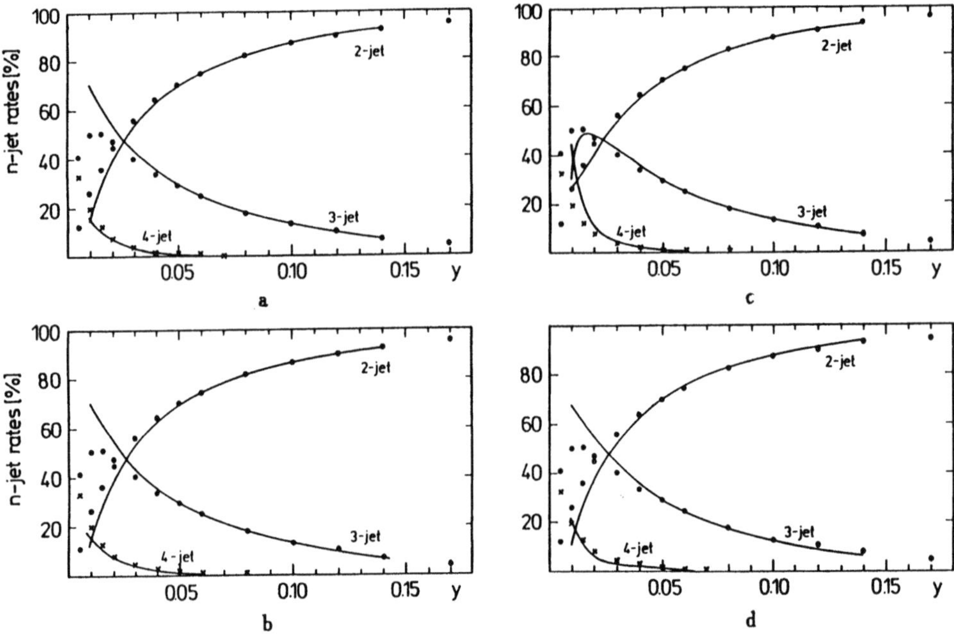

Fig. 4. Comparison between OPAL data and $O(\alpha_S^2)$ QCD predictions for n-jet production rates. The renormalization scale μ^2 is computed [29] according to the optimization procedures in (a) [26], (b) [27], (c) [28] and (d) $\mu^2 = yQ^2$.

We regard the results described above as the failure of phenomenological attempts to justify the choice of values of μ^2 very different from Q^2 and we think we should offer our theoretical explanation of that.

Let us state more precisely the argument that the physical value for the renormalization scale is $\mu^2 \simeq Q^2$. In order to compare the $O(\alpha_S^2)$-expression (4) with data one has to assume that the perturbative expansion for R_n is convergent. If μ^2 is very different from Q^2, we know that the two loop result for R_n neglects large higher order logarithmic corrections of the type $(\ln Q^2/\mu^2)^k$ (for instance the $O(\alpha_S^3(\mu)\ln^2 Q^2/\mu^2)$ term in eq. (4)). Therefore, modulo fortuitous cancellations, when $\mu^2 \neq Q^2$ one expects the perturbative expansion for R_n is badly convergent and one cannot safely compare the fixed order truncation with data. And vice versa, by studying the μ^2 dependence of the perturbative series in $\alpha_S(\mu)$ one can test its convergence: if the series is convergent, R_n should be a stable function of μ^2 for $\mu^2 \simeq Q^2$ and a rapidly varying function for μ^2 very different from Q^2.

It turns out that in second order $dR_n/d\ln\mu^2$ for $\mu^2 \simeq Q^2$ strongly depends on $y_{\rm cut}$

for small y_{cut} values. This means that the perturbative series in α_S is not convergent for $y_{\text{cut}} \ll 1$. The choices $\mu^2 \simeq y_{\text{cut}} Q^2$ or $\mu^2 \simeq \mu_{op}^2$ are just guesses to improve the convergence of the perturbative expansion. We think the only reliable procedure is to identify the origin of the terms spoiling the convergence and to compute them in higher orders.

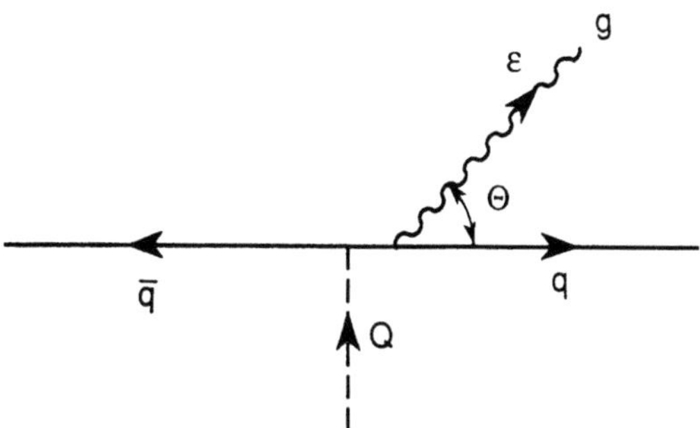

Fig. 5. Soft and collinear gluon emitted from the quark-antiquark pair.

Large coefficients in the α_S-expansion for R_n already appear in one loop order. From eqs. (4), (6) we have ($L \equiv \ln 1/y_{\text{cut}}$)

$$R_2 \simeq 1 - \frac{C_F \alpha_S}{\pi} L^2 \,, \qquad R_3 \simeq \frac{C_F \alpha_S}{\pi} L^2 \,, \quad (y_{\text{cut}} \ll 1) \,. \tag{8}$$

The double logarithmic term in eq. (8) comes from the bremsstrahlung spectrum for emission of a soft (with energy fraction $\epsilon = \omega/E \ll 1$) and collinear (with angle $\theta \ll 1$) gluon from the $q\bar{q}$ pair (Fig. 5). In the soft and collinear limit the single gluon emission probability is

$$dw(1) = \frac{C_F \alpha_S}{\pi} \frac{d\epsilon}{\epsilon} \frac{d\theta^2}{\theta^2} \,, \tag{9}$$

which, integrated over the phase space for the JADE algorithm ($y_{qg} \simeq \epsilon\theta^2$), gives

$$\int dw(1)\, \Theta(\epsilon\theta^2 - y_{\text{cut}}) = \frac{C_F \alpha_S}{\pi} \int_0^1 \frac{d\epsilon}{\epsilon} \int_0^1 \frac{d\theta^2}{\theta^2}\, \Theta(\epsilon\theta^2 - y_{\text{cut}}) = \frac{C_F \alpha_S}{2\pi} L^2 \,. \tag{10}$$

The large contributions $\alpha_S L^2$ are responsible for (i) the unphysical shape of the n-jet rates computed in fixed order perturbation theory (for instance in one loop order $R_2 \to -\infty$ for $y_{\text{cut}} \to 0$) and (ii) the instability of the perturbative expansion with respect to renormalization scale variations [30].

At small values of y_{cut}, higher order terms in α_S are enhanced by powers of $\ln^2 y_{\text{cut}}$. In this kinematical region the real expansion parameter is the large effective coupling $\alpha_S L^2$ and therefore *any* finite-order perturbative calculation cannot give an accurate evaluation of the cross section. The logarithmic terms must be *resummed* to all orders in α_S before a reliable prediction can be made. No choice of μ^2 in a fixed order result can account for such a resummation because changing μ^2 one can only introduce single logarithmic corrections $(\alpha_S L)^k$ in higher orders. A particular choice of μ^2 can at most simulate the double logarithmic resummation over a limited range of y_{cut} [30,31].

3. The small y_{cut} region

3.1 Exponentiation of Sudakov logarithms

The appearance of large double logarithmic terms is a common feature of any hard process in the semi-inclusive or Sudakov region, where emission of radiation is inhibited by the kinematics [32,33]. In the case of jet cross sections at $y_{\text{cut}} \ll 1$, the jet invariant mass is constrained to be so small as to allow only emission of gluons that are soft and collinear with respect to the parton generating the jet. The double logarithmic terms $\alpha_S \ln^2 y_{\text{cut}}$ are due to such soft and collinear gluons.

In recent years, following the pioneering calculations [34] of double logarithms, the programme of resummation of Sudakov logarithms has been succesfully carried out to next-to-leading order for transverse [35] and longitudinal momentum distributions [36,37] in hadron-initiated processes, as well as for quantities like the energy-energy correlation [38] and the thrust [30] and heavy jet mass distribution [31] in e^+e^- annihilation. All these successful resummations depend on the fact that the large logarithmic corrections to the relevant quantity *exponentiate*. By this we mean that terms of the form $\alpha_S^n \ln^m y_{\text{cut}}$ with $m \leq 2n$ can be combined into an exponential function of less singular terms, in fact with $m \leq n + 1$. Once exponentiation has been established, the resummation of leading and next-to-leading logarithms to all orders is reduced to a much simpler, finite-order calculation of the relevant exponent.

Let us recall the physical basis for exponentiation starting from the simpler case of Quantum Electrodynamics (QED). In QED photons are not charged and this implies that multiple soft photon emission is an independent process. Thus the probability $dw(1, \cdots, n)$ for n-soft photon emission *factorizes* in the product of single photon factors $dw(i)$ in eq. (9)

$$dw(1, \ldots, n) = \frac{1}{n!} \prod_{i=1}^{n} dw(i) = \frac{1}{n!} \prod_{i=1}^{n} \frac{\alpha}{\pi} \frac{d\epsilon_i}{\epsilon_i} \frac{d\theta_i^2}{\theta_i^2} . \tag{11}$$

Starting from the factorized result (11) one can get the corresponding cross sections contribution by integration over the relevant phase-space $\Theta(1, \cdots, n; y_{\text{cut}})$. Therefore the factorization of multiphoton amplitudes leads to the exponentiation of the cross section if and only if the phase space constraint factorizes in the soft limit, i.e.

$$\Theta(1, \cdots, n; y_{\text{cut}}) \simeq \prod_{i=1}^{n} \Theta(i; y_{\text{cut}}) , \tag{12}$$

$$1 + \sum_{n=1}^{\infty} \int dw(1, \cdots, n) \, \Theta(1, \cdots, n; y_{\text{cut}})$$

$$\simeq 1 + \sum_{n=1}^{\infty} \frac{1}{n!} \prod_{i=1}^{n} \int dw(i) \, \Theta(i; y_{\text{cut}}) = \exp \left[\int dw(i) \, \Theta(i; y_{\text{cut}}) \right] . \tag{13}$$

The property (11) comes from QED dynamics whilst (12) depends on the kinematic definition of the cross section. The cross section exponentiates in the Sudakov region *iff* its definition does not induce *kinematic correlations* among soft photons.

Let us come to the QCD case. The main difference with respect to QED is due to the fact that gluons have colour charge. Therefore they can radiate in cascade and soft gluon emission is no longer an independent process. Strong gluon correlations are enforced by the dynamics, multiple gluon emission is not factorized into single emission contributions and we can only expect some kind of generalized exponentiation.

Nevertheless a simple exponentiation structure is still valid for highly inclusive cross sections like a large class of two jet dominated quantities [34-38,30,31]. Let us consider for example the two loop non-factorized contributions in Fig. 6. If the two-jet cross section is defined in such a way that the integration over the soft gluon degrees of freedom is unconstrained, the real (Fig. 6a) and virtual (Fig. 6b) contribution *cancel* each other in the IR and collinear regions. The only logarithmic term which survives comes from the ultraviolet region for the virtual diagram. It leads to the running of the QCD coupling $\alpha_S \to \alpha_S(k_\perp)$ with an argument as given by the transverse momentum $k_\perp$ of the emitted gluon [39].

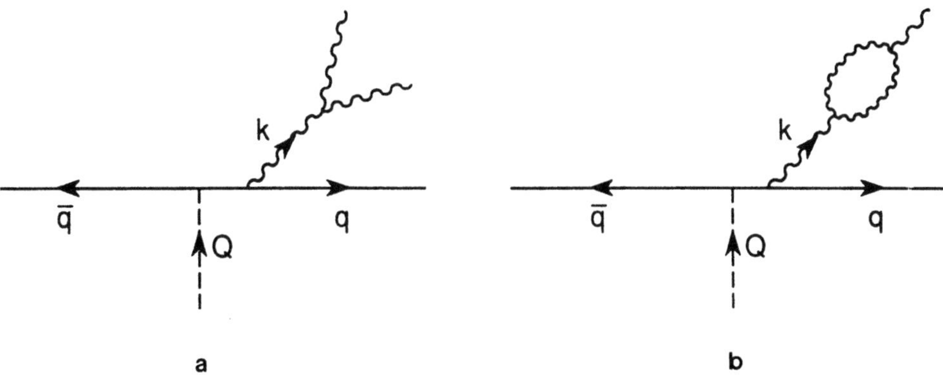

Fig. 6. Soft gluon dynamics correlations in two loop order: (a) real and (b) virtual emission contributions.

The cancellation described so far is preserved to higher orders due to *unitarity* and *gauge invariance* properties [40] and leads to the factorization of the one gluon emission probability (9) with $\alpha_S \to \alpha_S(k_\perp)$ $(k_\perp \sim \epsilon\theta Q)$. Exponentiation then follows in the absence of kinematic correlations among soft partons, related to the actual definition of the cross section.

3.2 Leading and next-to-leading logarithms for the JADE algorithm

In order to check whether exponentiation works for jet cross sections defined according to the JADE algorithm, one can start computing n-jet rates to $O(\alpha_S^2)$. In the small y_{cut} region one can write $(L \equiv \ln 1/y_{\text{cut}})$

$$
\begin{aligned}
R_n = \delta_{2n} \;\; &+ \;\; \frac{\alpha_S(Q)}{\pi} \left[A_2^{(n)} L^2 + A_1^{(n)} L + O(1) \right] \\
&+ \;\; \left(\frac{\alpha_S(Q)}{\pi} \right)^2 \left[B_4^{(n)} L^4 + B_3^{(n)} L^3 + O(L^2) \right]
\end{aligned}
\tag{14}
$$

We have computed the leading $(A_2^{(n)}, B_4^{(n)})$ and next-to-leading $(A_1^{(n)}, B_3^{(n)})$ logarithmic coefficients for the JADE algorithm in the E-recombination scheme and the results [41] are reported in Table 2 (the C_F^2 contributions for $B_4^{(n)}$ were first computed by Brown and Stirling [42]).

One has the surprising result that the leading double logarithmic contributions do not exponentiate even in the abelian limit $C_A \to 0$ [42]. In Table 2 we have split the C_F^2 contributions for the two loop coefficient $B_4^{(n)}$ into two terms. The first one in parentheses is the term corresponding to the exponentiation and the second one leads to the violation of exponentiation.

According to our discussion in Sec. 3.1, these results imply that the JADE algorithm leads to non-factorized phase space which, in the E-recombination scheme, produces *attractive* kinematic correlations among soft gluons. For the two jet cross section these correlations appear in the kinematic configuration in which the $q\bar{q}$ pair emits

Table 2. Leading and next-to-leading two loop coefficients for the JADE algorithm in the E-recombination scheme.

n	$A_2^{(n)}$	$A_1^{(n)}$	$B_4^{(n)}$	$B_3^{(n)}$
2	$-C_F$	$+\frac{3}{2}C_F$	$(\frac{1}{2} - \frac{1}{12})C_F^2$	$-\frac{3}{2}C_F^2 - \pi\beta_0 C_F$
3	$+C_F$	$-\frac{3}{2}C_F$	$(-1 + \frac{5}{24})C_F^2 - \frac{1}{24}C_F C_A$	$+3C_F^2 + \frac{1}{4}C_F C_A + \frac{7}{6}\pi\beta_0 C_F$
4	0	0	$(\frac{1}{2} - \frac{1}{8})C_F^2 + \frac{1}{24}C_F C_A$	$-\frac{3}{2}C_F^2 - \frac{1}{4}C_F C_A - \frac{1}{6}\pi\beta_0 C_F$

Table 3. Leading and next-to-leading two loop coefficients for the JADE algorithm in the massless recombination schemes.

n	$A_2^{(n)}$	$A_1^{(n)}$	$B_4^{(n)}$	$B_3^{(n)}$
2	$-C_F$	$+\frac{3}{2}C_F$	$(\frac{1}{2} + \frac{1}{24})C_F^2$	$-\frac{3}{2}C_F^2 - \pi\beta_0 C_F$
3	$+C_F$	$-\frac{3}{2}C_F$	$(-1 + \frac{1}{12})C_F^2 - \frac{1}{24}C_F C_A$	$+3C_F^2 + \frac{1}{4}C_F C_A + \frac{7}{6}\pi\beta_0 C_F$
4	0	0	$(\frac{1}{2} - \frac{1}{8})C_F^2 + \frac{1}{24}C_F C_A$	$-\frac{3}{2}C_F^2 - \frac{1}{4}C_F C_A - \frac{1}{6}\pi\beta_0 C_F$

Table 4. Leading and next-to-leading two loop coefficients for the $k_\perp$-algorithm in any (E, E0, P, P0) recombination scheme.

n	$A_2^{(n)}$	$A_1^{(n)}$	$B_4^{(n)}$	$B_3^{(n)}$
2	$-\frac{1}{2}C_F$	$+\frac{3}{2}C_F$	$+\frac{1}{8}C_F^2$	$-\frac{3}{4}C_F^2 - \frac{1}{3}\pi\beta_0 C_F$
3	$+\frac{1}{2}C_F$	$-\frac{3}{2}C_F$	$-\frac{1}{4}C_F^2 - \frac{1}{48}C_F C_A$	$+\frac{3}{2}C_F^2 + \frac{1}{8}C_F C_A + \frac{1}{2}\pi\beta_0 C_F$
4	0	0	$+\frac{1}{8}C_F^2 + \frac{1}{48}C_F C_A$	$-\frac{3}{4}C_F^2 - \frac{1}{8}C_F C_A - \frac{1}{6}\pi\beta_0 C_F$

Table 5. Leading and next-to-leading two loop coefficients for the Geneva algorithm.

n	$A_2^{(n)}$	$A_1^{(n)}$	$B_4^{(n)}$	$B_3^{(n)}$
2	$-C_F$	$+\frac{3}{2}C_F$	$\frac{1}{2}C_F^2$	$-\frac{3}{2}C_F^2 - \pi\beta_0 C_F$
3	$+C_F$	$-\frac{3}{2}C_F$	$-C_F^2 - (\frac{1}{24} + \frac{1}{8})C_F C_A$	$+3C_F^2 + \frac{1}{4}C_F C_A + \frac{7}{6}\pi\beta_0 C_F$
4	0	0	$\frac{1}{2}C_F^2 + (\frac{1}{24} + \frac{1}{8})C_F C_A$	$-\frac{3}{2}C_F^2 - \frac{1}{4}C_F C_A - \frac{1}{6}\pi\beta_0 C_F$

two back-to-back ($\theta_1^2, \theta_2^2 \ll 1$) and soft ($\epsilon_1, \epsilon_2 \ll 1$) gluons (Fig. 7). The corresponding factorized phase space is

$$\epsilon_1 \theta_1^2, \; \epsilon_2 \theta_2^2 < y_{\text{cut}} \; , \tag{15}$$

but, following the E-recombination procedure, the sub-region

$$\epsilon_1 \theta_1^2, \; \epsilon_2 \theta_2^2 > \epsilon_1 \epsilon_2 \; ; \quad \epsilon_1, \epsilon_2 > y_{\text{cut}} \tag{16}$$

has to be subtracted from the two jet region and included in the three jet rate. We see that the E-scheme gives rise in the region of eqs. (15), (16) to an "anomalous" three jet configuration in which a jet is given by two very soft partons with a large relative angle. One may say that the JADE algorithm counts the number of lumps instead of jets. [†].

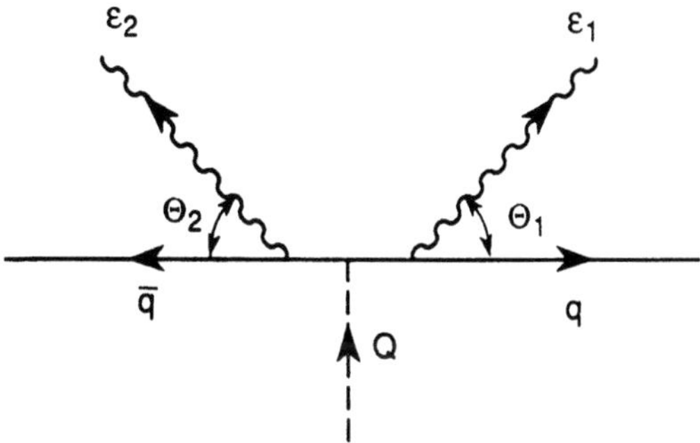

Fig. 7. Two back-to-back gluons emitted from the quark-antiquark pair.

The attractive kinematic correlations induced by the E-recombination scheme imply the following consequences.

- In two loop order, R_2 and R_4 decrease and R_3 increases with respect to the result for independent emission.

- The kinematic correlations strongly increase with the number of final state partons and therefore even the resummation of the leading $\ln y_{\text{cut}}$ terms to all orders appears hopeless.

- The recombination procedure produces a lot of "anomalous" jets: soft and large angle jets.

- Since, according to QCD preconfinement, hadronization is supposed to be a local process in energy and angle, we expect (and we do have) large hadronization corrections.

In order to check whether these features are just a pathology of the E-scheme, we have evaluated the jet rates (14) for the other recombination schemes. We found [41] that exponentiation is still violated and that leading and next-to-leading terms are the same for any massless recombination scheme (JADE, E0, P, P0). The results are reported in Table 3. Some comments are in order.

Firstly, the next-to-leading coefficients $B_3^{(n)}$ are equal to the ones for the E-scheme. This is due to the fact that the kinematic correlations related to the jet definition affect only the soft gluon phase space region.

Secondly, JADE, E0, P and P0-schemes have the same leading behaviour because the corresponding pseudoparticles are massless. That provides a first explanation of the different size of hadronization corrections for the E-scheme and the massless recombination schemes (cf. Sec. 2.2).

[†] We owe Nick Brown this nice picture.

33

Finally, soft gluon kinematic correlations for the 2 and 3 jet rates in the massless schemes are weaker than for the E-scheme (compare the second contribution in parentheses for $B_4^{(n)}$ in Tables 2 and 3). That supports our explanation of the smaller hadronization effects observed for massless schemes. Note also that for massless schemes, the attractive gluon correlations only reduce R_4, whilst both R_2 and R_3 are enhanced.

On balance, our conclusions about the JADE algorithm in the E-scheme are not changed very much by our results for the other schemes. The only point is that massless recombination procedures slightly reduce some bad features of the algorithm.

4. A new clustering algorithm: the $k_\perp$-algorithm

4.1 Definition

We believe that the analysis of the JADE algorithm carried out in Secs. 2 and 3 sufficiently motivates the need of a different jet definition. The main goal is to define a clustering procedure which destroys soft gluon kinematic correlations in order to make the theoretical calculation possible for $y_{\rm cut} \ll 1$ and to reduce hadronization effects.

We are looking for a jet algorithm whose dynamics features follow as close as possible QCD theory. In the naive parton model, hadronic jets have a *cylindrical shape*. By this we mean that they are produced with a total energy increasing with the centre of mass energy and with a limited fixed transverse momentum $< k_\perp >$ (of order of some hundred MeV) with respect to the jet axis. The QCD jets have instead an almost *conical shape*: parton multiplication due to the fragmentation of the parent parton leads to hadron jets whose mean transverse momentum increases with the jet energy [32, 33,43]. Thus it is natural to introduce a jet algorithm[§] in which the transverse momentum replaces the invariant mass of the original JADE algorithm as the jet resolution variable. To emphasize this change of variable, we call the new algorithm the $k_\perp$-*algorithm* [13].

The $k_\perp$-algorithm is defined in a similar way to the JADE algorithm, following the iterative procedure 1)-4) in Sec. 2.1 and just replacing the invariant mass by the transverse momentum as the jet resolution variable. Therefore, the test variable y_{kl} at the step 2) becomes

$$y_{kl} = \frac{2(1 - \cos \theta_{kl})}{Q^2} \, \min(E_k^2, E_l^2) = 2 \min(\epsilon_k^2, \epsilon_l^2)\,(1 - \cos \theta_{kl})\,. \qquad (17)$$

For the recombination procedure one can define several schemes (E, E0, P, P0) as in the JADE algorithm.

The main features of the $k_\perp$-algorithm (in any recombination scheme) we expect are (i) small hadronization effects (positive preliminary results have been already reported at this meeting [3]) and (ii) the possibility of resumming large $\ln y_{\rm cut}$ terms to all orders in perturbation theory. In fact, since the test variable (17) is diagonal with respect to parton energies, it is a much more local energy-angle resolution variable than the invariant mass (which depends on both parton energies). For instance, it is trivial to check that the back-to-back configuration in Fig. 7 leads only to two jets in the factorized phase space region $\epsilon_1^2\theta_1^2, \epsilon_2^2\theta_2^2 < y_{\rm cut}$.

[§]This algorithm was discussed at the Durham Workshop on JET Studies at LEP and HERA, December 1990, and is sometimes referred to as the Durham algorithm.

The n-jet rates

$$R_n(y_{\text{cut}}) = \sum_{m \geq n} \frac{1}{\sigma_{\text{TOT}}} \int d\sigma_m^{ex} \, \Theta_{n\text{-jet}}(y_{\text{cut}}) \tag{18}$$

are obtained by integration of the m-parton exclusive cross sections $d\sigma_m^{ex}$ ($m \geq n$) over the n-jet phase space $\Theta_{n\text{-jet}}(y_{\text{cut}})$ as given by the clustering procedure. We have computed the n-jet fractions using the coherent branching formalism [44,45,33] to evaluate the exclusive cross sections and working out the n-jet phase space to next-to-leading accuracy.

Using the $k_\perp$-algorithm we are able to resum the leading $\alpha_S^m L^{2m}$ and next-to-leading $\alpha_S^m L^{2m-1}$ logarithmic contributions to all orders in α_S for any number of jets[¶]. The result [13] can be simply stated in terms of the generating function $\phi(Q, Q_0; u)$ for jet rates ($Q_0^2 \equiv y_{\text{cut}} Q^2$)

$$R_n(y_{\text{cut}} = Q_0^2/Q^2) = \frac{1}{n!} \left(\frac{\partial}{\partial u} \right)^n \phi(Q, Q_0; u)\big|_{u=0} \,. \tag{19}$$

Defining the following emission probabilities

$$
\begin{aligned}
\Gamma_q(Q, q) &= \frac{2 C_F}{\pi} \frac{\alpha_S(q)}{q} \left(\log \frac{Q}{q} - \frac{3}{4} \right) , \\
\Gamma_g(Q, q) &= \frac{2 C_A}{\pi} \frac{\alpha_S(q)}{q} \left(\log \frac{Q}{q} - \frac{11}{12} \right) , \\
\Gamma_f(q) &= \frac{N_f}{3\pi} \frac{\alpha_S(q)}{q} ,
\end{aligned}
\tag{20}
$$

the generating function ϕ in eq. (19) is given by

$$\phi(Q, Q_0; u) = u^2 \exp\left(2 \int_{Q_0}^Q dq \, \Gamma_q(Q, q) \left[\phi_g(q, Q_0; u) - 1 \right] \right) , \tag{21}$$

where the gluon functional ϕ_g satisfies the following implicit equation

$$
\phi_g(Q, Q_0; u) = u \exp\left(\int_{Q_0}^Q dq \left\{ \Gamma_g(Q, q) \left[\phi_g(q, Q_0; u) - 1 \right] - \Gamma_f(q) \right\} \right)
$$
$$
\cdot \left\{ 1 + u \int_{Q_0}^Q dq \, \Gamma_f(q) \exp\left(\int_{Q_0}^q dq' \left\{ [2\Gamma_q(q, q') - \Gamma_g(q, q')] \right. \right. \right.
$$
$$
\cdot \left. \left. \left. [\phi_g(q', Q_0; u) - 1] + \Gamma_f(q') \right\} \right) \right\} . \tag{22}
$$

Solving eq. (22) as a power series in the jet label u, one can compute the jet rates R_n in eq. (19) for an arbitrary number of jets.

We refer to [13,41] for details of the calculations. Here, we want just to note that we are able to resum leading and next-to-leading $\ln y_{\text{cut}}$ terms because the $k_\perp$-resolution variable in eq. (17), which is diagonal with respect to parton energies, does not induce soft gluon kinematic correlations in the clustering procedure. Moreover, although various recombination schemes (E, E0, P, P0) can differ significantly at finite y_{cut}, they all give the same leading and next-to-leading logarithms at small y_{cut} [41].

[¶] The resummation of leading logarithms for the two-jet rate has been also considered in Ref. [46]

Therefore the formulae (21),(22) as well as the ones we shall give below are valid for any of the common recombination schemes.

The n-jet fractions one gets from eqs. (19)-(22) are (for $n \leq 4$)

$$R_2^{(e^+e^-)} = [\Delta_q(Q)]^2 \ , \tag{23}$$

$$R_3^{(e^+e^-)} = 2\,[\Delta_q(Q)]^2 \int_{Q_0}^{Q} dq\,\Gamma_q(Q,q)\Delta_g(q) \ , \tag{24}$$

$$\begin{aligned}
R_4^{(e^+e^-)} \ = \ & 2\,[\Delta_q(Q)]^2 \left\{ [\int_{Q_0}^{Q} dq\,\Gamma_q(Q,q)\Delta_g(q)]^2 \right. \\
& + \ \int_{Q_0}^{Q} dq\,\Gamma_q(Q,q)\Delta_g(q) \int_{Q_0}^{q} dq'\,\Gamma_g(q,q')\Delta_g(q') \\
& + \left. \int_{Q_0}^{Q} dq\,\Gamma_q(Q,q)\Delta_g(q) \int_{Q_0}^{q} dq'\,\Gamma_f(q')\Delta_f(q') \right\} \ , \tag{25}
\end{aligned}$$

where we have introduced the quark and gluon Sudakov form factors

$$\Delta_q(Q) \ = \ \exp\left(-\int_{Q_0}^{Q} dq\Gamma_q(Q,q)\right) \ , \tag{26}$$

$$\Delta_g(Q) \ = \ \exp\left(-\int_{Q_0}^{Q} dq[\Gamma_g(Q,q)+\Gamma_f(q)]\right) \ , \tag{27}$$

which depend implicitly on y_{cut} via $Q_0 = Q\sqrt{y_{\mathrm{cut}}}$, and we have defined

$$\Delta_f(q) = [\Delta_q(q)]^2/\Delta_g(q) \ . \tag{28}$$

The relevant two loop coefficients in eq. (14) for the $k_\perp$-algorithm are given in Table 4.

The results in eqs. (23)-(28) are in agreement with the general discussion in Sec. 3.1. We see that a simple exponentiation of $\ln y_{\mathrm{cut}}$ terms is valid for the two jet fraction R_2. The jet rates R_n with $n > 2$ satisfy simple exponentiation in the pseudo-abelian limit $\Gamma_g, \Gamma_f \to 0$

$$R_n = \frac{\Delta_q^2(Q)}{(n-2)!}\left[2\int_{Q_0}^{Q} dq\Gamma_q(Q,q)\right]^{n-2} + (\Gamma_g \text{ and } \Gamma_f \text{ terms}) \ , \tag{29}$$

whilst in the QCD case they follow a generalized exponentiation structure.

5. Discussion and outlook

In this paper we have presented a QCD theory based discussion of jet topology and jet algorithms. We have mainly focused on the small y_{cut} or Sudakov region. From our analysis we argued that soft gluon kinematic correlations induced by the JADE clustering procedure may be responsible for large hadronization corrections and for bad convergence properties of the QCD perturbative expansion (see also Appendix A for a different jet definition). In Sec. 4 we have discussed a new jet algorithm, the $k_\perp$-algorithm, which does not suffer from the above problems.

Using the $k_\perp$-algorithm, large higher order corrections at small values of y_{cut} can be easily evaluated. We were able to resum leading and next-to-leading logarithms of y_{cut} to all orders in perturbation theory for any number of jets. The results in eqs. (23)-(25) may be combined with the exact fixed order results, after subtraction of the terms that have been exponentiated, to obtain predictions for up to four jets over the full range of y_{cut}.

A question we have discussed at length is the renormalization scale dependence. The coefficients of the leading and next-to-leading logarithmic contributions resummed in eqs. (23)-(25) are renormalization group invariant. Therefore, we cannot properly discuss the renormalization scale dependence in the $k_\perp$-algorithm until the matching with fixed-order results and a smooth extrapolation between the region of small and large y_{cut} have been performed [41]. Anyway, according to our discussion in Sec. 2.3 and to our experience [30,31], the renormalization scale problem is not a problem at all! Once a given quantity has been correctly computed resumming classes of logarithmic higher order corrections, its expansion in this "improved perturbation theory" should be convergent for values of the renormalization scale μ^2 of the order of the physical scale of the process.

The $k_\perp$-algorithm can also be used to compute inclusive and differential jet cross sections as well as being applicable to hadron collision processes. Calculations are in progress.

We hope experimental data for jets within the $k_\perp$-algorithm will be soon available.

Acknowledgements. The results reported in this talk have been obtained in collaboration with Yu.L. Dokshitzer, M. Olsson, G. Turnock and B.R. Webber. Useful discussions with S. Bethke, M. Ciafaloni, G. Marchesini, D.E. Soper, W.J. Stirling and L. Trentadue are gratefully acknowledged. We would especially like to acknowledge many valuable and stimulating conversations with Nick Brown, to whose memory we respectfully dedicate this paper. It is a pleasure to thank Yuri Dokshitzer and Luisa Cifarelli for organizing this pleasant and efficient Workshop at the *Ettore Majorana Centre for Scientific Culture*.

Appendix A - The Geneva algorithm

A different jet algorithm called the Geneva or E_n-algorithm [14,3] has been recently proposed to overcome the difficulty of resumming $\ln y_{\text{cut}}$ terms in the JADE algorithm. In this clustering algorithm the jet resolution variable y_{kl} is defined by

$$y_{kl} = \frac{2E_k E_l}{(E_k + E_l)^2}(1 - \cos\theta_{kl}) = \frac{2\epsilon_k \epsilon_l}{(\epsilon_k + \epsilon_l)^2}(1 - \cos\theta_{kl}) \tag{30}$$

and the particle are recombined according to the E-scheme (although other recombination procedure can be used as well).

The test variable (30) is essentially the two parton invariant mass times a rescaling factor $1/(\epsilon_k + \epsilon_l)^2$. This rescaling factor acts as a repulsive potential counteracting the attractive soft gluon correlations arising in the JADE algorithm. In order to have a better understanding of this effect we have performed some calculations for the small y_{cut} region.

Let us start by considering a pseudo-abelian model in which "gluons" cannot radiate in cascade gg or $q\bar{q}$ real pairs. This model is similar to QED apart from the fact that the coupling constant runs as in QCD. For this model we were able to compute the generating function $\tilde{\phi}(Q, Q_0; u)$ of n-jet rates in the Geneva algorithm

$$\tilde{R}_n(y_{\text{cut}} = Q_0^2/Q^2) = \frac{1}{n!}\left(\frac{\partial}{\partial u}\right)^n \tilde{\phi}(Q, Q_0; u)\Big|_{u=0} . \tag{31}$$

For the generating function $\tilde{\phi}$ to leading and next-to-leading logarithmic accuracy we find [41]

$$\tilde{\phi}(Q,Q_0;u) = u^2 \exp\left\{(u-1)\left[8\int_{Q_0/Q}^1 \frac{dx}{x}\int_{xQ_0}^{x^2Q}\frac{dq}{q}\alpha_S(q) - 3\int_{Q_0}^Q \frac{dq}{q}\alpha_S(q)\right]\right\} \ . \qquad (32)$$

The generating function (32) leads to a Poisson distribution and hence the corresponding n-jet rates exponentiate in this pseudo-abelian limit.

Let us come back to the QCD case. We have evaluated the two jet fraction R_2 resumming leading and next-to-leading $\ln y_{\rm cut}$ terms to all orders in α_S and we found the exponentiated expression

$$R_2 = \exp\left\{-\left[8\int_{Q_0/Q}^1 \frac{dx}{x}\int_{xQ_0}^{x^2Q}\frac{dq}{q}\alpha_S(q) - 3\int_{Q_0}^Q \frac{dq}{q}\alpha_S(q)\right]\right\} \ . \qquad (33)$$

The results (32,33) confirm that the rescaling factor $1/(\epsilon_k + \epsilon_l)^2$ in eq. (30) is able to cancel the attractive kinematic correlations for the two jet cross section and for any jet rate in the abelian limit. However it looks surprising that the jet resolution variable (30), depending on both parton energies, may lead to a soft gluon phase space which is completely uncorrelated. A full two loop calculation confirms that kinematic correlation are still there. The result [41] for the two loop coefficients of eq. (14) in the Geneva algorithm is given in Table 5.

The abelian C_F^2 contributions to $B_4^{(n)}$ exponentiate, whilst the non-abelian terms $C_F C_A$ differ from those obtained for the JADE algorithm (these latter are reported in Table 5 as the first contribution to $B_4^{(n)}$ inside the parentheses). This difference is due to kinematic configurations like the one in Fig. 8 in which two soft gluons $(1 > \epsilon_1 > \epsilon_2)$ are radiated in cascade $(1 > \theta_1^2 > \theta_2^2)$ at large angle $(\epsilon_1\theta_1^2 > y_{\rm cut})$. In the JADE algorithm this configuration leads to a 4-jet event fraction $\{\bar{q},q,g_1,g_2\}$ and to a 3-jet event fraction $\{\bar{q},q,(g_1,g_2)\}$. In the Geneva algorithm the rescaling factor $1/(\epsilon_k + \epsilon_l)^2$ in the test variable (30) induces a strong effective repulsion between the soft gluons g_1 and g_2 to produce a non-local 3-jet recombination $\{\bar{q},(q,g_2),g_1\}$. This repulsive correlation enhances the 4-jet cross section and reduces the 3-jet cross section with respect to the JADE algorithm.

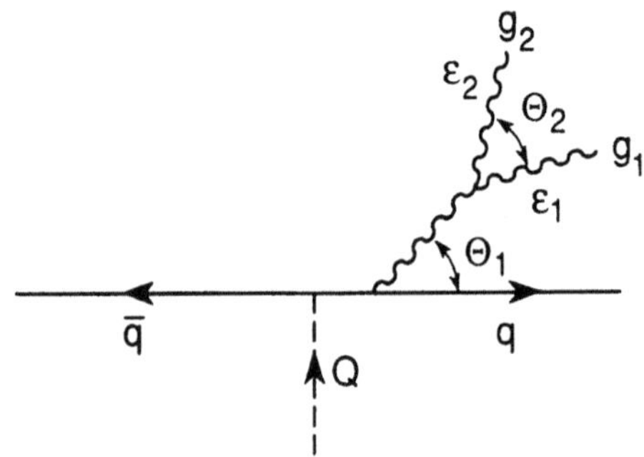

Fig. 8 - Two soft gluons radiated in cascade.

We conclude that the clustering procedure of the Geneva algorithm introduces *repulsive kinematic correlations* among soft gluons, thus leading to analogous problems to the ones arising in the JADE algorithm, viz.

- the resummation of the $\ln y_{\mathrm{cut}}$ terms for R_n with $n \geq 3$ is (probably) hopeless;

- the jet recombination pattern is non-local in energy-angle and sizeable hadronization corrections are expected (preliminary results presented at this meeting [3] confirm this expectation).

References

1. G. Hanson et al., Phys. Rev. Lett. 35 (1975) 1609.

2. S.L. Wu, Phys. Rep. 107 (1984) 59;
   B. Naroska, Phys. Rep. 148 (1987) 67;
   S. Bethke, CERN preprint PPE/91-36.

3. S. Bethke, these proceedings.

4. UA2 Collaboration, J. Alitti et al., CERN preprint PPE/90-188;
   B.L. Flaugher, FERMILAB-Conf.-90/159-E.

5. M. Mangano, these proceedings.

6. V. Khoze, these proceedings.

7. D. Amati and G. Veneziano, Phys. Lett. 83B (1979) 87;
   A. Bassetto, M. Ciafaloni and G. Marchesini, Phys. Lett. 83B (1979) 207;
   G. Marchesini, L. Trentadue and G. Veneziano, Nucl. Phys. B181 (1980) 335.

8. Ya.I. Azimov, Yu.L. Dokshitzer, V.A. Khoze and S.I. Troyan, Phys. Lett. 165B (1985) 147; Zeit. Phys. C27 (1985) 65.

9. G. Sterman and S. Weinberg, Phys. Rev. Lett. 39 (1977) 1436.

10. T. Kinoshita, J. Math. Phys. 3 (1962) 650;
    T.D. Lee and M. Nauenberg, Phys. Rev. 133 (1964) 1549.

11. B. Flaugher and K. Meier, FERMILAB-Conf.-90/248-E.

12. JADE Collaboration, W. Bartel et al., Zeit. Phys. C33 (1986) 23; S. Bethke et al., Phys. Lett. 213B (1988) 235.

13. S. Catani, Yu.L. Dokshitzer, M. Olsson, G. Turnock and B.R. Webber, Cambridge preprint Cavendish–HEP–91/5 to be published in Phys. Lett. B.

14. S. Bethke, Z. Kunszt, D.E. Soper and W.J. Stirling, CERN preprint TH.6222-91.

15. Z. Kunszt, P. Nason, G. Marchesini and B.R. Webber, in 'Z Physics at LEP 1', CERN 89-08, vol. 1, p. 373.

16. T. Sjöstrand, Computer Phys. Comm. 39 (1984) 347;
    M. Bengtsson and T. Sjöstrand, Computer Phys. Comm. 43 (1987) 367.

17. B.R. Webber, Nucl. Phys. B238 (1984) 492;
    G. Marchesini and B.R. Webber, Nucl. Phys. B310 (1988) 461;
    HERWIG Monte Carlo: G. Abbiendi, I.G. Knowles, G. Marchesini, M.H. Seymour, L. Stanco and B.R. Webber, Cambridge preprint Cavendish-HEP-90/26 (1990), to be published in Computer Phys. Comm.

18. U. Petterson, Lund preprint LU TP 88-5 (1988);
    L. Lönnblad and U. Petterson, Lund preprint LU TP 88-15 (1988);
    L. Lönnblad, Lund preprint LU TP 89-10 (1989).

19. ALEPH Collaboration, D. Decamp et al., Phys. Lett. 255B (1991) 623; CERN
    preprint PPE/90-196;
    DELPHI Collaboration, P. Abreu et al., Phys. Lett. 247B (1990) 167;
    L3 Collaboration, B. Adeva et al., Phys. Lett. 248B (1990) 464;
    OPAL Collaboration, M.Z. Akrawy et al., Phys. Lett. 235B (1990) 389; Phys. Lett.
    252B (1990) 159; Zeit. Phys. C47 (1990) 505; Zeit. Phys. C49 (1991) 375.

20. G. Kramer and B. Lampe, J. Math. Phys. 28 (1987) 945; Zeit. Phys. C34 (1987)
    497; Zeit. Phys. C39 (1988) 101; Zeit. Phys. C42 (1989) 504 (E); Fortschr. Phys.
    37 (1989) 161.

21. R.K. Ellis, D.A. Ross and A.E. Terrano, Nucl. Phys. B178 (1981) 421.

22. Mark II Collaboration, S. Komamiya et al., Phys. Rev. Lett. 64 (1990) 987.

23. TASSO Collaboration, W. Braunschweig et al., Phys. Lett. 214B (1988) 286.

24. AMY Collaboration, I. Park et al., Phys. Rev. Lett. 62 (1989) 1713.

25. S. Bethke, Zeit. Phys. C43 (1989) 331.

26. P.M. Stevenson, Phys. Rev. D 23 (1981) 2916; Nucl. Phys. B203 (1982) 472; Nucl.
    Phys. B231 (1984) 65.

27. G. Grunberg, Phys. Lett. 95B (1980) 70; Phys. Rev. D 29 (1984) 2315.

28. S.J. Brodsky, G.P. Lepage and P.B. Mackenzie, Phys. Rev. D 28 (1983) 228.

29. G. Kramer and B. Lampe, CERN preprint TH.5810-90.

30. S. Catani, G. Turnock, B.R. Webber and L. Trentadue, Phys. Lett. 263B (1991)
    491.

31. S. Catani, G. Turnock and B.R. Webber, CERN preprint TH.6231-91 to be pub-
    lished in Phys. Lett. B.

32. Yu.L. Dokshitzer, D.I. Dyakonov and S.I. Troyan, Phys. Rep. 58 (1980) 270.

33. A. Bassetto, M. Ciafaloni and G. Marchesini, Phys. Rep. 100 (1983) 202.

34. Yu.L. Dokshitzer, and D.I. Dyakonov, Phys. Lett. 84B (1979) 234;
    G. Parisi and R. Petronzio, Nucl. Phys. B154 (1979) 427;
    A. Bassetto, M. Ciafaloni and G. Marchesini, Nucl. Phys. B163 (1980) 477;
    G. Curci and M. Greco, Phys. Lett. 92B (1980) 175.

35. J. Kodaira and L. Trentadue, Phys. Lett. 112B (1982) 66;
    C.T.H. Davies, J. Stirling and B.R. Webber, Nucl. Phys. B256 (1985) 413;
    J.C. Collins, D.E. Soper and G. Sterman, Nucl. Phys. B250 (1985) 199;
    S. Catani, E. d'Emilio and L. Trentadue, Phys. Lett. 211B (1988) 335.

36. S. Catani and L. Trentadue, Phys. Lett. 217B (1989) 539; Nucl. Phys. B327 (1989)
    353; Nucl. Phys. B353 (1991) 183.

37. G. Sterman, Nucl. Phys. B281 (1987) 310.

38. J. Kodaira and L. Trentadue, Phys. Lett. 123B (1982) 335; preprint SLAC-PUB-2934 (1982).

39. D. Amati, A. Bassetto, M. Ciafaloni, G. Marchesini and G. Veneziano, Nucl. Phys. B173 (1980) 429.

40. S. Catani and M. Ciafaloni, Nucl. Phys. B236 (1984) 61; Nucl. Phys. B249 (1985) 301.

41. S. Catani, Yu.L. Dokshitzer, M. Olsson, G. Turnock and B.R. Webber, in preparation.

42. N. Brown and W.J. Stirling, Phys. Lett. 252B (1990) 657.

43. Yu.L. Dokshitzer, V.A. Khoze, A.H. Mueller and S.I. Troyan, *Basics of Perturbative QCD*, Editions Frontieres, 1991.

44. A.H. Mueller, Phys. Lett. 104B (1981) 161;
B.I. Ermolaev and V.S. Fadin, JETP Lett. 33 (1981) 285.

45. Yu.L. Dokshitzer, V.S. Fadin and V.A. Khoze, Zeit. Phys. C15 (1982) 325; Zeit. Phys. C18 (1983) 37:
Yu.L. Dokshitzer and S.I. Troyan, Leningrad report LNPI-922 (1984).

46. N. Brown and W.J. Stirling, preprint RAL-91-049.

CHROMODYNAMICS OF JETS TODAY AND THE DAY AFTER TOMORROW

V. A. Khoze

INFN Eloisatron Project and World Laboratory, Spb
and Leningrad Institute for Nuclear Physics, Gatchina, USSR

ABSTRACT

We review some selected topics in QCD physics at supercolliders. We concentrate on the colour-related effects and the prospects of using them for studying the new physical phenomena.

1. INTRODUCTION

This talk covers some selected topics in QCD physics of jets in the TeV range. We discuss some applications of the analytical perturbative approach (APA)[1-5] and attempt to demonstrate the present level of its credibility and maturity. We concentrate on the applications of the results of this approach to the physics at Supercolliders. Noteworthy to mention that only the standard, the so-called "back to the earth", problems are discussed here. The new theoretical ideas and their possible manifestations are covered in the talks given by V. Gribov[6] and V^2. Khoze[7].

2. JET PHYSICS AT SUPERHIGH ENERGIES AS SEEN THROUGH THE EYES OF LEP

LEP proves to be an exceedingly rich source of data for QCD tests[8-9] . In particular, the detailed experimental study of the inclusive distributions of charged particles, π^0 and K^0 has been performed. The first data on the inclusive K^* and Λ spectra appeared[10]. The new results from TRISTAN on the identified hadron spectra have also become available[11]. The data on particle distributions strongly support the ideas of the APA. In some sense one could consider LEP as the testing ground for the applications of the APA technology for physics at future supercolliders.

In ref. [12] a number of new results in the APA description of particle spectra in QCD jets has been obtained.

Remind the reader that when measuring the inclusive distributions in the process $e^+e^- \to$ hadrons one studies the properties of an individual quark jet. So the inclusive spectrum is the sum of two q-jet distributions

$$\frac{1}{\sigma}\frac{d\sigma}{dl} = 2\ \overline{D}_q^h\ (l,\ Y,\ \lambda) \tag{1}$$

where

$$l = \frac{1}{x}\ ,\quad Y = \ln\frac{E}{Q_0}\ ,$$

$$x = \frac{E_h}{E}\ ,\quad \lambda = \ln\frac{Q_0}{\Lambda}\ ,$$

here $2E = W$ is the total c.m.s. energy, Q_0 is the cutoff parameter in QCD cascades, which sets a formal boundary between the perturbative and nonperturbative phases of jet evolution.

When confronting with the experimental data the partonic distributions, calculated in the framework of the APA technique[1, 12], are multiplied by the overall normalization factors K^h. These factors connect the number of hadrons of the given species and partons

$$\overline{D}_q^h = K^h\ \overline{D}_q \tag{2}$$

Their values as well as the cutoff parameters Q_0 ($Q_0 \equiv Q_0\ (m_h,\ J_h^{PC})$) should be determined phenomenologically.

In the context of the APA the distributions of the "massless" hadrons ($\pi^\pm$, π^0) are described by the simplified version of the general analytical formula where $Q_0 = \Lambda$ (the so called limiting spectrum), see [12,13]. As it is shown experimentally[8, 9, 11] the limiting formulae work surprisingly well even for moderate energies.

According to the APA ideology to model the distributions of "massive" hadrons (K, P, Λ ...) the partonic spectra truncated at the different cutoff values Q_0 ($Q_0(m_h) > \Lambda$) can be used.

The effective parameter Λ_{eff} could be found from the fit to the π-spectra using limiting distribution. Once determined this quantity should remain fixed when fitting the data for different particle species using the truncated parton distributions.

Up to now there were no reasons to update the value

$$\Lambda_{eff} = 150\ \text{MeV}$$

found from the first MLLA analysis[15] of the PEP/PETRA data.

When the limiting distribution is applied for comparison with the data on all charged particle production the effective value of Λ appears to be much larger than Λ_{eff} (see Refs. [8, 9, 13]). $\Lambda_{ch} \simeq 1.6\Lambda_{eff} \simeq 250\ \text{MeV}$.

In Refs. [14, 15] the expressions for the truncated parton distributions have been presented in terms of the contour integral of the confluent hypergeometric functions. However the obtained formulae are not well suited for the numerical calculations of the truncated spectra.

Luckily, in the most interesting region, not too far from the maximum of the spectrum,

$$|\delta| \leq 1 \quad , \quad \delta = \frac{1 - <l>}{\sigma} \quad , \quad \sigma^2 = <l^2> - <l>^2 \tag{3}$$

one can follow the route proposed in Ref. [16] for the limiting case (see for details Ref. [12]). The idea is to encode the effects of the modified leading logarithmic approximation (MLLA)[14, 17] in terms of a few analytically calculated shape parameters by means of a distorted Gaussian

$$\bar{D}\,(l, Y, \lambda) = \frac{N(Y, \lambda)}{\sigma\sqrt{2\pi}}\ \exp\left[\frac{1}{8}k - \frac{1}{2}s\delta - \frac{1}{4}(2 + k)\delta^2 + \frac{1}{6}s\delta^3 + \frac{1}{24}k\delta^4\right] \tag{4}$$

with s and k the skewness and the kurtosis of the distribution defined as:

$$s \equiv \frac{<(l - <l>)^3>}{\sigma^3} = k_3 \ , \tag{5}$$

$$k \equiv \frac{<(l - <l>)^4>}{\sigma^4} - 3 = k_4 \ .$$

The multiplicity $N(Y, \lambda)$ can be presented in a compact form in terms of the modified Bessel functions $I_\nu(z)$ and Mackdonald functions $K_\nu(z)$, see Refs. [14, 15].

$$N(Y, \lambda) = z_1 \left(\frac{z_2}{z_1}\right)^B [I_{B+1}(z_1)\,K_B(z_2) + K_{B+1}(z_1)\,I_B(z_2)] \tag{6}$$

Here

$$z_1 = \left(\frac{16N_c}{b}(Y + \lambda)\right)^{\frac{1}{2}} \ , \quad z_2 = \left(\frac{16N_c}{b}\lambda\right)^{\frac{1}{2}} \ . \tag{7}$$

In Ref. [12] the analytical procedure is described in detail for calculation of the first four momenta

$$l_k \equiv <l^k>, \quad k = 1....4 \tag{8}$$

of the truncated spectra. These momenta allow one to construct dispersion σ, skewness and kurtosis of the distribution (4).

For illustration we present below the explicit expression for the simplest case of $l_1 = <l>$. Neglecting the terms $\sim 0\,(1/N^2)$ one arrives at

$$l_1 = <l> = (Y + \lambda)\left[\frac{1}{2} + \frac{B}{z_1}\frac{I_{B+2}(z_1)}{I_{B+1}(z_1)}\right] - \lambda\left[\frac{1}{2} + \frac{B+1}{z_2}\frac{K_{B-1}(z_2)}{K_B(z_2)}\right] \tag{9}$$

Eq. (9) shows an influence of finite Q_0 ($\lambda \neq 0$) on $<l>$. Stiffening of parton distribution originating from truncated cascades is predicted to be energy independent. This fact can be used to measure effective Q_0 values by comparative study of the energy evolution of the peak position of massive hadron spectra.

The remarkable fact is that the results of the straightforward numerical calculations of the truncated parton distributions which solve the MLLA evolution equations appear to be reproduced well enough by the approximate formula (4) with the analytically calculated shape parameters, see Ref. [12].

Remind the reader that due to coherent suppression of soft gluon multiplication (angular ordering) the parton spectra exhibit the famous "hump-backed" structure (see Refs. [1-3]).

In the limiting case one gets the simply testable prediction revealing the net subleading MLLA effects in parton cascades[13]

$$l_{max} = Y\left[\frac{1}{2} + \sqrt{\frac{C}{Y}} - \frac{C}{Y} + O(Y^{-\frac{3}{2}})\right] \tag{10}$$

with

$$C = \frac{a^2}{16N_c b} = 0.2915\ (0.3513) \qquad\qquad \text{for } n_f = 3(5) \tag{11}$$

$$a = \frac{11}{3}N_c + \frac{2n_f}{3N_c^2}\ , \quad b = \frac{11}{3}N_c - \frac{2}{3}n_f\ ,$$

$$l_{max} - <l> \approx \frac{1}{2}\frac{3a}{16Nc} = 0.351\ (0.355) \qquad\qquad \text{for } n_f = 3(5) \tag{12}$$

The LEP results wonderfully confirm QCD cascading picture of multiple hadroproduction. Especially spectacular is the energy evolution of the peak position, shown in Fig. 1. To combine the data on π^0's with the results for all charged particle production we replotted the latter on an effective energy scale $(\sqrt{s})_{ch.part} = W(\Lambda_{eff}/\Lambda_{ch})$. Fig. 1 nicely confirms the universality of the energy dependence of the peak position.

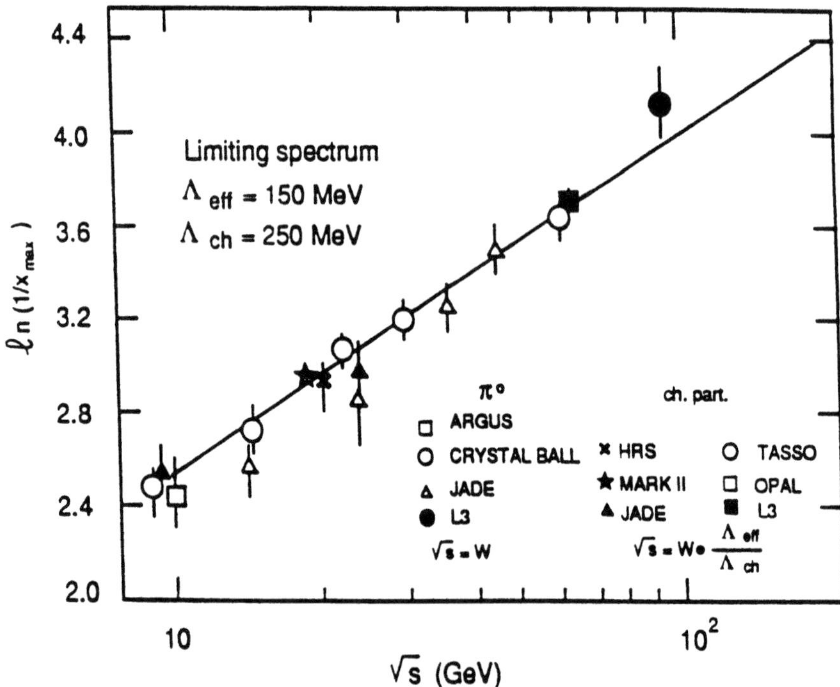

Fig. 1. Energy evolution of the peak position compared to the MLLA limiting result. The data on π^0 production are presented as a function of $\sqrt{s} = W = 2E$ at $\Lambda_{eff} = 150$ MeV. The data on all charged particles are replotted as a function of $\sqrt{s} = W(\Lambda_{eff}/\Lambda_{ch})$.

Turning to $Q_0 > \Lambda$ case the main effect one observes[12] is the constant shift of the peak due to Eq. (9), see Fig. 2. Let us emphasize that in a rather broad energy region the slopes of the curves corresponding to different values of Q_0 are the same. It would be interesting to check whether such a universality holds true for distributions of π, K, p, Λ etc.

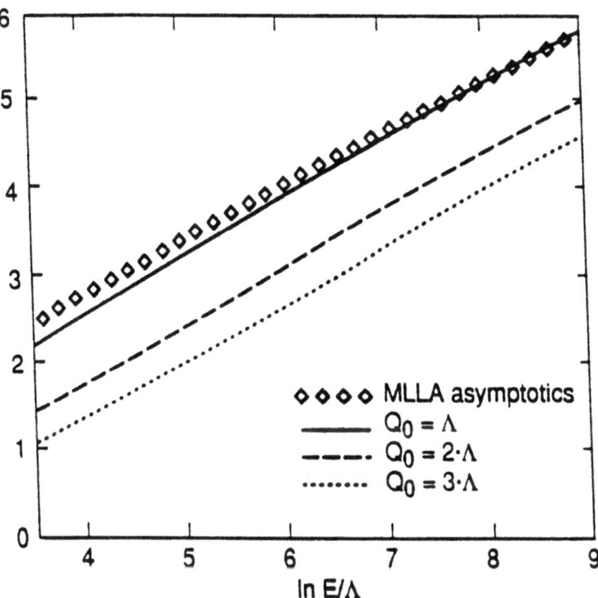

Fig. 2.   Energy evolution of the peak position for different values of Q_0[12]. The limiting case (solid line) is compared to the MLLA asymptotical formula (10).

To demonstrate how the APA picture works in the case of massive particles we presented in Fig. 3 the new OPAL data[8] on the K^0 production together with the spectrum of partons from the truncated MLLA cascades and its distorted Gaussian approximation. Fit to these data gives the values of the only free parameters $Q_0\,(m_K) = 300$ MeV, $K^{K^0} = 0.38$.

One may expect even higher credibility of the MLLA results at larger jet energies E_{jet} where the subasymptotic corrections are less important.

Fig. 4 illustrates the APA predictions of the π and K distributions in a quark jet at $E_{jet} = 1$ TeV together with the results at the Z^0 [1]. In this figure one clearly sees the rise of the maximum with the jet energy and an increase of the hump height, reflecting the rise of multiplicity.

The dynamics of the multiplicity increase is shown in Fig. 5 [1] $(N^h_{e^+e^-} = 2N^h_q)$

Finally let us remind that to explore the coherent origin of the hump-backed particle spectrum and in an attempt to study the depletion in its soft part for jets produced in hadronic collisions, it proves to be important to look at particles restricted to lie within a particular opening angle with respect to the jet[18]. For example, one might consider the energy distribution of particles accompanying the production of an energetic particle and lying within an opening angle θ about the direction of the trigger particle momentum.

---

[1] I am indebted to Yu. L. Dokshitzer for supplying me with these curves.

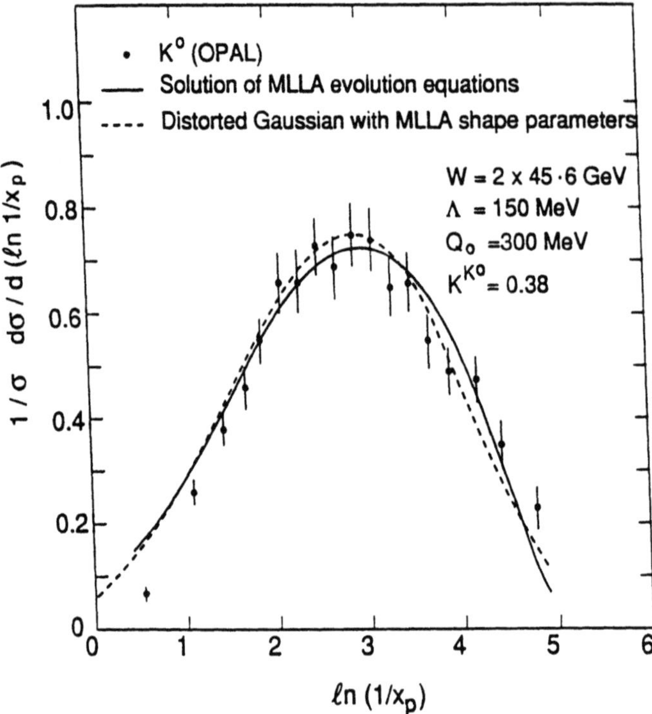

Fig. 3. ln(1/x_p) distribution of K^o as compared to the numerical results of solution of MLLA Evolution Equation for truncated cascades (solid line) and to the distorted Gaussian formula (4) with the MLLA shape parameters (dashed curve). Phase space effects are not taken into account.

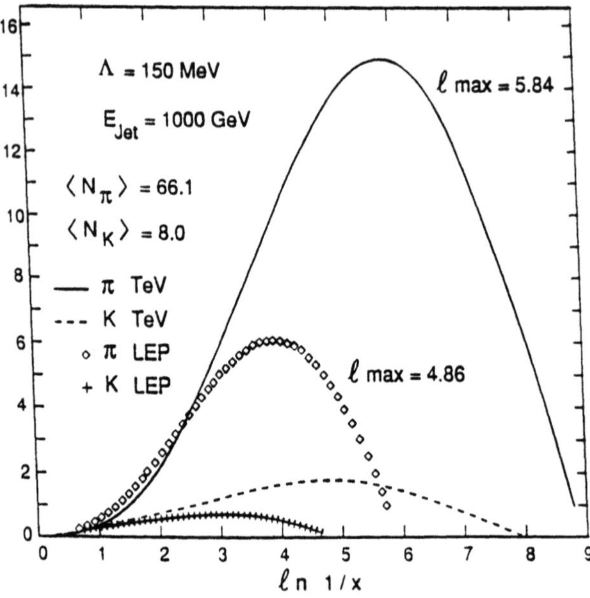

Fig. 4. Energy evolution of the spectra of π and K.

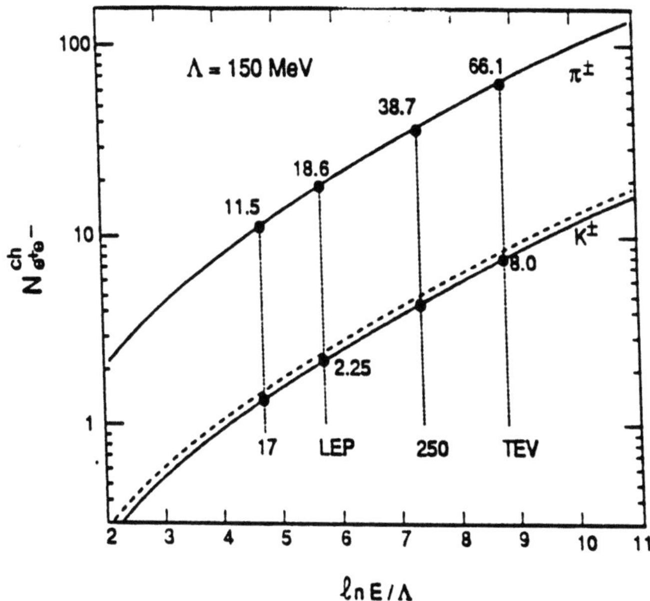

Fig. 5.   The energy rise of particle multiplicities in e^+e^- collisions.

Parton cascades in these situations will populate mainly the region

$$\frac{m_h}{\sin\frac{\theta}{2}} < E_h < E$$

The maximum of the distribution, in E_h, is now forced to larger energies. The first results of the ALEPH collaboration at the Z^0 look very promising[19]. The angular cut θ, is expecially useful for jets produced in hadronic collisions, since one is able to eliminate much of the soft background.

3. SPECIFIC PROPERTIES OF HEAVY QUARK JETS

Hadron jets accompanying heavy quarks $Q = c, b, ...$ differ from u-, d-, s-, and gluon-initiated jets because of the suppression of bremsstrahlung off massive "charge" in the forward direction. This phenomenon, very well known from QCD, leads to the depopulation of the angular region $\theta < \theta_0 = m_Q/E_Q$ - the Dead Cone. This has a number of consequences described in [13]. Here we illustrate some of them.

Leading particle effect. The inclusive spectrum of Q is commonly believed to be an Infrared Stable QCD prediction, since the larger quark mass $m_Q \gg \Lambda$ seems to cut off collinear divergences in PT gluon radiation. This conclusion stems from the standard LLA logic when one looks for contributions

$$\frac{\alpha_s}{\pi} \ln \frac{E}{\Lambda} \sim 1 \quad , \quad \frac{\alpha_s}{\pi} \ln \frac{m_Q}{\Lambda} \sim 1,$$

to $\overline{D}(x; E, m_Q)$ at $x \sim 1$ and systematically neglects corrections of the order of α_s. The LLA result for the valence Q distribution reads

$$\overline{D}_{val}(x, E, m_Q) = \int\limits_{(\gamma)} \frac{dj}{2\pi i} \, x^{-j} \, \exp[\Phi_F^F(j) \, \Delta\xi] , \tag{13a}$$

$$\Phi_F^F(j) = \int\limits_0^1 dz[(1 - z)^{j-1} - 1] \, \Phi(z) \quad , \quad \Phi(z) \equiv 2C_F \frac{1 + (1 - z)^2}{z} ; \tag{13b}$$

$$\Delta\xi = \xi(W^2) - \xi(m_Q^2) \equiv \frac{1}{b} \ln\left(\frac{\ln(W^2/\Lambda^2)}{\ln(m_Q^2/\Lambda^2)}\right) , \tag{13c}$$

where $W = 2E$ and the contour (γ) in (13a) lies to right of all singularities: $\mathrm{Re}j > 0$. As one can see from the approximate expression

$$\overline{D}_Q^Q(x; W, m_Q) \propto (1 - x)^{-1 + 4C_F\Delta\xi} \frac{1}{\Gamma(4C_F\Delta\xi)} , \tag{14}$$

the spectrum for small values of the PT-parameter $\Delta\xi$ (which is really the case) peaks near $x = 1$. This undermines the very LLA approach since typical values of scaled momenta $<z>$, carried by bremsstrahlung gluons, appear to be small and a new large parameter $\alpha_s \ln<z> \sim 1$ enters the game[13]. An improved approximation which fully takes into account these soft Sudakov-type effects has to be applied[13]. Roughly speaking, this makes the $\Delta\xi$ parameter in Eq. (13a) z-dependent:

$$\Phi_F^F(j) \, \Delta\xi \Rightarrow \int\limits_0^1 dz\left[(1 - z)^{j-1} - 1\right] \Phi(z)\{(zW^2) - \xi(z^2 m_Q^2)\} \tag{15}$$

As a result, the shape of the spectrum changes drastically at large x because of the Sudakov suppression and $D_Q^Q(x, E, m_Q)$ becomes infrared sensitive at values of x parametrically close to one: $(1 - x) < \mu/m_Q$ where $\mu \geq \Lambda$ is the characteristic hadronization scale.

Fig. 6 shows the inclusive spectra of b-quarks at $E = 100$ and 1000 GeV. Comparison of the LLA and MLLA spectra demonstrates that the improved approach leads to noticeably softer Q-spectra with maxima in the APA- controlled momenta region.

Spectra of light hadrons from heavy quark jets are predicted to be depopulated in the region of hard momenta compared to light -q jets. This is a consequence of the Dead Cone phenomenon[13, 20]. Notice that the particle yield in the soft momentum region remains unaffected.

4. STUDIES OF THE FINAL STATES CONTAINING RAPIDITY GAPS

At high energies an intimate connection is expected between the hadronic event structure and the underlying production mechanism (see Refs. [1-3, 20]). In some sense the portrait of the event could be considered as a real partonometer making traceable the basic interaction processes. In particular the detailed features of the parton system, such as the flow of colour quantum numbers, influence significantly the distribution of colour-singlet hadrons in the final state.

Due to this a very important physical phenomenon could manifest itself, namely, the fact that the multihadron production in hard processes is in accordance with the so called radiophysics of colour flows[2, 3, 18].

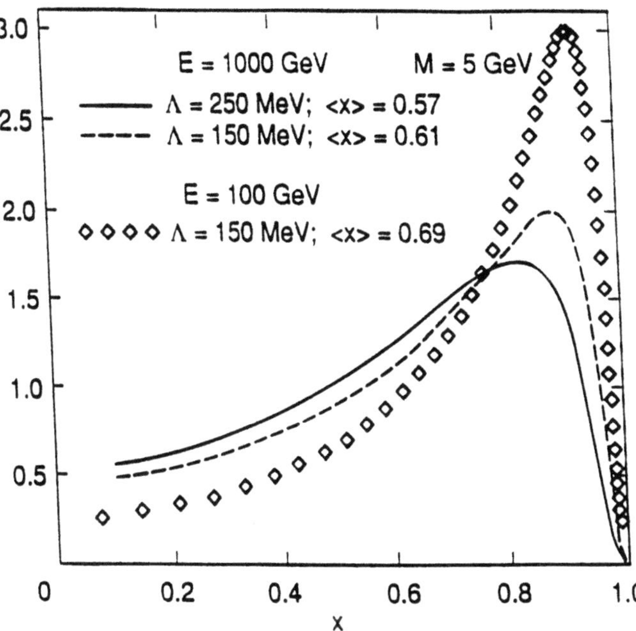

Fig. 6.   Inclusive x distribution of b-quark at E = 100, 1000 GeV.

The first and (still best) example of the colour-related phenomena is the so-called string[21] (or drag[22]) effect in the $q\bar{q}g$ events of e^+e^- annihilation.

Originally this effect was predicted as a result of a Lorentz boost exerted by a gluon on the nonperturbative string stretched between quarks[21]. In the perturbative scenario it is a result of interference among the gluon waves radiated from the $q\bar{q}g$ emitter. As it was demonstrated in Ref. [22] the treatment of the structure of final states given by string picture and by the QCD radiophysics are basically equivalent up to some subtle effects which we shall not discuss here.

The e^+e^- data on $q\bar{q}g$ events (e.g. Refs. [8, 9]) strongly support the predicted drag of the interjet particles in the direction of the gluon jet. They demonstrated that the wide angle particles do not belong to any particular jet (parton) but their emission properties depend upon the colour topology of the overall jet ensemble. These observations have beautifully demonstrated the connection between colour and hadronic flows.

The rich diversity of the interference drag effects has been discussed for hadronic collisions, see e.g. Refs. [3, 18, 20, 23]. One of the simplest examples is prompt γ (or W, or Z) production at large $P_\perp$[3, 18, 23, 24]. Here the accompanying particle distribution reveals some peculiar features. Thus the particle production proves to be largest between the directions for the incoming gluon and the outgoing quark, but approximately 3 times smaller between the directions of the incoming quark and outgoing quark.

The colour-related effects could become a phenomenon of large potential value as a new additional tool for discriminating between hard processes. Because of them some spectacular signatures arise in the structure of the final hadronic state. For example if in the t-channel of the underlying partonic process colour is not transferred, the emission of accompanying particles at large angles should be strongly suppressed (rapidity gap signature[20, 25]).

Recently the new Full-Acceptance Detector (FAD) was proposed[25] for the studies of the structure of individual hadronic events in phase space at the SSC. It looks essentially as two full-acceptance 20 TeV fixed target spectrometers face-to-face. Such a device should cover all of the phase space (pseudorapidities up to ±12) reasonably uniformly.

One of the main targets of this detector is the physics of event structures containing "rapidity gap". Special attention is paid here to the physics of diffractive processes (pomeron physics) which definitely suffered from the lack of attention in the programs of supercolliders.

Remind the reader that in the Reggeon field theory diffraction dissociation of a hadron proceeds by the t-channel exchange of the colourless vector object-Pomeron(P). Therefore the rapidity gap signature could be well applied for studies of a vast area of strong interaction phenomena (some of which are at the moment out of theoretical control).

Fig. 7 borrowed from Ref. [25] illustrates the event topologies for various diffraction dissociation processes. The coordinates of a particle are described by the familiar lego variables: azimuthal angle Φ and (pseudo-) rapidity η. The minimum width for a significant rapidity gap is about 2 to 3 units, in order to reduce Poisson fluctuations in multiplicity.

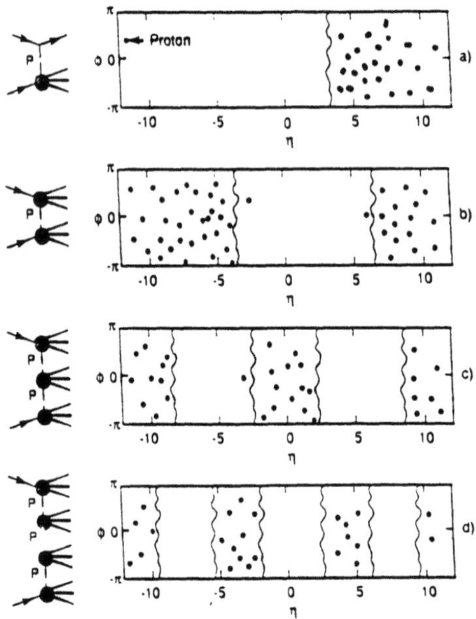

Fig. 7. Event topologies for various diffraction dissociation processes: (a) single diffraction, (b) double diffraction, (c) Pomeron-Pomeron absorption (triple diffraction), (d) diffraction dissociation of Pomerons via Pomeron exchange (quadruple diffraction).

There is enough space at the SSC for 3 rapidity gaps in a single event, taking into account that a region between gaps must also be at least 2 to 3 units wide. So one can study not only pp collisions, but also diffraction excitation of Pomerons via P-exchange. The corresponding cross sections are expected to be quite large.

One of the most important goals of supercolliders is the Higgs physics. We should briefly discuss here the possibility to discriminate two competing main Higgs production mechanisms (see Fig. 8) using a rapidity gap signature. This example was first discussed in[20], see also[3] and (independently) first included in a full-fledged event simulator framework in the PYTHIA MC program[26].

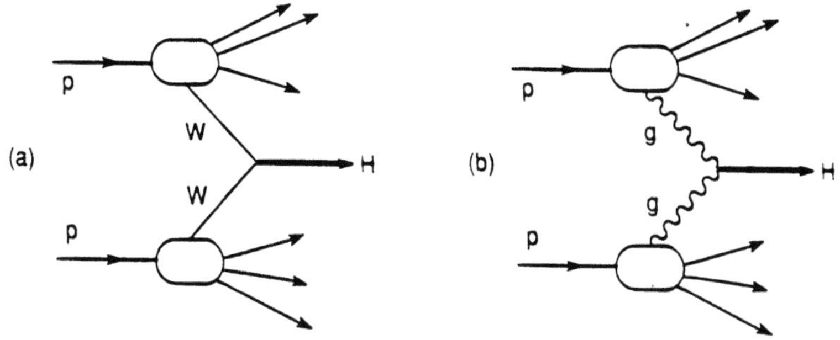

Fig. 8.   Hadronic Higgs boson production via a) WW fusion and b) gg fusion.

With the help of PYTHIA one can try to give a realistic appraisal of the size of expected difference between $WW \to H$ and $gg \to H$ (see Ref. [27] for details).

Let us assume that the heavy Higgs is observed via the gold-plated channel $H \to ZZ \to 4$ leptons (to avoid contamination of the rapidity gap from the Higgs decay products).

In Ref. [27] events have been generated with PYTHIA for 200 and 500 GeV Higgs masses at 16, 40 and 100 TeV pp collisions. Each run consisted of 10^4 $WW \to H$ events and equally many $gg \to H$ ones, with curves normalized per event of the given type. The EHLQ set 1 structure functions[28] have been used. Although the absolute cross-sections, and even the relative $WW \to H / gg \to H$ event composition, would be somewhat different with another choice of structure functions, the properties of the events themselves are but little affected.

Fig. 9 demonstrates the result in terms of the charged rapidity distribution at $m_H = 500$ GeV and $\sqrt{s} = 100$ TeV. At central rapidities the expected difference between the two scenarios is clearly seen. As is shown in [27] the size of the dip in the $WW \to H$ channel is increased for heavier Higgs masses. Further, the rapidity distance between the two peaks is increased when the CM energy, $\sqrt{s}$, is increased.

However, we must confess that the scenario described above looks too optimistic. A number of complications could arise in a realistic experimental setup which, without special care, could mask the effect. These difficulties and the possible ways to overcome them are analysed in [27].

Another interesting example is the production of a single top by the Wg fusion mechanism as shown in Fig.10.

The accompanying particle distribution here should consist of two parts: the standard plateau generated by the fragmentation of b-quark and the gluon exchange plus quasi-diffractive contribution corresponding to the W exchange. The contribution of the lower part of the diagram of Fig. 10 reminds the structure of the target fragmentation region in DIS. There are three sources of accompanying radiation: fragmentation of the b-quark, fragmentation of the gluon rungs and the t-channel emission due to colour exchange. With decrease of x_B[2] the rapidity distribution should become more dense since the role of colour octet t-channel exchange increases (see Refs. [20, 18]).

---

2) Here x_B means Bjorken variable. In the next section we denote it as x.

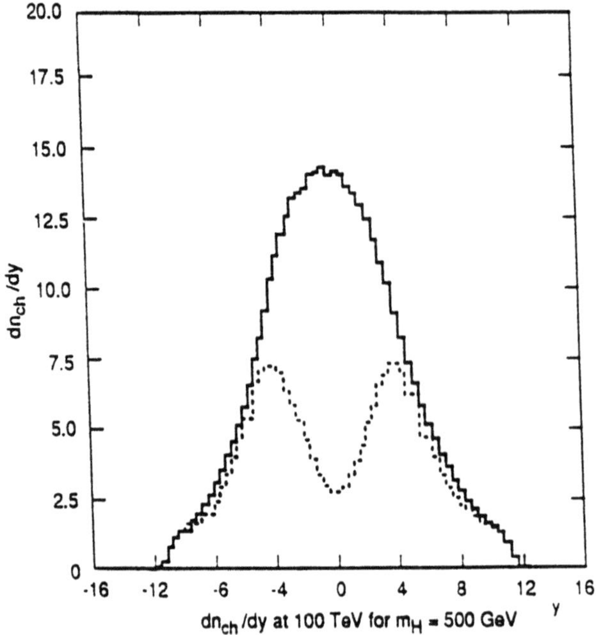

Fig. 9. Charged particle rapidity distribution for two mechanisms of Higgs production in pp collisions at 100 TeV for m_H = 500 GeV. The full line corresponds to the gg fusion, the dashed one to the WW fusion.

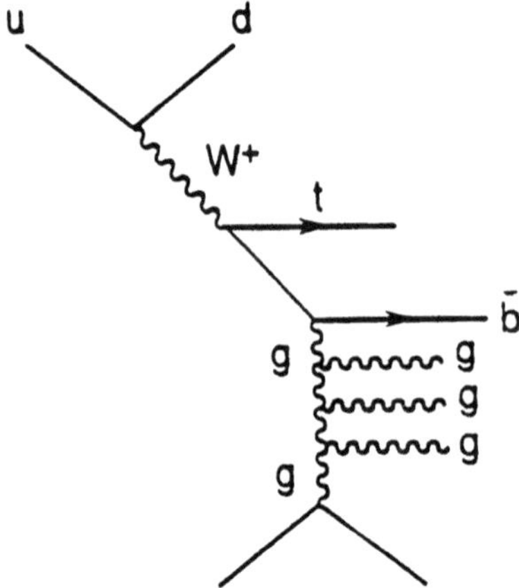

Fig. 10. Single top production via Wg fusion mechanism.

5. ON THE STRUCTURE OF THE FINAL STATES IN DIS

We shall briefly discuss here the manifestations of colour coherence in lepton-hadron deep inelastic scattering (DIS), see Ref. [29] for details. It should from the onset be mentioned that we concentrate mainly on the qualitative features of the final states. One can expect that the detailed study of their structure in a realistic experimental setup will be performed in the framework of the approach developed in Ref. [4, 5].

When discussing the structure of the final state of a DIS process with momentum transfer $-Q^2 \gg \Lambda^2$ and a fixed value of Bjorken variable x, one must take two phenomena into account: the dissociation of the initial parton fluctuation, the coherence of which was destroyed by the "removal" of a virtual quark (target fragmentation), and the evolution of the struck quark (current fragmentation). As well known, these two fragmentation regions are best separated kinematically in the Breit frame ($Q_0 = 0$, $2x\vec{p} = -\vec{Q}$). Here the process looks like the abrupt spatial spreading of two colour states: 3 (the struck quark q) and $\bar{3}$ (the disturbed proton), moving in the opposite directions.

Not the least role in the justification of the celebrated Feynman hypothesis about a universal hadronic plateau in DIS (see Fig.11a) has been assigned to the argument that the continuously distributed hadrons connecting the two fragmentation regions are necessary to compensate the fractional charges of quarks-partons. At the same time, some serious doubts have been expressed about the possibility to organize dynamically such a state if one proceeds from the natural idea to consider successive decays of outgoing partons as an only source of multiple production of final particles. The problem has been formulated most clearly by Gribov in 1973[30]. It was shown that in a DIS process the offsprings of the fragmentation of an initially prepared fluctuation had to leave the rapidity interval $0 < \ln\omega < \ln Q$ depopulated as shown in Fig.11b (ω being the energy of registered hadron h).

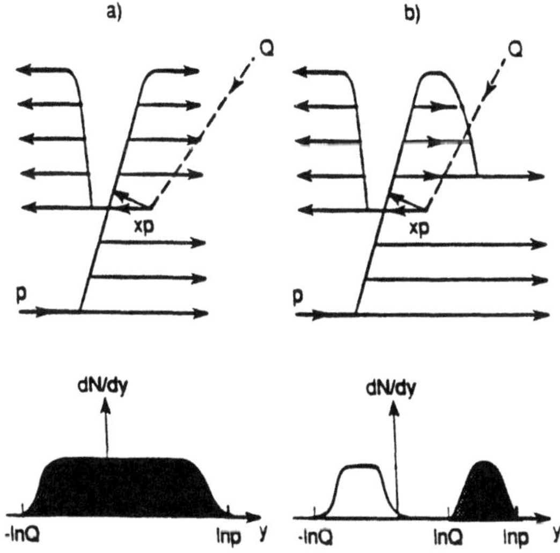

Fig. 11. Structure of the hadron plateau in DIS process according to Feynman (a)
and Gribov (b) (shaded area shows the target fragmentation region),
$y = \ln\omega$.

At the basis of this conclusion an analysis lays of the space-time picture of the process development in the framework of a field-theoretical approach to the description of the wave function of the target hadron as a coherent system of partons. The absence of hadrons with momenta $\omega \ll Q$ in the target fragmentation in the Gribov picture can be simply explained by noticing that the coherence of the partonic wave function in this region remains, in fact, not destroyed by hard knocking out a parton with momentum $xp \approx Q/2$, as a consequence of which the upper (low momenta) part of the partonic fluctuation in Fig.11b "collapses" as if there were no scattering at all.

The experimental observation of a uniform flat plateau (Fig. 11a), which was interpreted as a proof of the correctness of the identification of partons with quarks, did not eliminate the Feynman-Gribov puzzle[3]), but only sharpened it in fact, since it required one to point out a clear physical mechanism responsible for the filling of the "Gribov's gap", that would not come into conflict with quantum mechanics.

From the present-day standpoints the foundations for the Gribov phenomenon have only been strengthened, since QCD indeed corroborated the treatment of the structure of a relativistic hadron in terms of field fluctuations composed of quasi-real quarks and gluons - the partons of QCD.

The way in which this puzzle is resolved in the QCD context[29] appears to be eclectic in a certain sense. On the one hand, here the Gribov phenomenon does occur indeed: the energy spectrum of the offsprings of the parton dissociation − elements of the "ladder" fluctuations determining the DIS cross-section − proves to be concentrated mainly in the region of large energies $Q < \omega < P$. On the other hand, there is a specific mechanism responsible for bridging the regions of target and current fragmentation in the spirit of the Feynman picture. This role is taken on by coherent bremsstrahlung of soft gluons which is insensitive to details of the structure of partonic wave function of the target hadron, being governed exclusively by the value of the total colour charge transferred in the scattering process.

Because of the fact that at small x's the "sea" contribution dominates, the scattering amplitude corresponding to the gluon colour transfer in the t-channel makes the accompanying radiation in the target fragmentation region approximately twice as intensive as that in the fragmentation of current.

In the target fragmentation there arises the hump-backed particle distribution evolving with $\ln Q^2$ and $\ln(1/x)$. The resulting spectrum can be represented (see Ref. [29]) as a sum of three terms: the contribution related to the upper quark cell in Fig. 12 (I), the contribution accounting for effective emissions off the vertical gluon lines (II), the contribution combining the relatively soft gluons off the lower part of the diagram in Fig. 12 with fragments of ladder rungs (III). Fig. 13 shows the structure of these terms for certain specific values of $\ln(1/x)$ and $\ln(Q/\Lambda)$. The contributions II and III, which are negligibly small for $x \sim 1$, increase with decrease of x.

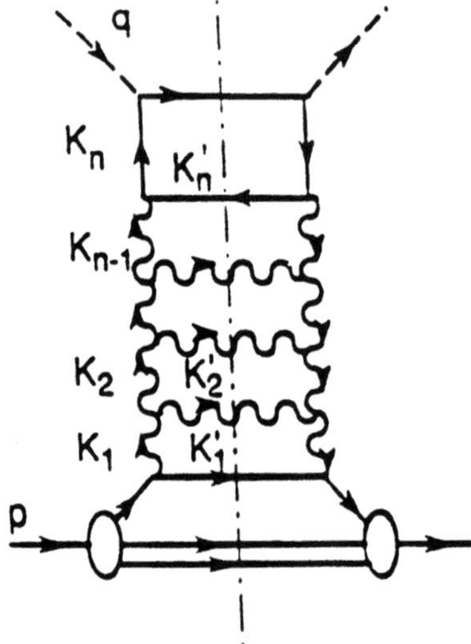

Fig. 12.   Many-rung parton ladder determining the DIS structure functions.

---

[3]) we are not sure that Feynman himself ever heard of such a term

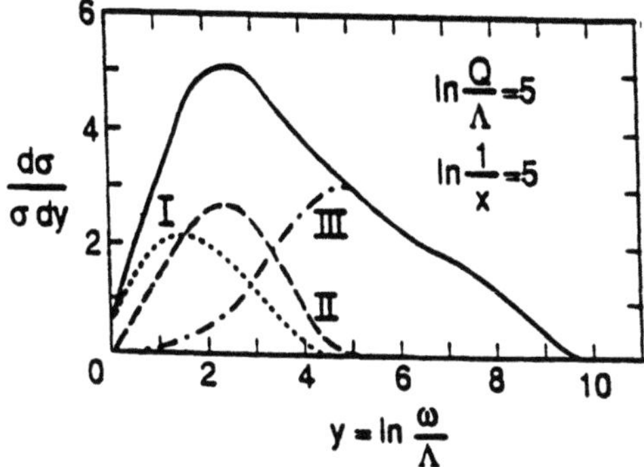

Fig. 13. Contributions to the energy spectrum of particles from the region of fragmentation of the target for $\ln(Q/\Lambda) = 5$ and $\ln(1/x) = 5$. The dotted line is the quark contribution I, the dashed line is the coherent t-channel emission II, the dashed-dotted line corresponds to the fragmentation of the structure partons III, and the solid line is the total spectrum.

As in e^+e^- annihilation, coherence in DIS leads to hardening of the energy spectra: the yield of particles with finite (in the Breit) frame, energies ($\ln(\omega/\Lambda) \sim 1$), remains practically constant with increase of the hardness of the process.

6. CONCLUSION

This review continues a series of talks given recently by the members[4] of the international QCD community (Columbia University - Leningrad - Italy - Cambridge $\equiv$ CLIC). One of the main aims of the present talk is to illustrate the recent developments in the APA and to emphasize the prospects of using the colour-related phenomena for studying physics at supercolliders.

7. ACKNOWLEDGEMENTS

It is a pleasure for me to thank my friends Luisa Cifarelli and Yuri Dokshitzer for having arranged a wonderful week of enjoyable physics and sun.

I am very much indebted to Yu. Dokshitzer, T. Sjöstrand and S. Troyan for their pleasant collaboration.

I am very much obliged to Despina Hatzifotiadou for helping me prepare the talk.

REFERENCES

1. For reviews see A. Bassetto, M. Ciafaloni and G. Marchesini, Phys. Rep. C100 (1983) 201;
    Yu. L. Dokshitzer, V. A. Khoze, A. H. Mueller and S. I. Troyan, Rev. Mod. Phys. 60 (1988) 373;

---

[4] The names can be found among the authors of Ref. [1-5] and in the references therein.

Yu. L. Dokshitzer, V. A. Khoze and S. I. Troyan, in Perturbative QCD, ed. A. H. Mueller (World Scientific, Singapore, 1989), p. 241.

2. Yu. L Dokshitzer, Talks given at the International Schools of Subnuclear Physics, Erice, 1989, 1990, 1991.

3. V. A. Khoze, Talk given at the 12th Workshop: "New Technologies for Supercolliders" Erice, Sept. 1990. UND-HEP-90-BIG-07

4. G. Marchesini, Talk at this Workshop.

5. B. Webber, Talk at this Workshop.

6. V. N. Gribov, Talk at this Workshop.

7. V. V. Khoze, Talk at this Workshop.

8. S. Bethke, Talk at this Workshop and OPAL Collab. G. Alexander et al., Phys. Lett. B264(1991)467.

9. T. Hebbeker, Plenary talk at the Int. Lepton-Photon Symposium and Europhysics Conference on High Energy Physics, Geneva, July-August 1991.

10. DELPHI collab. contribution to the Int. Lepton-photon symposium and Europhysics Conference on High Energy Physics, Geneva, July-August 1991.

11. R. Itoh, Talk given at the Int. Lepton-Photon Symposium and Europhysics Conference on High Energy Physics, Geneva, July-August 1991.

12. Yu. L. Dokshitzer, V. A. Khoze and S. I. Troyan, LU-TP91-12, 1991, to be published in Int. Journ. Mod. Phys. A.

13. Yu. L. Dokshitzer, V. A. Khoze and S. I. Troyan, Contribution to the Workshop on Jet Studies at LEP and HERA, Dec.10-15, 1990, Durham, England. DTP-91/04.

14. Yu. L. Dokshitzer, and S. I. Troyan. Proceedings of the XIX Winter School of the LNPI, volume 1, page 144. Leningrad, 1984; Yu. L. Dokshitzer, and S. I. Troyan. preprint LNPI-922, 1984.

15. Ya. I. Azimov, Yu. L. Dokshitzer, V. A. Khoze and S. I. Troyan. Z. Phys., C27:65, 1985; Z. Phys., C31:213, 1986.

16. C. P. Fong and B. R. Webber, Phys. Lett. B229: 289, 1989; Nucl. Phys. B355: 54,1991.

17. A. H. Mueller. Nucl. Phys., B213:85, Erratum quoted ibid., B241(1984)141.

18. Yu. L. Dokshitzer, V. A. Khoze A. H. MuelLer and S. I. Troyan, in Ref.1.

19. G. Cowan, Talk at the QCD Workshop, Lund, May 1991.

20. Yu. L. Dokshitzer, V. A. Khoze and I. Troyan, in Proceedings of the 6th Int. Conference on Physics in Collisions 1986, ed. M. Derrick (World Scientific, Singapore, 1987) p. 365.

21. B. Andersson, G. Gustafson, and T. Sjöstrand, Phys. Lett. 94B (1980) 211;

22. Ya. I. Azimov et al., Phys. Lett. B165 (1985)147; Yad. Fiz. 43 (1986) 149.

23. R. K. Ellis, G. Marchesini, and B. R. Webber, Nucl. Phys. B286 (1987) 643.

24. B. Andersson, H. U. Bengtsson and G. Gustafson, LU-TP-82-10.

25. J. D. Bjorken, SLAC-PUB-5545, 1991.

26. H-U Bengtsson and T. Sjöstrand, Computer Physics Commun. 46 (1987) 43.

27. Yu. L. Dokshitzer, V. A. Khoze and T. Sjöstrand, CERN TH. 6269/91.

28. E. Eichten, I. Hinchliffe, K. Lane and C. Quigg, Rev. Mod. Phys. 56 (1984) 579; 58 (1986) 1065.

29. L. V. Gribov, Yu. L. Dokshitzer, S. I. Troyan, and V. A. Khoze. Sov. Phys. JETP, 68:1303, 1988.

30. V. N Gribov. Proceedings of the VIII Winter School of the LNPI, volume 2, page 5. Leninggrad, 1973.

HIGH ENERGY FACTORIZATION AND HEAVY FLAVOUR PRODUCTION

Marcello Ciafaloni

Dipartimento di Fisica, Università di Firenze and I.N.F.N., Sezione di Firenze
L.go E. Fermi, 2
Firenze 50125, Italy

ABSTRACT

A form of high-energy factorization has been recently proposed in order to find the small x behaviour of heavy mass production in QCD. This factorization in $k_\perp$-dependent, and applies to various kinds of single-$k_\perp$ and double-$k_\perp$ processes. Here I briefly review the method and its applications to heavy flavour production, where large K-factors are found and computed.

1. INTRODUCTION

Production of heavy flavour (as b,t) and in general of heavy objects (as W and Higgs bosons) is of utmost interest for the new generation of colliders. Such processes occur at rather large values of $k_\perp = 0(M)$ because of the large mass M involved, and are proportional to some power of $\alpha_S(M^2) \ll 1$. Hence, they can be viewed as hard probes of the partons in the hadron and are treated perturbatively in QCD.

To be definite let me consider a single -$k_\perp$ process (Fig.1a) and the M^2-dependence of the cross section moments σ_N in the energy variable $\rho \equiv 4M^2/S$. According to the renormalization group factorization theorem we can write

$$4 M^2 \sigma_N \left(M^2, Q_o^2 \right) = C_N \left(\alpha_S(M^2) \right) \exp\left[\int_{\log \frac{Q_o^2}{\Lambda^2}}^{\log \frac{M^2}{\Lambda^2}} dt \; \gamma_N \left(\alpha_S(t) \right) \right] \quad ,$$

$$(1)$$

where γ_N is the anomalous dimension matrix in the structure function and C_N is the coefficient function, usually computed in fixed order perturbation theory.

However, with increasing energy, the heavy mass hard process enters a two-scale regime $S \gg M^2 \gg \Lambda^2$, where large factors of $\log \rho$ due to gluon exchange affect the QCD perturbative coefficients. In the ρ–moment space of Eq.(1) this means that the QCD expansion parameter is not α_S but α_S/N for both γ_N and C_N. For istance, for lowest order heavy flavour photoproduction, one has[1]

QCD at 200 TeV, Edited by L. Cifarelli
and Y. Dokshitzer, Plenum Press, New York, 1992

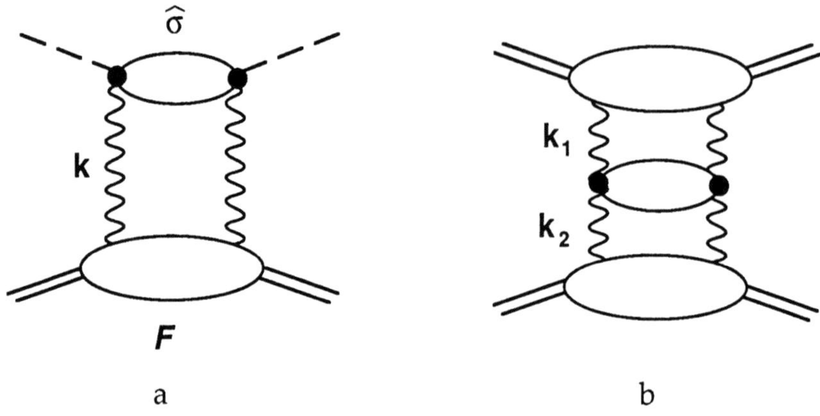

Fig.1

Factorization scheme for (a) single -$k_\perp$ and (b) double $k_\perp$ high energy hard processes. Wavy lines denote (Regge) gluon exchange.

$$\gamma_N = \frac{3}{\pi}\,\frac{\alpha_s}{N} + 2\,\zeta(3)\left(\frac{3\alpha_s}{\pi N}\right)^4 + \cdots$$

$$C_N = \frac{28\,\pi}{9}\,\alpha\,\alpha_s\,e_Q^2\left(1 + \frac{41}{21}\,\frac{3\,\alpha_s}{\pi N} + \cdots\right)$$

$$(2)$$

The problem then arises of how to evaluate such large terms to all orders.

The idea put forward by Catani, Hautmann and myself[2,3] is to replace the collinear (or parton pole) factorization of Eq.(1) by the high-energy (or gluon Regge pole) factorization in Figs.(1a,b). The latter reduces to the former for $S>M^2 \gg k_\perp^2$, but holds also for $S \gg k_\perp^2 \sim M^2$, i.e. for any value of $k_\perp/M$. Since the factors $(\alpha_s/N)^n$ in the coefficient function are generated precisely by the $\underline{k_\perp\text{-integration}}$ from the ones in the anomalous dimension, we are able to evaluate and resum them.

2 HIGH ENERGY FACTORIZATION AND RESUMMATION

Let me start illustrating single-$k_\perp$ processes (Fig.1a) At high energies, hard vertex and structure function terms factorize in a $k_\perp$-dependent way, due to (Regge) gluon exchange, as follows

$$4M^2\sigma\left(\varrho, M^2, Q_o^2\right) = \int d^2k\,\frac{dz}{z}\,\mathcal{F}\left(z, \underline{k}, Q_o\right)\,\hat{\sigma}\left(\frac{\varrho}{z}, \frac{k^2}{zS}\right)\,.$$

$$(3)$$

Here $\hat{\sigma}$ is the basic high-energy cross section defined by the physical process of Fig.(2a), and computed as function of $k_\perp$ by coupling the lowest order $\gamma g \to Q\bar{Q}$ absorptive part to eikonal partons. On the other hand, $\mathcal{F}(x,\underline{k})$ is the unintegrated gluon structure function, defined as the probability for a gluon to have energy fraction x associated to a total emitted transverse momentum $\underline{k}$.

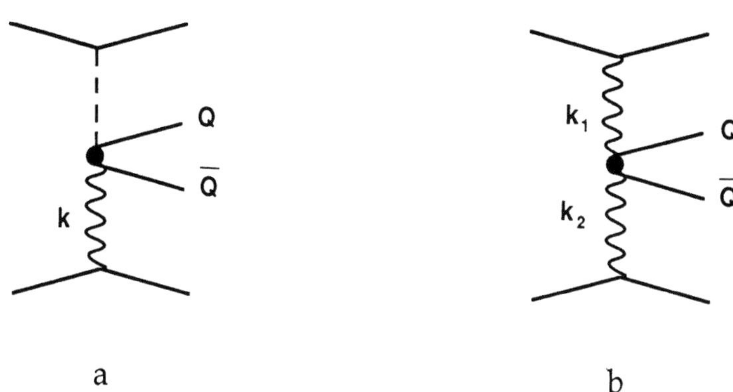

Fig.2

High-energy processes defining (a)the γ-g and (b) the g-g off-shell Born cross section $\hat{\sigma}$. External partons are coupled with eikonal vertices.

The full structure function is obtained by $\underline{k}$-integration

$$G\left(x, M^2, Q_0^2\right) \equiv \int_{Q_0^2}^{M^2} d^2k \; \mathcal{F}\left(x, \underline{k}, Q_0\right)$$

(4)

and has a well-known[4,5] small x behaviour.

The main point is that, if Eq.(3) is coupled to the customary renormalization group behaviour for $\mathcal{F}(k_\perp)$ then it yields automatically a <u>resummed</u> coefficient function. Assume in fact, for definiteness, that $\mathcal{F}(k_\perp)$ has a well defined anomalous dimension, so that, in moment space,

$$\mathcal{F}_N\left(k_\perp^2, Q_0^2\right) \equiv \int \frac{dx}{x} \, x^N \mathcal{F}\left(x, \underline{k}, Q_0\right) = \frac{\gamma_N\left(\alpha_s(M^2)\right)}{\pi \, k_\perp^2} \left(\frac{k_\perp^2}{M^2}\right)^{\gamma_N\left(\alpha_s(M^2)\right)} G_N\left(M^2, Q_0^2\right),$$

(5)

where the α_s coupling has been frozen at M^2 because $k_\perp^2/M^2 = O(1)$. By inserting Eq.(5) into Eq.(3) we get immediately

$$4 \, M^2 \sigma_N\left(M^2, Q_0^2\right) = h_N\left(\gamma_N(\alpha_s(M^2))\right) \; G_N\left(M^2, Q_0^2\right),$$

(6)

where $\gamma(\alpha_s/N) \equiv \gamma_N$ is the gluon anomalous dimension of Eq.(2) and h_N, the coefficient function, is given by the $k_\perp$-transform

$$\gamma^{-1} h_N(\gamma) \equiv \int_0^\infty \frac{dk^2}{k^2} \left(\frac{k^2}{M^2}\right)^\gamma \hat{\sigma}_N^{\gamma g}\left(\frac{k^2}{4M^2}\right).$$

(7)

We note in h_N the γ-dependent weight of the $k_\perp$-integration. Its explicit expression can be obtained from the perturbative computation of $\hat{\sigma}$. In the case of $Q\bar{Q}$ photoproduction, we obtain, for N=0,[2]

$$h_{N=0}(\gamma) = \frac{4\pi}{3} \alpha \, \alpha_s \, e_Q^2 \cdot$$

$$\cdot \frac{7-5\gamma}{3-2\gamma} \; B(1-\gamma, 1-\gamma) \; B(1+\gamma, 1-\gamma) \quad .$$

$$(8)$$

The resummation effect is incorporated in Eq.(6) through the α_S/N dependence of γ_N (known from earlier work[4,5]) and the γ-dependence of h_N (known from Eq.8), which is an increasing function of γ for $0\leq\gamma<1$.

A result similar to Eq.(6) has been recently obtained by other authors[6].

One can thus understand the energy dependence of σ/G, which follows from Eq.(6) by inverse Mellin transform. As the energy increases, the effective anomalous dimension and $h(\gamma)$ increase also, and saturate at $\gamma=1/2$ at energies so large that $\alpha_S \log(S/4M^2)>\log(Q^2/Q^2)$ according to the FLK[5] evolution[3].

The asymptotic limit is

$$\frac{4M^2 \sigma(\rho, M^2/Q_o^2)}{G(\rho, M^2/Q_o^2)} \xrightarrow[\rho\to 0]{} h_{\bar{N}}\left(\frac{1}{2}\right)$$

$$\xrightarrow[\bar{N}\to 0]{} \frac{3\pi^3}{2} \alpha \, \alpha_s \, e_Q^2 = \frac{27}{56} \pi^2 h(0) \quad . \tag{9}$$

Note that this asymptotic "K-factor" is rather large (about 5 times $h(o)$) and not much dependent on the absorptive phenomena[7] which are expected to affect in the same way cross section and structure function, thus cancelling in the ratio.

Care should be taken however of the $\bar{N}$-dependence of Eq.9, where the effective asymptotic moment $\bar{N}$ is expected to be of order $\bar{N}\sim\alpha_S$. In principle, this is a subleading effect, because h_N has a smooth $N \to 0$ limit. However, in practice $\alpha_S(M^2)$ is not very small and, in the FLK evolution, $\bar{N}=4\alpha_S(M^2)\log 2$ is sizeable compared to the asymptotic value $\gamma_N=1/2$. Since $\sigma_N \sim (k^2)^{-N-1}$ for $k_\perp \gg M$, the parameter $(\gamma_N-\bar{N})$ governs the large $k_\perp$ part of the integral (7), and thus the size of the coefficient function. If $\bar{N}\ll\gamma_N$, the estimate in Eq.(9) holds, but a numerical evaluation is needed, depending on the given energy and on the gluon structure function in the hadron.

3. LEPTOPRODUCTION AND HADROPRODUCTION PROCESSES

A similar treatment holds for the lepto-production process, which contains the additional scale Q^2 due to the off-shell photon. The small y spectrum is given in terms of the cross section $\sigma^{(2)}=\sigma_T+ \sigma_L$, defined as follows ($y\equiv(E_e-E'_e)/E_e$)

$$d\sigma^{(\ell)}_{y\ll 1} = \frac{\alpha}{\pi} \frac{dQ^2}{Q^2} \frac{dy}{y} \sigma^{(2)}\left(\frac{\rho}{y}, \frac{Q^2}{4M^2}, M^2, Q_o^2\right) \Theta\left(Q^2 - \frac{m^2 y^2}{(1-y)^2}\right) \quad .$$

$$(10)$$

High-energy factorization can then be used to predict the ratio

$$R_N \equiv \frac{\sigma_N^{(2)}(Q^2/4M^2, M^2, Q_0^2)}{\sigma_N(M^2, Q_0^2)} = \frac{1}{h_N(\gamma_N)} \int_{\frac{1}{2}-i\infty}^{\frac{1}{2}+i\infty} \frac{d\gamma}{2\pi i\, \gamma} \left(\frac{Q^2}{M^2}\right)^{-\gamma} h_N^{(2)}(\gamma, \gamma_N) \ ,$$

(11)

where the h-function is now given by a double $k_\perp$-transform of the $\gamma g \to Q\bar{Q}$ off-shell cross section

$$(\gamma_1 \gamma_2)^{-1} h_N(\gamma_1, \gamma_2) \equiv \int \frac{d^2 k_1}{\pi k_1^2} \frac{d^2 k_2}{\pi k_2^2} \left(\frac{k_1^2}{M^2}\right)^{\gamma_1} \left(\frac{k_2^2}{M^2}\right)^{\gamma_2} \hat{\sigma}_N^{\gamma g}\left(\frac{k_1}{M}, \frac{k_2}{M}\right)$$

(12)

and has the explicit expression(3)

$$h_N^{(2)}(\gamma_1, \gamma_2) = \pi^2 \, 4^{\gamma_1 + \gamma_2 - 1} \alpha_S \, e_Q^2 \left(7 - 5(\gamma_1 + \gamma_2) + 3\gamma_1 \gamma_2\right) \ .$$

$$\cdot \ \Gamma(1 - \gamma_1 - \gamma_2) \ \frac{B(1 + \gamma_1, 1 - \gamma_1)}{\Gamma\left(\frac{5}{2} - \gamma_1\right)} \ \frac{B(1 + \gamma_2, 1 - \gamma_2)}{\Gamma\left(\frac{5}{2} - \gamma_2\right)} \ .$$

(13)

In Eq.(10), the variable γ_2 related to the gluon transverse momentum takes the anomalous dimension value $\gamma_2 = \gamma(\alpha_S/N)$, while γ_1 yields, instead the photon Q^2-dependence. A simple analysis(3) shows that $R_N(Q^2/M^2)$, starting from $R_N = 1$ for $Q^2 \ll 4M^2$, decreases with increasing Q^2 and reaches, for $Q^2 \gg 4M^2$, the asymptotic form

$$R_N \simeq \frac{3}{7} \left(\frac{4M^2}{Q^2}\right)^{N+1} \frac{1}{\gamma_N} \left[\left(\frac{Q^2}{4M^2}\right)^{\gamma_N} - 1\right] \ ,$$

(14)

needed to perform the expected change of scale $M^2 \to Q^2$.

Similar results hold for the complete leptoproduction spectrum. Since the total cross section is dominated by the small y form in Eq.(10), we obtain from Eq.(11), integrated up to some value $\bar{Q}^2 \ll M^2$, the result

$$\frac{\sigma_N^{(\ell)}(M^2, Q_0^2; Q^2 > \bar{Q}^2)}{\sigma_N(M^2, Q_0^2)} = \frac{\alpha}{\pi N} \left[\log \frac{4M^2}{\bar{Q}^2} + \frac{1}{h_N(\gamma_N)} \frac{\partial h_N(\gamma, \gamma_N)}{\partial \gamma}\bigg/_{\gamma=0}\right] \ .$$

(15)

Note the rather large factor coming from the Q^2- and y-integration, which becomes even larger if the full phase space $Q^2 > m^2 y^2$ is covered.

The heavy flavour hadroproduction process is a typical double-$k_\perp$ process (Fig.1b) and is qualitatively different from leptoproduction because the hard subprocess $g(k_1)+g(k_2)\to Q\bar{Q}$ involves two gluons. Correspondingly, the high-energy factorization formula involves two structure functions, and has the form

$$4M^2 \sigma^{(h)}(\rho, M^2, Q_0^2) = \int d^2k_1\, dz_1\, d^2k_2\, dz_2\, (z_1 z_2)^{-1}$$

$$\cdot \mathcal{F}(k_1, z_1)\, \mathcal{F}(k_2, z_2)\, \hat{\sigma}^{gg}\left(\frac{\rho}{z_1 z_2}, \frac{k_1}{M}, \frac{k_2}{M}\right) , \tag{16}$$

where $\hat{\sigma}^{gg}$ is now is now the (non-abelian) hard vertex cross section of Fig.(2b).

In moment space, Eq.(14) reads

$$4M^2 \sigma_N^{(h)}(M^2, Q_0^2) = h_N^{(h)}(\gamma_N, \gamma_N)\, G_N^2(M^2, Q_0^2) , \tag{17}$$

where the h-function is defined as in Eq.(12) with $\hat{\sigma}^{\gamma g}$ replaced by $\hat{\sigma}^{gg}$. Unfortunately, the computation of the non-abelian part of h has not been completed yet. We know the small γ behaviour

$$h^{(h)}(\gamma_1, \gamma_2) = 2\pi \alpha_s^2 \left[\frac{7}{36} C_F - \frac{11}{360} N_c + (\gamma_1 + \gamma_2)\left(\frac{41}{108} C_F - \frac{101}{5400} N_c\right) + \cdots \right] , \tag{18}$$

which allows comparison with low order perturbation theory.

We have also noticed a peculiar feature occurring in the extreme energy regime $\alpha_s \log(S/4M^2) \gg \log(Q^2/Q_0^2)$. Since in the hadroproduction case both γ_1 and γ_2 become large $(\gamma_1, \gamma_2 \to 1/2)$, the h-function becomes sensitive to the large $k_\perp$ behaviour of σ, showing a singularity, close to $\gamma_1 + \gamma_2 = 1$, of the form[3]

$$h_N^{(h)} \simeq \frac{8\pi}{3} \alpha_s^2 \frac{N_c}{N_c^2 - 1} \left(N + 1 - \gamma_1 - \gamma_2\right)^{-2} \left(1 - \gamma_1 - \gamma_2\right)^{-1} . \tag{19}$$

Hence, the ratio $\sigma^{(h)}/\sigma^2$ becomes anomalously large, increasing logarithmically with energy.

The physical origin for the singular behaviour (19) lies in the large $k_\perp$ form of the non-abelian part of the Born cross section, roughly $\sigma^{gg} \sim (k_1 + k_2)^{-2} \log(k_1 + k_2)^2$. This $k_\perp^{-2}$ decrease is a sign of the compositeness of the $Q\bar{Q}$ probe, capable to damp the $k_\perp$-integrals.

However, with increasing anomalous dimensions these integrals provide larger K-factors and eventually, for $\gamma_1 = \gamma_2 = 1/2$, they start diverging, thus causing the logarithmic enhancement mentioned before.

In conclusion, it appears that the heavy flavour production cross sections stay large in the small ρ regime, and are dominated by gluon exchange. The resummed K-factors can be reliably computed, sometimes analytically, by the high-energy factorization method.
Some word of caution is needed for the hadroproduction case, where the extreme energy regime is somewhat delicate, because large values of $k_\perp = 0(\sqrt{S})$ become important. But on the whole , this approach seems well suited to a more detailed phenomenological analysis and to experimental checks.

References

1) P. Nason, S. Dawson and R.K. Ellis, Nucl.Phys.B312 (1989) 551.
2) S. Catani, M. Ciafaloni and F. Hautmann, Phys. Lett. 242B (1990) 97.
3) S. Catani, M. Ciafaloni and F. Hautmann, Nucl.Phys., B 366 (1991) 135.
4) E.A. Kuraev, L.N. Lipatov and V.S. Fadin, Sov.Phys. JETP 45 (1977) 199.
5) A. Bassetto, M. Ciafaloni and G. Marchesini, Phys. Rep.100 (1983) 201.
6) J.C. Collins and R.K. Ellis, Fermilab. PUB 91/22.
   See also E.M. Levin, M.G. Ryskin, Yu.M. Shabelski and A.G. Shuvaev, DESY preprint 91-065.
7) L.V. Gribov, E.M. Levin and M.G. Ryskin, Phys. Rep. 100 (1983)1.

HEAVY QUARK PRODUCTION IN NUCLEON COLLISIONS

Yuly Shabelski

Leningrad Nuclear Phyiscs Institute Gatchina

Leningrad 188380, USSR

ABSTRACT

We compare the results of calculations of heavy quark production cross section in leading log QCD to those parton model. The formulae of parton model are obtained from QCD via several subsequent simplifications. The cross section of b-quark production in LLA QCD is predicted to be $2 - 3$ times larger than the parton predictions.

The presented results were obtained together with E.M.Levin, M.G.Ryskin and A.G.Shuvaev.

In this paper we compare two approaches to the production of the heavy quarks in high energy hadron-hadron collisions. One of them employs the parton model, another one is based on the theory of semihard processes [1]. The heavy pair production is described in the lowest order of QCD as an elementary process given by the sum of three graphs in Fig.1 in both approaches. In the parton model all particles involved are assumed to be on mass shell and the cross section is averaged over two transverse polarizations of the gluons. The virtualities q^2 of the initial partons are taken into account through their densities. The latter are calculated in leading logarithm approximation (LLA) collecting the terms of the form $(\alpha_s \ln q^2)^n$ and resulting in the well-known Gribov-Lipatov-Altarelli-Parisi evolution equation. The probabilistic picture of noninteracting partons underlies this way of proceeding.

The theory of semihard processes deals with the region where the values of Bjorken variables x are very small. It is just the region that dominates in the heavy quark production at high energies $\sqrt{s}$ since the characteristic value of x is $x \sim 2m_T/\sqrt{s}$ ($m_T^2 = m^2 + p_T^2$ is transverse mass of the quark). The drastical growth of the parton density in this domain makes the effects of parton-parton interaction very significant. For the correct description of these phenomena it is necessary to sum up in Feynman diagrams not only the terms of the form $(\alpha_s \ln q^2)^n$ but also the terms $(\alpha_s \ln 1/x)^n$ and $(\alpha_s \ln q^2 \ln 1/x)^n$. Another problem that appeared at $x \sim 0$ is that of the screening (absorption) corrections which stop the growth of the cross section and restore the unitary. As a result the gluon structure function $xG(x, q^2)$ becomes proportional to $q^2 R^2$ at relatively small virtuality $q^2 \leq q_0^2(x)$ and the cross section is $\sigma \sim (1/q^2)xG(x, q^2) \sim R^2$.

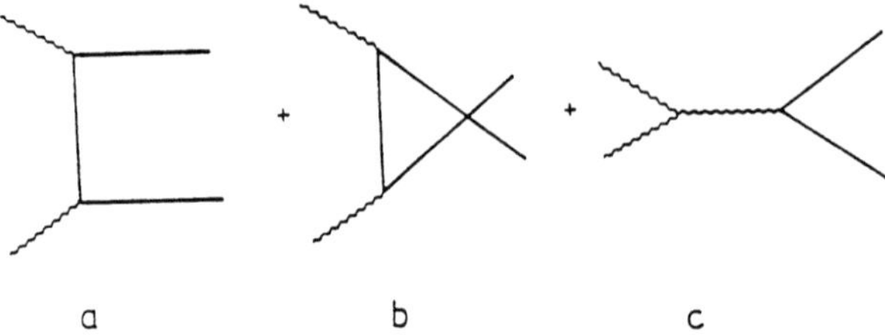

Fig.1. The diagrams, taking into account in calculation of heavy quark production via gluon-gluon collision.

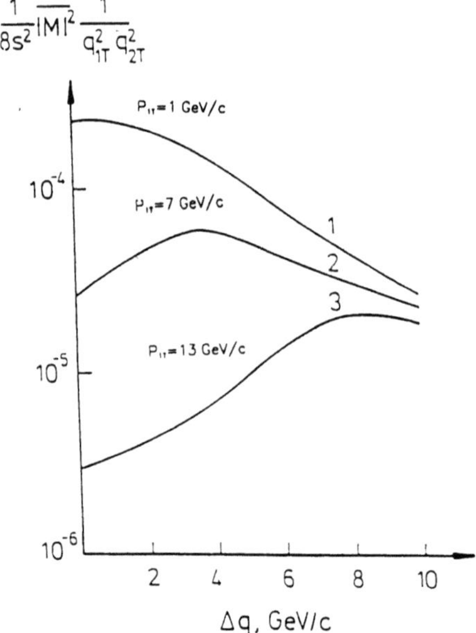

Fig.2. Dependences of the matrix element square of $gg \to \bar{Q}Q$ reaction on the gluon transverse momenta Δq at different values of the heavy quark transverse momentum p_{1T}. The values of another variables are $\sqrt{s} = 1.8\,TeV$, $y_1^* = 0.5$, $y_2^* = 0$, $m_Q = 4.7\,GeV$.

Here $R = const$ is a new phenomenological parameter with dimension of (mass)$^{-1}$ and the value of $q_0(x)$ can be considered as a new typical transverse momentum of partons in the parton cascade of the hadron which leads to natural infrared cut-off in semihard processes (see ref. [1] for details).

The main contribution to the cross section at small x is known to come from gluons. Their distribution over x and transverse momenta q_T in hadron is given in semihard theory by function $\varphi(x, q^2)$. It differs from the usual function $G(x, q^2)$:

$$xG(x, q^2) = \frac{1}{4\sqrt{2}\,\pi^3} \int_0^{q^2} \varphi(x, q_1^2) dq_1^2 \tag{1}$$

Such definition of $\varphi(x, q^2)$ makes possible to treat correctly the effects arising from gluons virtualities. The exact expression for this function can be obtained as a solution of the evolution equation which, contrary to the parton model case, is nonlinear due to interactions between the partons in small x region.

The differential cross section of heavy quark production has the form

$$\frac{d\sigma}{dy_1^* dy_2^* d^2 p_{1T} d^2 p_{2T}} = \frac{1}{(2\pi)^8} \frac{1}{(s)^2} \int d^2 q_{1T} d^2 q_{2T} \delta(q_{1T} + q_{2T} - p_{1T} - p_{2T})$$

$$\times \frac{\alpha_s(q_1^2)}{q_1^2} \frac{\alpha_s(q_2^2)}{q_2^2} \varphi(q_1^2, y) \varphi(q_2^2, x) |M|^2. \tag{2}$$

Here $s = 2p_A p_B$, $y_{1,2}^*$ are the quarks rapidities in c.m.s.,

$$\begin{aligned}
x_1 &= \tfrac{m_{1T}}{\sqrt{s'}} e^{-y_1^*}, & x_2 &= \tfrac{m_{2T}}{\sqrt{s'}} e^{-y_2^*}, & x &= x_1 + x_2 \\
y_1 &= \tfrac{m_{1T}}{\sqrt{s'}} e^{y_1^*}, & y_2 &= \tfrac{m_{2T}}{\sqrt{s'}} e^{y_2^*}, & y &= y_1 + y_2.
\end{aligned} \tag{3}$$

$q_{1,2T}$ are the gluons' transverse momenta. $|M|^2$ is the square of the matrix element. The explicit expression for $|M|^2$ is rather bulky and we do not present it here (one can find it in ref. [2]).

There is a certain limit case in which our formulas can be transformed into parton model ones. Introducing the polar coordinates

$$d^2 q_{1T} = \frac{1}{2} dq_{1T}^2 d\theta_1 \tag{4}$$

we obtain

$$\int_0^{2\pi} d\theta_1 q_{1T}^\mu q_{1T}^\nu = \pi\, q_{1T}^2 \delta_T^{\mu\nu} \tag{5}$$

$$\int d\theta_1 \int d\theta_2 |M|^2 = 2\pi^2 \frac{q_{1T}^2 q_{2T}^2}{(x\,y)^2} |M_{part}|^2. \tag{6}$$

Here M_{part} is just the matrix element in the parton model since the result is the same as that calculated for the real (mass shell) gluons and averaged over transverse polarizations. Then we obtain the cross section (2) in the form

$$\frac{d\sigma}{dy_1^* dy_2^* d^2 p_{1T}} = \int |M|^2 \frac{d\theta_1 d\theta_2}{2\pi^2 q_1^2 q_2^2 (s)^2} \int \frac{\alpha_s(q_1^2)\varphi(y, q_{1T}^2)}{4\sqrt{2}\pi^3} \frac{\alpha_s(q_2^2)\varphi(x, q_{2T}^2)}{4\sqrt{2}\pi^3} dq_{1T}^2 dq_{2T}^2$$

$$= |M_{part}|^2 \frac{1}{(\hat{s})^2} \int \frac{\alpha_s(q_1^2)\varphi(y, q_{1T}^2)}{4\sqrt{2}\,\pi^3} \frac{\alpha_s(q_2^2)\varphi(x, q_{2T}^2)}{4\sqrt{2}\,\pi^3} dq_{1T}^2 dq_{2T}^2 \tag{7}$$

where $\hat{s} = xys$ is the mass square of $\bar{Q}Q$ pair.

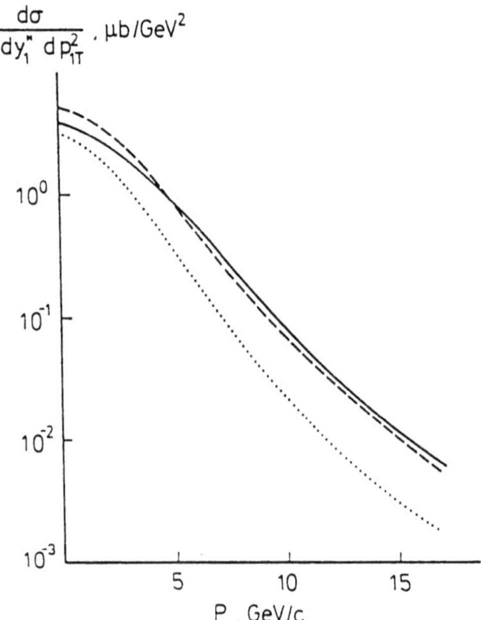

Fig.3. The calculated dependences of heavy quark production cross section on its transverse momenta in the LLA QCD, eq.(2) (solid curve), in the approximations of eq. (7) (dashed curve) and eq.(8) (dotted curves). The values of another variables are $\sqrt{s} = 1.8 \; TeV$, $m_Q = 4.7 \; GeV$, $y_1^* = 0.5$.

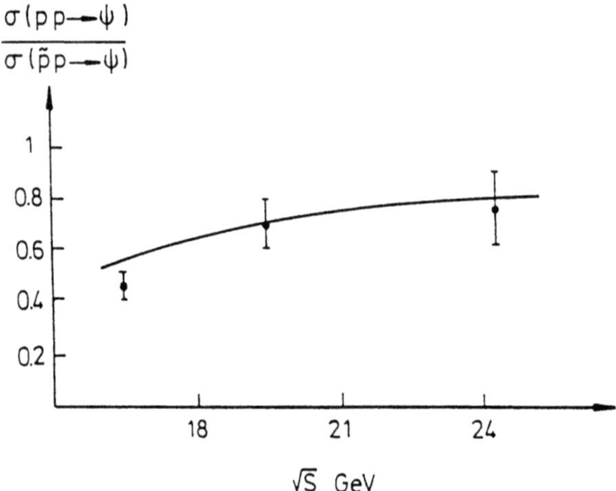

Fig.4. Comparison of calculated values of $\sigma(pp \longrightarrow J/\Psi)/\Psi(\bar{p}p \longrightarrow J/\Psi)$ ratios with the data of refs.[5,6].

The difference between eqs. (2) and (7) is caused by change of the matrix element square $|M|^2$ in (2) by $q_{1T}^2 q_{2T}^2 [M/(q_{1T}^2 q_{2T}^2)]|_{q_{1T}=q_{2T}=0}$. In order to show this we present in Fig. 2 the values of $|M|^2$ averaged over gluon polarizations, as the functions of $q_{1T} = q_{2T} = \Delta q$ at three values of b-quark ($m_b = 4.7\,GeV$) transverse momenta, $\sqrt{s} = 1.8\,TeV$, $y_1^* = 0.5$, $y_2^* = 0$. Here we integrate $|M|^2$ over transverse momentum of the second b-quark. The values of $\Delta q \geq 10\,GeV/c$ don't contribute really to eq.(2) because of fast decrease of the gluon distributions. The value of $|M|^2$ decrease with Δq at small p_{1T}. Thus eq. (7) overestimates the cross section in this region. But with the growth of p_{1T} the minimum of $|M|^2$ appears at $\Delta q = 0$ and the maximum of $|M|^2$ moves away (it is in the region $\Delta q \simeq p_{1T}/2$). The numerical calculations show that the cross sections given by eqs.(2) and (7) coincide approximately at $p_{1T} = 5 - 7\,GeV/c$. At larger values of p_t the cross section of eq.(7) becomes smaller. The difference is, however, not so large and does not vary with p_{1T} (see fig.3). The reason is that the maximum value of $|M|^2$ at high p_T is achieved at large values of Δq which give very small contribution to the cross section. The most important is the different p_T dependence of the two eqs. (2) and (7). For example, the cross section ratios at $p_{1T} = 0$ to $p_{1T} = 10\,GeV/c$ are about 1.5 times different.

The next step to the well-known parton model is changing of the argument in α_s QCD constant. As it was discussed earlier (see, e.g., refs.[3, 4]) the theory can not fix the value of this argument eq. (6) corresponds to the direct calculation of the diagrams of Fig.1 and here it seems to be the most natural to use $\alpha_s(q_1^2) \cdot \alpha_s(q_2^2)$. The essential values of q_1^2 and q_2^2 in eq. (2) are of the order of $q_0^2(x)$, which is equal to several GeV^2 and increase very slowly with initial energy [1]. Thus the essential values of α_s are of the order of $\alpha_s \sim 0.2$. In the parton model the value of mass of heavy quark is used usually as an argument of α_s, with the possible variation of numerical factor. In this case the results of calculations of heavy quark production cross section or these of the distributions $d\sigma/dp_{1T}^2$ should differ by the numerical factor (which can depends on the value of initial energy). However, it seems to be more natural to use the transverse mass of heavy quark, $m_T^2 = m_Q^2 + p_{1T}^2$.

The results of calculations of cross section

$$\frac{d\sigma}{dy_1^* dy_2^* d^2 p_{1T}} = |M_{part}|^2 \left[\frac{\alpha_s(4m_T^2)}{\hat{s}}\right]^2 \int \frac{\varphi(y, q_{1T}^2)}{4\sqrt{2}\,\pi^3} \frac{\varphi(x, q_{2T}^2)}{4\sqrt{2}\,\pi^3} dq_{1T}^2 dq_{2T}^2 \qquad (8)$$

are shown in fig. 3 by dotted curve.

The factorization of gluon distributions in eq. (8) is one more distinguish from parton model. It connects with accounting for the dependence of x and y distributions on the transverse momentum, see eqs.(3). Nevertheless, if we rewrite eq.(8) in the form

$$\frac{d\sigma}{dy_1^* dy_2^* d^2 p_{2T}} = |M_{part}|^2 \left[\frac{\alpha_s(4m_T^2)}{\hat{s}}\right]^2 \int \frac{\varphi(y, q_{1T}^2)}{4\sqrt{2}\,\pi^3} dq_{1T}^2 \int \frac{\varphi(x, q_{2T}^2)}{4\sqrt{2}\,\pi^3} dq_{2T}^2$$

$$= |M_{part}|^2 \left[\frac{\alpha_s(4m_T^2)}{\hat{s}}\right]^2 xG(x, 4m_T^2) \cdot yG(y, 4m_T^2) \qquad (9)$$

the distributions $d\sigma/dp_{1T}^2$ change at small transverse momenta and the differences depend on the initial energy. The cross section of b-quark production, $d\sigma/dp_{1T}^2$ increase in the region $p_{1T}^2 \ll m_b^2$ about 20% at $\sqrt{s} = 1.8\,TeV$ and by $1.5-2$ times at $\sqrt{s} = 18\,TeV$.

In practical calculations we need in the explicit form of gluon distribution fuction $\varphi(x, q_T^2)$. It was obtained in ref. [1] that $\varphi(x, q_T^2) \sim q_T^{-2}$ at $q_T^2 \longrightarrow \infty$. At not so large q_T^2 it is necessary to account for preexponent factor, so the approximation $\varphi(x, q_T^2) \sim q_T^{-4}$

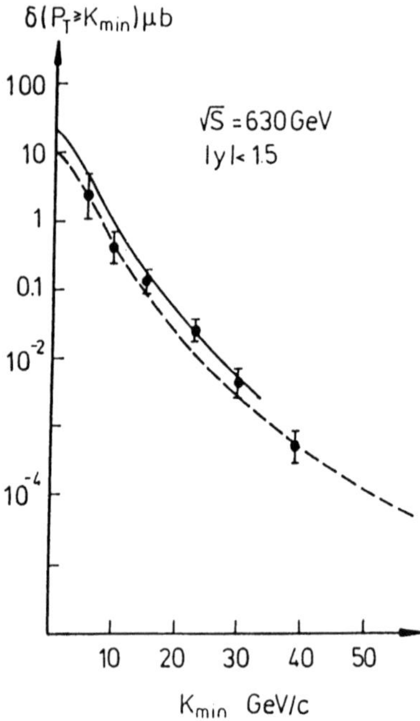

Fig.5. Calculated cross section of b-quark production in $\bar{p}p$ collisions at $\sqrt{s}$ = 630 GeV in LLA QCD (solid curve) and parton model (dashed curve).

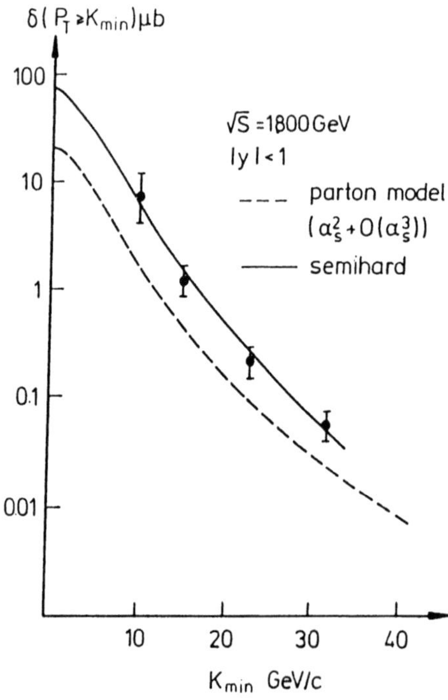

Fig.6. Calculated cross section of b-quark production in $\bar{p}p$ collisions at $\sqrt{s}$ =1800 GeV in LLA QCD (solid curve) and parton model (dashed curve).

seems to be better. As a function of $x\varphi(x, q_T^2)$ should decrease at $x \longrightarrow 1$ at least as $\sim (1-x)^3$. The value of $\varphi(x, q_T^2)$ at $x \longrightarrow 0$ and large q_T^2 should determine also the cross section of high p_T jet production that allow one to normalize $\varphi(x, q_T^2)$ in this region. It is possible to select the most consistent parametrization with the help of the data on J/Ψ production in pp and $\bar{p}p$ collisions. In the first case the gluons and sea antiquarks contribute only, whereas in the case of $\bar{p}p$ collisions there is the valence quark- antiquark contribution also. So we can normalize our gluon distribution to the valence quark sructure functions.

The result of calculation of the ratio

$$R = \frac{\sigma(pp \longrightarrow J/\Psi)}{\sigma(\bar{p}p \longrightarrow J/\Psi)} \tag{10}$$

is show together with existing experimental data [5, 6] in Fig.4. We used Duke-Owens quark structure function [7] (set 1) and the gluon distribution in the form

$$\varphi(x, q_T^2) = \varphi(1-x)^3 \frac{0.05}{0.05 + x} f(x, q_T^2), \tag{11}$$

$$f(x, q_T^2) = \left\{ \begin{array}{ll} 1 & q_T^2 < q_0^2(x) \\ [q_0^2(x)/q_T^2]^2 & q_T^2 > q_0^2(x) \end{array} \right. \tag{12}$$

$$q_0^2(x) = Q_0^2 + \Lambda^2 exp(3.56\sqrt{\ln x/x_0}) \tag{13}$$

$\varphi_0 = 170\ mb$, $Q_0^2 = 2\ GeV^2$, $\Lambda = 52\ McV$, $x_0 = 1/3$. The reasonable agreement allows us to use this parametrization in the further calculations.

In Figs. 5 and 6 we compare the results of our calculations [2] (solid curves) with new experimental data present in last Rencontres de Moriond. Parton model predictions are shown by dashed curves. The $S\bar{p}pS$ collider data at $\sqrt{s} = 630\ GeV$ are in agreement with the parton model calculations. The results of calculations of eq. (2) don't contradict the data also. The Tevatron-collider data at $\sqrt{s} = 1800\ GeV$ lay significantly higher than the parton model curve and agree with our predictions. It can be connected with the specific behaviour of the gluon distribution function at small x.

REFERENCES

[1] Gribov L.V., Levin E.M.,Ryskin M.G. Phys.Rep., 1983, v.100, p.1.

[2] Levin E.M. et al.Preprint LNPI-1643, Leningrad (1990).

[3] Nason P., Dawson S., Ellis R.K. Nucl.Phys., 1988, v.B303, p.607.

[4] Altarelli G. et al. Nucl.Phys., 1988, v.B308, p.724.

[5] Morel C. et al. UA6-Coll.Preprint CERN-PPE/90-127 (1990).

[6] Badier J. et al. Z.Pyhs., 1983, v.C20, p.101.

[7] Duke D., Owens J. Phys.Rev. 1984, v.D30, p.49.

RESULTS FROM THE L3 EXPERIMENT AT LEP

Pierre Lecomte

Swiss Federal Institute of Technology
ETH Zürich, Switzerland

INTRODUCTION

The large Electron-Positron collider LEP at CERN was commissioned in Summer 1989 and has produced three quarter of a million Z^0 as of this writing (July 1991). The data were collected by four detectors known as ALEPH, DELPHI, L3 and OPAL. We describe here some of the results obtained sofar by the L3 experiment concerning precision tests of the standard model, $B\bar{B}$ mixing, determination of the strong coupling constant, Higgs and other particle searches. Since the collaboration has produced 32 publications to date, what follows will be a brief survey of the main results with reference to the corresponding publications for details of the analysis procedure.

THE L3 EXPERIMENT

The L3 experiment is the largest of the four LEP detectors and it differs from the other three in that it is optimized for high resolution measurement of electrons, photons and muons. The incentive for this optimization comes from the past discoveries of J, Υ, W and Z^0 through leptonic decay channels. the magnitude of the L3 effort can be gauged from a few numbers: it took eight years, two hundred millions swiss francs and one thousand one hundred man-years of technical support to design and build the 8500t, 16m high detector which 479 scientists belonging to 44 institutions from 13 countries are presently operating[1].

All the L3 subdetectors are installed within a 7800t conventional solenoidal magnet which provides a 0.5T field parallel to the beam. The relatively low field in a very large volume was selected to optimize muon momentum resolution, which improves linearly with the field, but quadratically with the track length. The muon detector uses precision drift chambers and extends radially from 5.6m down to 2.4m away from the interaction point; it has achieved a dimuon mass resolution of better than 1.5% at the Z^0. A 305t hadron calorimeter made of proportional chambers sandwiched between depleted uranium plates provides a hadron energy resolution of $(\frac{55}{\sqrt{E}} + 5)\%$ and a segmentation $\Delta\theta=2.5°, \Delta\phi=3.5°$. Inside the hadron calorimeter, radially between 0.8m and 0.5m, approximately 11000 crystals of bismuth germanate form an electromagnetic calorimeter with a resolution better than 2% above 2GeV and of 5% at 0.1GeV. A central

tracker using the time expansion principle surrounds the beam pipe and
provides 58μm single wire accuracy as well as 640μm double track resolution. A luminosity monitor, scintillators, triggering and acquisition
systems complete the detector. Side and end views of the L3 detector are
shown in Fig.1 and 2 respectively, whereas the design, construction, calibration and performance of the L3 subdetectors are described in Ref.2.

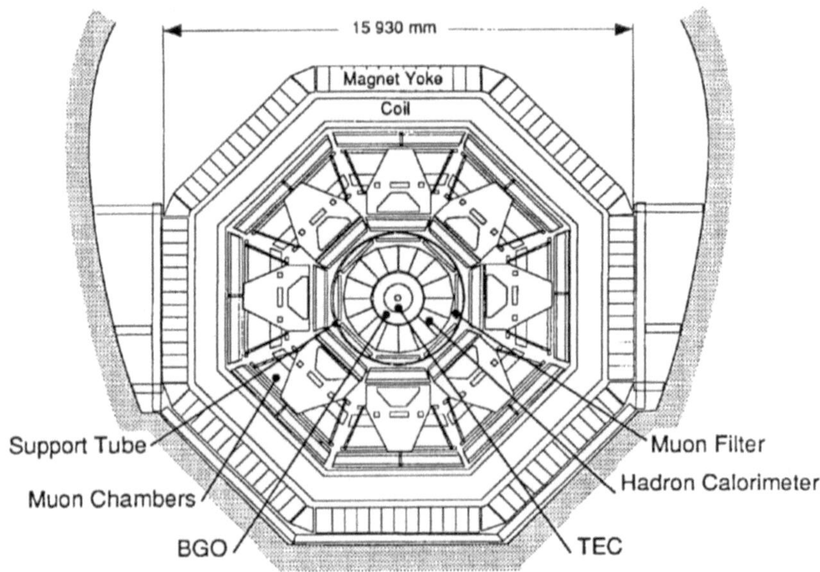

Fig.1 End view of the L3 detector: from the interaction point outwards,
there are the Time Expansion Chamber (TEC), BGO Electromagnetic
Calorimeter, depleted uranium Hadron Calorimeter with additional
absorber (Muon Filter), Muon Detector, the 1000t aluminum coil and
the magnet return yoke.

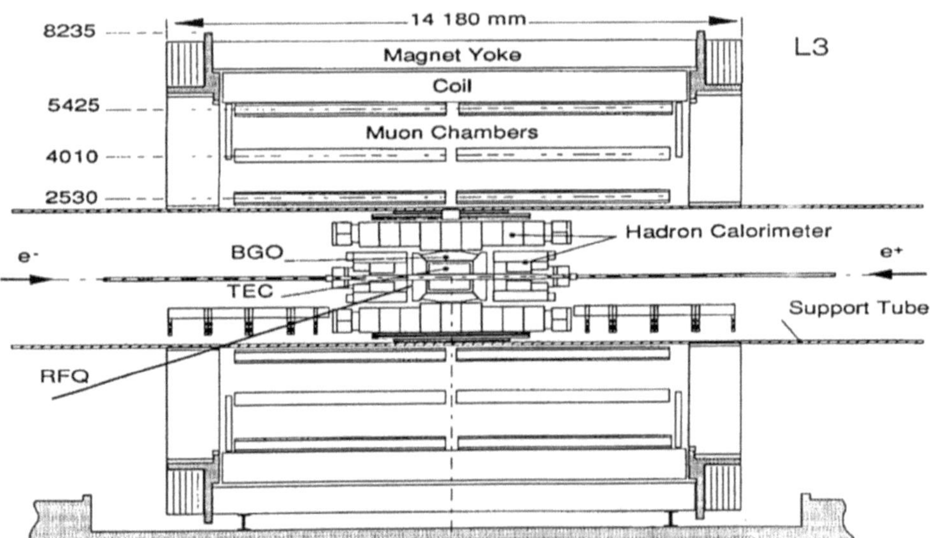

Fig.2 Side view of the L3 detector: all the subdetectors as well as the
LEP machine elements are supported independently of the magnet by
a 32m long, 4.45m diameter stainless steel/carbon steel tube dynamically positioned to 20μm accuracy.

We have used for this analysis[3-9] approximately 115000 hadronic and 10000 leptonic Z^0 decays and have studied $e^+e^-\rightarrow$hadrons, $e^+e^-\rightarrow e^+e^-(\gamma)$, $e^+e^-\rightarrow\mu^+\mu^-(\gamma)$, and $e^+e^-\rightarrow\tau^+\tau^-(\gamma)$ in the energy range 88.2GeV $\leq \sqrt{s} \leq$ 94.2GeV. The fitted cross-sections are shown in Fig.3a to 3d; table 1 summarizes the results obtained from fits to all our measured cross-sections, with and without lepton universality. The number of light neutrino species was obtained from the measured ratio of the Z^0 invisible and leptonic decay widths, using

$$N_\nu = \frac{\Gamma_{invis}}{\Gamma_{lept}} \bullet \left(\frac{\Gamma_{lept}}{\Gamma_\nu}\right)_{S.M.} = \frac{\Gamma_{invis}}{\Gamma_{lept}} \bullet 0.502 \quad \text{for } M_{Z^0}=91.181\text{GeV}.$$

The vector and axial vector neutral current coupling constants of charged leptons to the Z^0 were determined from a simultaneous fit to cross-sections and to the measured forward-backward asymmetry of the leptonic channels, assuming lepton universality; the sign is inferred from other experiments (Fig.4). We obtain ρ_{eff} and $\text{Sin}^2\theta_w$ using the relations $\overline{g}_A = -0.5\sqrt{\rho_{eff}}$ and $\overline{g}_V = \overline{g}_A\left(1-4\text{Sin}^2\theta_w\right)$. The vector and axial vector couplings of the electron are obtained from a fit to the electron and hadron data, the latter providing a precise determination of the Z^0 mass and width[3], the sign being again inferred from other experiments[10].

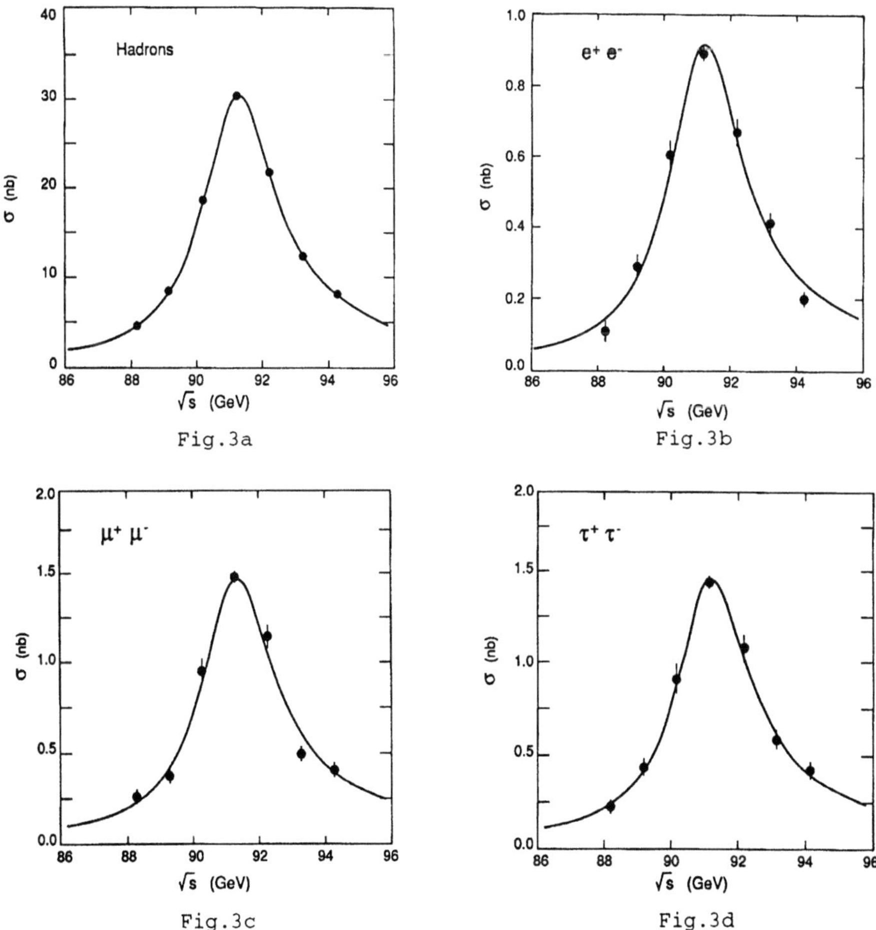

Fig.3a

Fig.3b

Fig.3c

Fig.3d

Fig.3 Hadronic and leptonic cross-sections as a function of $\sqrt{s}$.

Table 1. Electroweak parameters measured by the L3 experiment. See text and Ref.3 for details.

Parameter	Value	Errors	Unit
M_z	91.181	±0.010±0.02(LEP)	[GeV]
Γ_z	2501	±17	[MeV]
Γ_{had}	1742	±19	[MeV]
Γ_{lept}	83.6	±0.8	[MeV]
Γ_e	83.3	±1.1	[MeV]
$\Gamma\mu$	84.5	±2.0	[MeV]
$\Gamma\tau$	84.0	±2.7	[MeV]
Γ_{invis}	508	±17	[MeV]
N_ν	3.05	æ0.10	
$\overline{g}_A$	-0.500	±0.003	
$\overline{g}_V$	-0.046	$^{+0.015}_{-0.012}$	
$Sin^2\theta_W$	0.227	$^{+0.008}_{-0.006}$	
ρ_{eff}	1.000	±0.011	
$\overline{g}_A^e$	-0.501	$^{+0.004}_{-0.003}$	
$\overline{g}_V^e$	-0.008	$^{+0.060}_{-0.044}$	

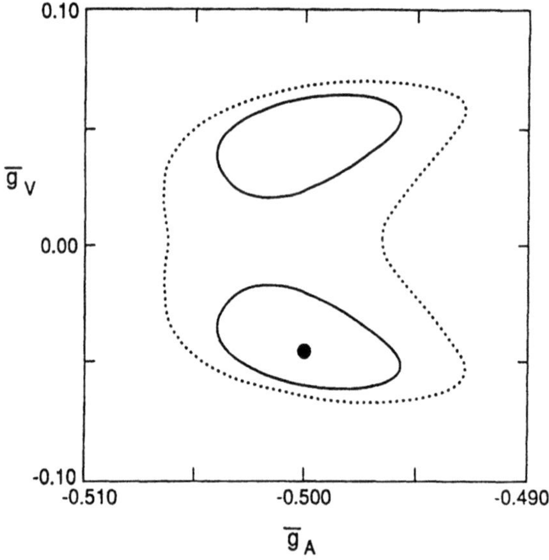

Fig.4 Vector and axial vector coupling constants obtained from a fit to the leptonic and hadronic cross-sections and leptonic forward backward asymmetries. The dotted and solid lines are the 95% and 68% confidence level limits. The dot indicates the solution favored by neutrino experiments.

The top quark mass can be estimated within the framework of the standard model since the total and partial decay widths of the Z^0 and the leptonic forward-backward asymmetries depend on m_t. In our fit, we leave M_Z, m_t as free parameters and constrain $\alpha_s=0.115\pm0.009$. We repeat the fit for $50\mathrm{GeV} \leq M_H \leq 1000\mathrm{GeV}$ and quote the central value obtained for $M_H=300\mathrm{GeV}$, with an additional error taking the variation of M_H into account. We obtain $m_t=193\,{}^{+52}_{-69}\pm16(\mathrm{Higgs})$ GeV. Alternatively, we get $M_W=80.44\pm0.45\mathrm{GeV}$ or

in terms of $\mathrm{Sin}^2\theta_w=1-\dfrac{M_W^2}{M_Z^2} = 0.222\pm0.008$, these results being independent

of the Higgs mass between 50GeV and 1000GeV.

BOTTOM QUARK PHYSICS

The $Z^0{\to}b\bar{b}$ partial decay width $\Gamma_{b\bar{b}}$ is insensitive to the mass of the top quark and thus provides a significant test of the Standard Model. Bottom quarks events are identified through their semileptonic decay. We select inclusive muon and electron events and favor $Z^0{\to}b\bar{b}$ by taking advantage of the fact that leptons from b semileptonic decays have a large momentum and transverse momentum relative to the b direction. Our selection procedure is described in Ref.11 and the number of events passing our cuts is summarized in Table 2.

Table 2. Number of selected inclusive muon and
electron events

Event Type	Number of events
μ+hadrons	2621
e+hadrons	1085
$\mu\mu$+hadrons	83
ee+hadrons	26
eμ+hadrons	78

The muon and electron samples consists of 62.7% and 76.7% prompt b decays respectively. The b semileptonic branching ratio is obtained by two methods: a) from the ratio of the number of dilepton events to single lepton events, we obtain the first two numbers in Table 3; the third number corresponds to the addition of eμ events with a combined fit to the muon and electron events. b) We set $\Gamma_{b\bar{b}}$ = 378MeV, the Standard Model value, and perform a one parameter fit,i.e. we mainly rely on the number of single lepton events. We also show in Table 3 the value obtained for the combined fit to e and μ data.

Table 3. Semileptonic branching ratios

Parameter	Value	Error	Method
Br(b${\to}\mu$+X)	.113	±0.012(stat)±0.006(sys)	a)
Br(b${\to}$e+X)	.138	±0.032(stat)±0.008(sys)	a)
Br(b${\to}$lept+X)	.113	±0.010(stat)±0.006(sys)	a)
Br(b${\to}\mu$+X)	.123	±0.003(stat)±0.006(sys)	b)
Br(b${\to}$e+X)	.112	±0.004(stat)±0.008(sys)	b)
Br(b${\to}$lept+X)	.119	±0.003(stat)±0.006(sys)	b)

We perform a one parameter fit to the data, with $\Gamma_{b\bar{b}}$ as free parameter and obtain $\Gamma_{b\bar{b}}$ =394±9 [MeV] for b→μ+X, $\Gamma_{b\bar{b}}$ =370±12 [MeV] for b→e+X and $\Gamma_{b\bar{b}}$ =385±7(stat) ±11(sys) ±19(Br) [MeV] for the combined fit to electron and muon samples. Combining all errors, we obtain, in good agreement with the Standard Model value of 378±3 [MeV]:

$$\Gamma_{b\bar{b}} = 385 \pm 23 \ [\text{MeV}]$$

The forward-backward asymmetry $A_{b\bar{b}}$ of $b\bar{b}$ pairs from Z^0 decays is sensitive to the electroweak mixing angle. We use the same selection procedure described above, with slightly different cuts[12], to obtain a sample of inclusive lepton events and perform a fit to the p versus p_{perp} distributions for single lepton and dilepton events, assuming no $B^0\bar{B}^0$ mixing and then correct for the effect of mixing, using our measured value (see below) of χ_B=0.178 $^{+0.049}_{-0.040}$. We obtain:

$$A_{b\bar{b}} = 0.130 \ ^{+0.044}_{-0.042} \ (\text{stat})$$

Using the improved Born approximation, QED and QCD corrections as well as correcting for our effective center of mass energy[13], we get:

$$\text{Sin}^2\theta_w = 0.226 \pm 0.008$$

The charge of the lepton produced in a direct semileptonic decay of a b quark tags the quark charge, and thus mixing can result in like-sign dileptons. The probability that a hadron containing a b quark oscillates at decay time into a hadron containing a $\bar{b}$ is given by the mixing parameter χ_B which we obtain by fitting the p and p_{perp} spectra of dileptons from Z^0 decays. The signature for $B^0-\bar{B}^0$ mixing in our case is two like-sign leptons on opposite sides of the event. Background from $Z^0{\rightarrow}b\bar{b}$, where $b{\rightarrow}\ell^-$ and $\bar{b}{\rightarrow}\bar{c}{\rightarrow}\ell^-$ can be rejected because those leptons have large p and p_{perp}. The event selection and the various sources of contamination are discussed in Ref.14. Our result is:

$$\chi_B = 0.178 \ ^{+0.049}_{-0.040}(\text{stat}) \pm 0.02(\text{sys})$$

The signed impact parameter distribution of lepton tracks from semileptonic b decays has been used[15] to measure the lifetime of B hadrons produced at the Z^0. Because B hadrons are not fully reconstructed, we measure their average lifetime weighted by their production rate and semileptonic branching ratios. The B hadron direction is approximated by the event thrust axis. The impact parameter is positive if the lepton track intersects the event thrust axis in the flight direction of the B; negative values are due to the experimental resolution and to the approximation of using the thrust axis as the B direction. The average e^+e^- interaction point is measured for each fill. The measured beam spot size is σ_x = 196±5μm, σ_y = 24±5μm and multiple scattering contributes for $\sigma_{mult} = \dfrac{83\mu m}{p[\text{GeV}]}$. The measured impact parameter distribution has a mean of +176±20μm and is a convolution of the true distribution, which depends of the B lifetime and which we determine from Monte Carlo simulations, with the smearing due to the experimental resolution. The B hadron lifetime is then obtained from a maximum likelihood fit to the impact parameter distribution. An extensive discussion of the procedure used can be found in Ref.15; our result is:

$$\tau_B = 1.32 \pm 0.08(\text{stat}) \pm 0.09(\text{sys}) \ [\text{ps}]$$

In agreement with PEP, PETRA and other LEP results.

The Energy-Energy correlation EEC and its asymmetry AEEC allow a determination of the strong coupling constant[16]. We used in this analysis 83000 events collected in 1990 at $\sqrt{s}$=91.2GeV. Our selection cuts and analysis procedure are described in Ref.17. The EEC is defined as an histogram of all angles between any two particles i,j in hadronic events, weighted by the product of their energies and averaged over the number N of events:

$$EEC(\chi_{bin}) = \frac{1}{N\Delta_{bin}} \sum_{1}^{N} \sum_{i,j} \frac{E_i E_j}{\left(\sum_i E_i\right)^2} \delta_{bin}(\chi_{bin}-\chi_{ij})$$

Where $\delta_{bin}(\chi_{bin}-\chi_{ij})$=1 for angles χ_{ij} inside the bin χ_{bin}, 0 otherwise.

Most angles are close from 0° or 180° for 2-jet events. Events with hard gluon radiation appear in the central region; thus the integral of the EEC distribution in this central region is a measure of the strong coupling constant; since such events contribute asymmetrically to the EEC distribution, we define the asymmetry $AEEC(\chi)=EEC(180°-\chi)-EEC(\chi)$. The integral of this distribution in the range ≈30°-90° is a measure of α_s.

We apply bin to bin corrections to the EEC distribution to account for detector effects, hadronization and decay, initial and final state photon radiation. The unfolded AEEC distribution is derived from the corrected EEC; both distributions are then compared to second order QCD calculations. The first order term can be written in analytical form and we use the second order term from Ref.18, reflecting differences between calculations from other groups in an additional error. The measured and calculated distributions are shown in Fig.5. We obtain (see Ref.17 for a discussion of the errors):

$$\alpha_s(EEC) = 0.121 \pm 0.004 \ (exp) \pm 0.002 \ (hadr) \ ^{+0.009}_{-0.006} \ (scale) \pm 0.006 \ (theory)$$

$$\alpha_s(AEEC) = 0.115 \pm 0.004 \ (exp) \ ^{+0.007}_{-0.004} \ (hadr) \ ^{+0.002}_{-0.000} \ (scale) \pm 0.006 \ (theory)$$

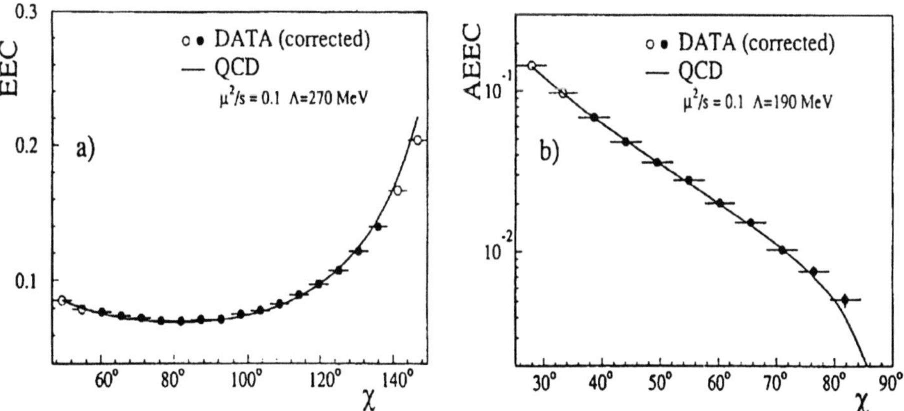

Fig.5 Comparison of EEC and AEEC distributions to second order QCD predictions for a renormalization scale f=0.1.

An earlier L3 result[19] based on 37000 Z^0 hadronic decays made use of jet multiplicities; the result is dominated by theoretical errors, particularly in the renormalisation scale:

$$\alpha_s(\sqrt{s}=91.22GeV) = 0.115 \pm 0.005(exp) \ ^{+0.012}_{-0.016}(theory)$$

We have looked[20] for narrow, high mass resonances Y produced in $e^+e^-\to Z^0\to Y\gamma$, where $Y\to e^+e^-$ or $Y\to\mu^+\mu^-$ or $Y\to$hadrons. In total respectively 150, 104 and 178 events satisfied our selection criteria for those channels; this corresponds to approximately 5.5pb^{-1} of integrated luminosity. We took advantage of the high precision of the photon energy measurement provided by the BGO electromagnetic calorimeter to search the mass spectrum of the system of particles recoiling against the photon for indications of resonances in the mass range 30-89GeV. We found no evidence for resonances with width smaller than our mass resolution and obtained the following upper limits on the product of the branching ratios, at the 95% confidence level:

$$BR(Z^0\to Y\gamma)\cdot BR(Y\to e^+e^-) < 2.8\cdot10^{-4}\ \text{for } 30\text{GeV} < M_Y < 89\text{GeV}$$

$$BR(Z^0\to Y\gamma)\cdot BR(Y\to\mu^+\mu^-) < 2.3\cdot10^{-4}\ \text{for } 30\text{GeV} < M_Y < 89\text{GeV}$$

$$BR(Z^0\to Y\gamma)\cdot BR(Y\to\text{hadrons}) < 4.7\cdot10^{-4}\ \text{for } 30\text{GeV} < M_Y < 86\text{GeV}$$

Searches for excited electrons and muons[21], for excited taus[22] and for excited neutrinos[23] have been carried out in Z^0 decays. An excited lepton is assumed to have spin 1/2 and to decay into an ordinary lepton and a photon with 100% branching ratio. The excited leptons can be produced in pairs ($e^+e^-\to\ell^*\ell^*$) or singly ($e^+e^-\to\ell\ell^*$), the second case reaching a mass region closer to the center of mass energy. The channels studied are listed in Table 4, together with the corresponding integrated luminosities and mass limits. We included the case when one electron escapes detection because the excited electron production is dominated by small angle scattering. Excited neutrinos are assumed to be of the Dirac type, thus having the same coupling as normal neutrinos to vector bosons.

Table 4. Measured mass limits, channels studied and corresponding integrated luminosities used in this analysis.

Channel	Integrated Luminosity [pb^{-1}]	Mass Limit (95% C.L.) [GeV]
$e^*\bar{e}^*\to e^+e^-\gamma\gamma$	1.0	$m_{e^*} > 45.0$
$ee^*\to e^+e^-\gamma$	"	Fig.6a
$ee^*\to(e^\pm)e^\mp\gamma$	"	Fig.6a
$\mu^*\bar{\mu}^*\to\mu^+\mu^-\gamma\gamma$	1.5	$m_{\mu^*} > 45.0$
$\mu\mu^*\to\mu^+\mu^-\gamma$	"	Fig.6b
$\tau^{*+}\tau^{*-}\to\tau^+\tau^-\gamma\gamma$	2.2	$m_{\tau^*} > 45.5$
$\tau\tau^*\to\tau^+\tau^-\gamma$	"	Fig.6c
$\nu\nu^*\to\nu e W^\pm\ \begin{array}{l}\to\nu e + 2\text{jets}\\ \to\nu e + \text{lept}^\pm\end{array}$	3.3	Fig.6d
$\nu\nu^*\to\nu\nu\gamma$	3.8	Fig.6d

Excited charged leptons are thus excluded for masses below $0.5M_Z$ and upper limits of couplings for masses up to almost M_Z have been obtained.

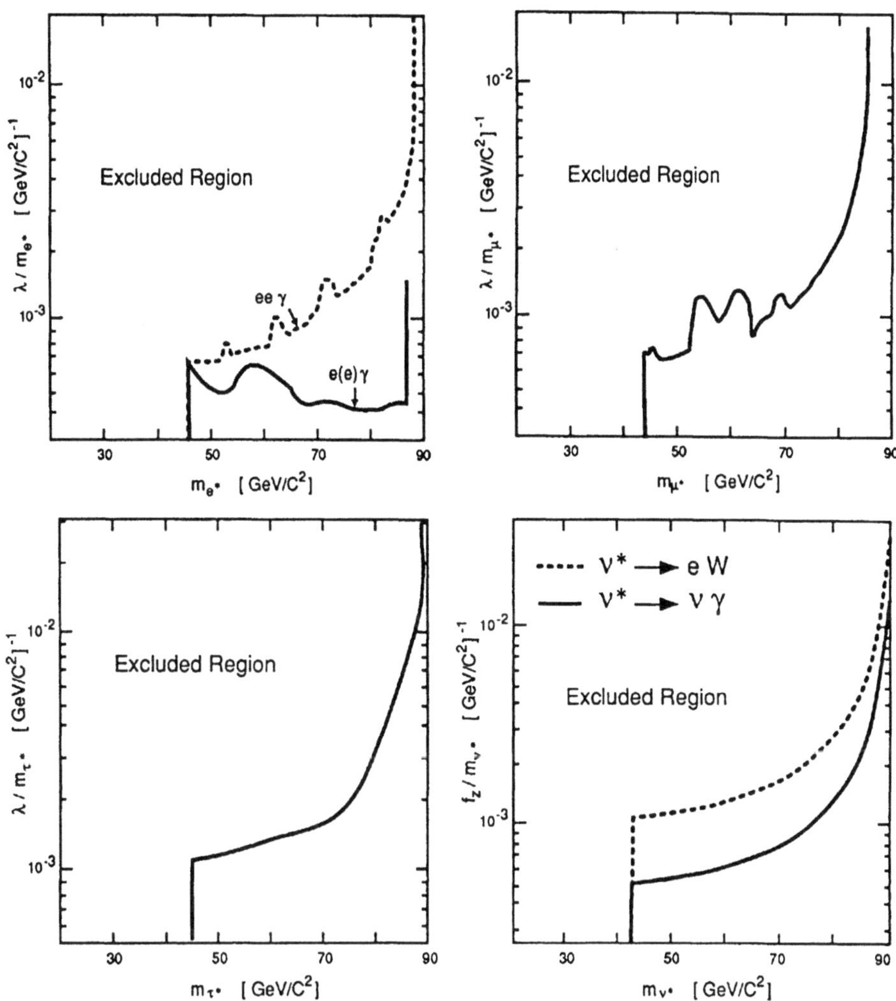

Fig.6 The 95% confidence level upper limits of the coupling constants as
a function of the mass of the excited leptons.

We have searched for new charged and neutral leptons[24]. A new type
of lepton would increase the total width of the Z^0, whereas the invisible
width of the Z^0 leads to a limit on heavy stable neutrinos. Using our measured values of M_Z, Γ_Z, Γ_{invis} we obtain two sets of limits (table 5),
with the following hypotheses:

a) M_{top}=150GeV, M_{Higgs}=100GeV, α_s=0.115 (this set of parameters is in good
agreement with our measurements)

b) M_{top}=90GeV, M_{Higgs}=1000GeV, α_s=0.110 (this set of parameters is still
compatible with our measurements and gives a lower Γ_Z)

We conducted a search for scalar leptoquarks[25], using 5.2pb^{-1} of
data, in the data samples $e^+e^-{\rightarrow}e^+e^-X$, $\mu^+\mu^-X$, $\tau^+\tau^-X$, $\nu\bar{\nu}X$. The signature of a
leptoquark is two leptons and two jets; the expected production rate is
estimated from Ref.26 and 27.

We exclude at the 95% confidence level the existence of scalar
leptoquarks with masses below 41 to 44GeV, depending on the charge assignment.

Table 5. Mass limits at 95% confidence level for new heavy leptons

Type of heavy lepton	Mass limit a) [GeV]	Mass Limit b) [GeV]
Sequential charged lepton (from Γ_Z)	27.9	14.2
Dirac neutrino (from Γ_Z)	43.2	38.8
Majorana neutrino (from Γ_Z)	35.4	30.0
Dirac neutrino (from Γ_{invis})	44.2	42.8
Majorana neutrino (from Γ_{invis})	37.6	34.8

We have conducted several searches for Higgs bosons:

a) In $Z^0 \to H^0 \mu^+ \mu^-$, $H^0 e^+ e^-$, $H^0 \nu \bar{\nu}$. Combining these three channels, we can exclude a minimal standard model Higgs boson in the mass range 2GeV $< M_{H^0} <$ 32GeV at 95% confidence level[28].

b) A light neutral scalar Higgs boson h^0 and the pseudoscalar Higgs boson A^0 of the minimal supersymmetric standard model. We consider $Z^0 \to h^0 A^0$, with $h^0 \to b\bar{b}$ and $A^0 \to b\bar{b}$, or with $h^0 \to \tau\bar{\tau}$ and $A^0 \to b\bar{b}$ or $A^0 \to \tau\bar{\tau}$, ie:

$$h^0 A^0 \to b\bar{b}b\bar{b}$$
$$h^0 A^0 \to \tau\bar{\tau}b\bar{b}$$
$$h^0 A^0 \to \tau\bar{\tau}\tau\bar{\tau}$$

We exclude[29] nearly the entire mass region up to 41.5GeV.

c) Pair-produced charged Higgs in $Z^0 \to H^+ H^-$, where :

$$H^+ H^- \to \tau^+ \nu \tau^- \bar{\nu}$$
$$H^+ H^- \to \tau^+ \nu cs$$
$$H^+ H^- \to c\bar{s}cs$$

We found no indication of $H^+ H^-$ production and can exclude a charged Higgs of mass lower than 36.5GeV at 95% Confidence level[30].

d) A low mass neutral Higgs in $Z^0 \to H^0 \mu^+ \mu^-$ or $Z^0 \to H^0 e^+ e^-$, where the signature is two isolated high momentum leptons recoiling against a low multiplicity jet. We exclude a Minimal Standard Model Higgs boson with $M_{H_0} <$ 2GeV at 99% confidence level[31].

Combining some of these results with higher statistics analysis of a)[32], we can exclude a Minimal Standard Model Higgs boson at the 95% confidence level in the mass range:

$$0 < M_{H_0} < 41.8 \text{GeV}.$$

CONCLUSION

Because of the large number of Z^0 produced at LEP and because of the excellent resolution of L3 for e, μ, γ we were able to produce a wide array of results which in most cases are either new or improve significantly on previous experiments. Our thanks must go to the people who supported, designed and built LEP, as well as to all the institutions and people who encouraged us to design L3 and helped build it.

REFERENCES

1. The L3 Collaborators are:

B.Adeva, O.Adriani, M.Aguilar-Benitez, H.Akbari, J.Alcaraz, A.Alisio,
B.Alpat, G.Alverson, M.G.Alviggi, G.Ambrosi, Q.An, H.Anderhub, A.L.Ander-
son, V.P. Andreev, T. Angelov, L.Antonov, D.Antreasyan, P.Arce,
A.Arefiev, T.Azemoon, T.Aziz, P.V.K.S.Baba, P.Bagnaia, J.A.Bakken,
L.Baksay, R.C.Ball, S.Banerjee, J.Bao, R.Barillère, L.Barone,
R.Battiston, A.Bay, U.Becker, F.Behner, J.Behrens, S.Beingessner,
Gy.L.Bencze, J.Berdugo, P.Berges, B.Bertucci, B.L.Betev, A.Biland,
G.M.Bilei, R.Bizzarri, J.J.Blaising, P.Blömeke, B.Blumenfeld,
G.J.Bobbink, M.Bocciolini, R.Bock, A.Böhm, B.Borgia, D.Bourilkov,
M.Bourquin, D.Boutigny, B.Bouwens, J.G.Branson, I.C.Brock, F.Bruyant,
C.Buisson, A.Bujak, J.D.Burger, J.P.Burq, J.Busenitz, X.D.Cai, M.Capell,
F.Carbonara, P.Cardenal, M.Caria, F.Carminati, A.M.Cartacci, M.Cerrada,
F.Cesaroni, Y.H.Chang, U.K.Chaturvedi, M.Chemarin, A.Chen, C.Chen,
G.M.Chen, H.F.Chen, H.S.Chen, M.Chen, M.L.Chen, W.Y.Chen, G.Chiefari,
C.Y.Chien, M.Chmeissani, C.Civinini, I.Clare, R.Clare, H.O.Cohn,
G.Coignet, N.Colino, V.Commichau, G.Conforto, A.Contin, F.Crijns,
X.Y.Cui, T.S.Dai, R.D'Alessandro, R.de Asmundis, A.Degré, K.Deiters,
E.Dénes, P.Denes, F.DeNotaristefani, M.Dhina, D.DiBitonto, M.Diemoz,
H.R.Dimitrov, C.Dionisi, M.T.Dova, E.Drago, T.Driever, D.Duchesneau,
P.Duinker, I.Duran, H.El Mamouni, A.Engler, F.J.Eppling, F.C.Erné, P.Ex-
termann, R.Fabbretti, M.Fabre, S.Falciano, Q.Fan, S.J.Fan, O.Fackler,
J.Fay, T.Ferguson, G.Fernandez, F.Ferroni, H.Fesefeldt, E.Fiandrini,
J.Field, F.Filthaut, G.Finocchiaro, P.H.Fisher, G.Forconi, T.Foreman,
K.Freudenreich, W.Friebel, M.Fukushima, M.Gailloud, Yu.Galaktionov,
E.Gallo, S.N.Ganguli, P.Garcia-Abia, S.S.Gau, D.Gele, S.Gentile, M.Glaub-
man, S.Goldfarb, Z.F.Gong, E.Gonzalez, A.Gordeev, P.Göttlicher, D.Gou-
jon, G.Gratta, C.Grinnell, M.Gruenewald, M.Guanziroli, J.K.Guo, A.Gurtu,
H.R.Gustafson, L.J.Gutay, H.Haan, A.Hasan, D.Hauschild, C.F.He, T.Hebbe-
ker, M.Hebert, G.Herten, U.Herten, A.Hervé, K.Hilgers, H.Hofer,
H.Hoorani, L.S.Hsu, G.Hu, G.Q.Hu, B.Ille, M.M.Ilyas, V.Innocente,
H.Janssen, S.Jezequel, B.N.Jin, L.W.Jones, A.Kasser, R.A.Khan,
Yu.Kamyshkov, Y.Karyotakis, M.Kaur, S.Khokhar, V.Khoze, M.N.Kienzle-
Focacci, W.Kinnison, D.Kirkby, W.Kittel, A.Klimentov, A.C.König,
O.Kornadt, V.Koutsenko, R.W.Kraemer, T.Kramer, V.R.Krastev, W.Krenz,
J.Krizmanic, K.S.Kumar, V.Kumar, A.Kunin, V.Lalieu, G.Landi, K.Lanius,
D.Lanske, S.Lanzano, P.Lebrun, P.Lecomte, P.Lecoq, P.LeCoultre, D.Lee,
I.Leedom, J.M.Le Goff, L.Leistam, R.Leiste, M.Lenti, E.Leonardi,
J.Lettry, P.M.Levchenko, X.Leytens, C.Li, H.T.Li, J.F.Li, L.Li, P.J.Li,
X.G.Li, J.Y.Liao, Z.Y.Lin, F.L.Linde, B.Lindemann, D.Linnöhfer, R.Liu,
Y.Liu, W.Lohmann, E.Longo, Y.S.Lu, J.M.Lubbers, K.Lübelsmeyer, C.Luci,
D.Luckey, L.Ludovici, L.Luminari, W.G.Ma, M.MacDermott, R.Magahiz,
P.K.Malhotra, C.Mana, R.Malik, A.Malinin, C.Maña, G.Mantovani, D.N.Mao,
Y.F.Mao, M.Maolinbay, P.Marchesini, A.Marchionni, J.P.Martin, L.Martinez-
Laso, F.Marzano, G.G.G.Massaro, T.Matsuda, K.Mazumdar, P.McBride,
T.McMahon, D.McNally, Th.Meinholz, M.Merk, L.Merola, M.Meschini,
W.J.Metzger, Y.Mi, G.B.Mills, Y.Mir, G.Mirabelli, J.Mnich, M.Möller,
B.Monteleoni, G.Morand, R.Morand, S.Morganti, N.E.Moulai, R.Mount,
S.Müller,E.Nagy,M.Napolitano, H.Newman, C.Neyer, M.A.Niaz, L.Niessen,
W.D.Nowak,D.Pandoulas, M.Pauluzzi, F.Pauss, G.Passaleva, G.Paternoster,
S.Patricelli, Y.J.Pei, D.Perret-Gallix, J.Perrier, A.Pevsner, M.Pieri,
P.A.Piroué, F.Plasil, V.Plyaskin, M.Pohl, V.Pojidaev,N.Produit,
J.M.Qian, K.N.Qureshi, R.Raghavan, G.Rahal-Callot, G.Raven, P.Razis,
K.Read, D.Ren, Z.Ren, S.Reucroft, A.Ricker, T.Riemann, O.Rind,
C.Rippich, H.A.Rizvi, B.P.Roe, M.Röhner, S.Röhner, L.Romero, J.Rose,
S.Rosier-Lees, R.Rosmalen, Ph.Rosselet, A.Rubbia, J.A.Rubio, M.Rubio,
W.Ruckstuhl, H.Rykaczewski, M.Sachwitz, J.Salicio, J.M.Salicio, G.Sand-
ers, A.Santocchia, M.S.Sarakinos, G.Sartorelli, G.Sauvage, A.Savin,
V.Schegelsky, K.Schmiemann, D.Schmitz, P.Schmitz, M.Schneegans,

H.Schopper, D.J.Schotanus, S.Shotkin, H.J.Shreiber, R.Schulte,
S.Schulte, K.Schultze, J.Schütte, J.Schwenke, G.Schwering, C.Sciacca,
I.Scott, R.Sehgal, P.G.Seiler, J.C.Sens, L.Servoli, I.Sheer, D.Z.Shen,
V.Shevchenko, S.Shevchenko, V.Shevchenko, X.R.Shi, K.Shmakov, V.Shoutko,
E.Shumilov,N.Smirnov, E.Soderstrom, N.Sopczak, C.Spartiotis,
T.Spickermann, P.Spillantini, R.Starosta, M.Steuer, D.P.Stickland,
F.Sticozzi, W.Stoeffl, H.Stone, K.Strauch, B.C.Stringfellow, K.Sudhakar,
G.Sultanov, R.L.Sumner, L.Z.Sun, H.Suter, R.B.Sutton, J.D.Swain,
A.A.Syed, X.W.Tang, E.Tarkovsky, L.Taylor, C.Timmermans, Samuel
C.C.Ting, S.M.Ting, Y.P.Tong, F.Tonisch, M.Tonutti, S.C.Tonwar, J.Tóth,
G.Trowitzsch, C.Tully, K.L. Tung, J.Ulbricht, L.Urbàn, U.Uwer,
E.Valente, R.T.Van de Walle, I.Vetlitsky, G.Viertel, P.Vikas, U.Vikas,
M.Vivargent, H.Vogel, H.Vogt, G.Von Dardel, I.Vorobiev, A.A.Vorobyov,
An.A.Vorobyov, L.Vuilleumier, M.Wadhwa, W.Wallraff, C.R.Wang, G.H.Wang,
J.H.Wang, Q.F.Wang, X.L.Wang, Y.F.Wang, Z.Wang, Z.M.Wang, A.Weber,
J.Weber, R.Weill, T.J.Wenaus, J.Wenninger, M.White, C.Willmott,
F.Wittgenstein, D.Wright, R.J.Wu, S.L.Wu, S.X.Wu, Y.G.Wu, B.Wyslouch,
Y.Y.Xie, Y.D.Xu, Z.Z.Xu, Z.L.Xue, D.S.Yan, X.J.Yan, B.Z.Yang, C.G.Yang,
G.Yang, K.S.Yang, Q.Y.Yang, Z.Q.Yang C.H.Ye, J.B.Ye, Q.Ye, S.C.Yeh,
Z.W.Yin, Z.M.You, M.Yzerman, C.Zaccardelli, P.Zemp, M.Zeng, Y.Zeng,
D.H.Zhang, Z.P.Zhang, J.F.Zhou, R.Y.Zhu, H.L.Zhuang, A.Zichichi.

The L3 participating institutions are:

1. Physikalisches Institut, RWTH, Aachen, and 3. Physikalisches In-
stitut, RWTH, Aachen, Germany
National Institute for High Energy Physics, NIKHEF, Amsterdam; NIKHEF-H
and University of Nijmegen, The Netherlands
University of Michigan, Ann Arbor, USA
Laboratoire de Physique des Particules, LAPP, Annecy, France
Johns Hopkins University, Baltimore, USA
Institute of High Energy Physics, IHEP, Beijing, People's Republic of
China
INFN Sezione di Bologna, Italy
Tata Institute of Fundamental Research, Bombay, India
Northeastern University, Boston, USA
Central Research Institute for Physics of the Hungarian Academy of Sci-
ences, Budapest, Hungary
Harvard University, Cambridge, USA
Massachusetts Institute of Technology, Cambridge, USA
INFN Sezione di Firenze and University of Firenze, Italy
Leningrad Nuclear Physics Institute, Gatchina, Soviet Union
European Laboratory for Particle Physics, CERN, Geneva, Switzerland
World Laboratory, FBLJA Project, Geneva, Switzerland
University of Geneva, Geneva, Switzerland
Chinese University of Science and Technology, USTC, Hefei,People's Repub-
lic of China
University of Lausanne, Lausanne, Switzerland
Lawrence Livermore National Laboratory, California, USA
Los Alamos National Laboratory, Los Alamos, New Mexico, USA
Institut de Physique Nucléaire de Lyon, IN2P3-CNRS/Université Claude Ber-
nard, Villeurbanne, France
Center of Energy and Environmental Research, CIEMAT, Madrid, Spain
Institute of Theoretical and Experimental Physics, ITEP,Moscow, Soviet
Union
INFN-Sezione di Napoli and University of Naples, Italy
Oak Ridge National Laboratory, Oak ridge, Tennessee, USA
California Institute of Technology, Pasadena, USA
INFN Sezione di Perugia and Universita Degli Studi di Perugia, Italy
Carnegie Mellon University, Pittsburgh, USA
Princeton University, Princeton, USA
INFN-Sezione di Roma and University of Roma, "La Sapienza", Italy

University of California, San Diego, USA
Union College, Schenectady, USA
Shanghai Institute of Ceramics, SIC, Shanghai, People's Republic of China
Central Laboratory of Automation and Instrumentation, Sofia, Bulgaria
University of Alabama, Tuscaloosa, Alabama, USA
Purdue University, West Lafayette, Indiana, USA
Paul Scherrer Institut, PSI, Würenlingen, Switzerland
High Energy Physics Institute, Zeuthen-Berlin, Germany
Eidgenössische Technische Hochschule, ETH Zürich, Switzerland
University of Hamburg, Germany
National Science Council, Taiwan, China

2. The construction of the L3 experiment, Nucl.Instr.Meth. A289 (1990) 35

3. Measurement of electroweak parameters from hadronic and leptonic decays of the Z^0, Zeitschrift für Physik. C51 (1991) 179

4. A determination of the properties of the neutral intermediate vector boson Z^0, Phys.Lett. 231B (1989) 509

5. Measurement of Z^0 decays to hadrons and a precise determination of the number of neutrino species, Phys.Lett. 237B (1990) 136

6. A measurement of the Z^0 leptonic partial widths and the Vector and Axial Vector Coupling Constants, Phys.Lett. 238B (1990) 122

7. A determination of electroweak parameters from $Z^0 \Rightarrow \mu^+\mu^-$ (γ) Phys.Lett. 247B (1990) 473

8. A precision measurement of the number of neutrino species Phys.Lett. 249B (1990) 341

9. A determination of electroweak parameters from Z^0 decays into charged leptons, Phys.Lett. 250B (1990) 183

10. CHARM collaboration, J.Dorenbosch et al., Z.Phys. C41 (1989) 567
K.Abe et al., Phys.Rev.Lett. 62 (1989) 1709
CHARM II collaboration, D.Geigerat et al., Phys.Lett. B232 (1989) 539
F.Avignone et al., Phys.Rev. D16 (1977) 2383
U.Amaldi et al., Phys.Rev. D36 (1987) 1385

11. Measurement of $Z^0 \Rightarrow b\bar{b}$ decays and the semileptonic branching ratio Br(b $\Rightarrow$ lepton + X), Phys.Lett. 261B (1991) 177

12. A measurement of the $Z^0 \Rightarrow b\bar{b}$ forward-backward asymmetry, Phys.Lett. 252B (1990) 713
Measurement of $Z^0 \Rightarrow b\bar{b}$ Decay Properties, Phys.Lett. 241B (1990) 416

13. A.Djouali et al., Z.Phys. C46 (1990) 411

14. A measurement of B^0-$\bar{B}^0$ mixing in Z^0 decays, Phys.Lett. 252B (1990) 703

15. Measurement of the Lifetime of B Hadrons and a determination of |Vcb| Submitted to Phys.Lett.

16. P.M.Stevenson, Phys.Rev. D23 (1981) 2916

17. Determination of α_s from Energy-Energy correlations measured on the Z^0 resonance. Phys.Lett. 257B (1991) 469

18. Z.Kunszt and P.Nason, "Z physics at LEP 1", CERN Rep.89-08, vol 1, 373

19. Determination of α_s from jet multiplicities measured on the Z^0 resonance, Phys.Lett. 248B (1990) 464

20. Search for narrow high mass resonances in radiative decays of the Z^0 Phys.Lett. 262B (1991) 15

21. Mass limits for excited electrons and muons from Z^0 decay Phys.Lett. 247B (1990) 177

22. Search for excited Taus from Z^0 decays
Phys.Lett. 250B (1990) 203

23. Search for excited neutrinos from Z^0 decays
Phys.Lett. 252B (1990) 525

24. A search for heavy charged and neutral leptons from Z^0 decays
   Phys.Lett. 251B (1990) 321

25. Search for leptoquarks in Z^0 decays
   Phys.Lett. 261B (1991) 169

26. D.Schaile and P.M.Zerwas, CERN 87-07, Vol.II, p.251

27. J.L.Hewett and T.G.Rizzo, Phys. Rev. D36 (1987) 3367

28. Search for the neutral Higgs boson in Z^0 decay
   Phys.Lett. 248B (1990) 203

29. Search for the neutral Higgs bosons of the minimal supersymmetric
   standard model from Z^0 decays
   Phys.Lett. 251B (1990) 311

30. Search for the charged Higgs boson in Z^0 decay
   Phys.Lett. 252B (1990) 511

31. Search for a low mass neutral Higgs boson in Z^0 decay
   Phys.Lett. 252B (1990) 518

32. Search for the neutral Higgs boson
   Phys.Lett. 257B (1991) 450

STRUCTURE FUNCTIONS AT SMALL-X AND THE REGGE LIMIT IN QCD

J.Bartels

II. Institut für Theoretische Physik, Universität Hamburg

Hamburg, Germany

ABSTRACT

An review is given of our present understanding of the behavior of deep inelastic structure functions in the small-x region. Particular emphasis is given to the connection with the Regge limit and our current knowledge of this high energy limit in QCD.

1 Introduction

Deep Inelastic scattering of electrons on protons (DIS) (Fig.1) provides the classical test of our parton picture of hadrons and, hence, of QCD at short distances. With the advent of HERA (anticipated for the summer of 1991) as well as the advanced planning of the SSC, LHC and the ep-version LEP-LHC, we are entering new kinematical regions in both $Q^2 = -q^2$ and

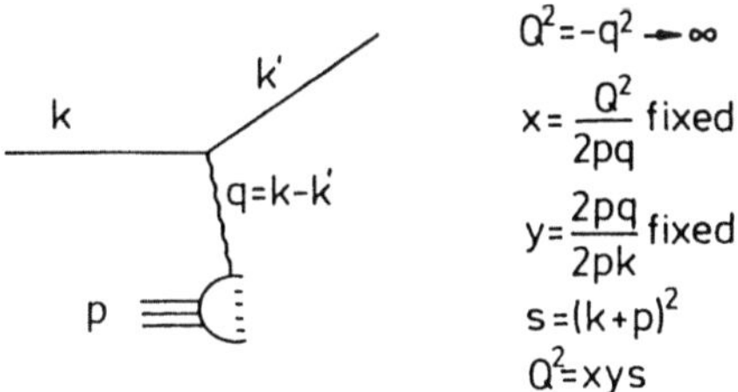

Figure 1. *Kinematics of deep inelastic scattering.*

Bjorken-$x = \frac{Q^2}{2pq}$ (Fig.2). Whereas the large-Q^2 region is expected to exhibit the logarithmic scaling violations predicted by QCD (unless deviations, indicative of new physics in the TeV region, show up), the other direction towards small-x leads to a new kinematic region ,which has not been explored so far and, hence, provides new insight into the dynamics of QCD. Furthermore, estimates indicate that numbers of events at HERA will accumulate at small x: it is therefore mandatory to have a thorough understanding of this kinematic regime. In this talk I will try to summarize, what kind of phenomena we are expecting to see and to what extent we are theoretically prepared. For many details I refer to [1].

QCD at 200 TeV, Edited by L. Cifarelli
and Y. Dokshitzer, Plenum Press, New York, 1992

Let me briefly recapitulate the set-up. With the variables $Q^2 = -q^2$ and $s = (k + p)^2$ (Fig.1) it is convenient to define, in addition to Bjorken-x, the variable y:

$$y = \frac{2pq}{2pk} \tag{1.1}$$

with

$$Q^2 = xys. \tag{1.2}$$

The Bjorken limit is defined as

$$Q^2 \to \infty, \ x\, fixed. \tag{1.3}$$

The momentum-transfer Q^2 defines the resolution by which the photon probes the short-distance structure of the nucleon, x is the momentum fraction carried by the struck parton inside the nucleon, and y describes, in the rest frame of the proton, the energy transfer from the incoming to the outgoing electron. For small x we have $\frac{1}{x} \approx \frac{s}{Q^2}$. The connection between different high energy limits is illustrated in Fig.3, in particular the Bjorken limit and the Regge limit $(s \to \infty, Q^2 \approx Q_0^2 \ fixed)$. The small-x region above the Bjorken limit is not far from the Regge limit: it is the continuation of the Regge limit towards large Q^2, and one therefore expects ultimately to face the same dynamics as in the Regee limit. For the discussion of the small-x limit of DIS scattering it is useful to use the variables

$$\xi = \ln\ln\frac{Q^2}{\Lambda^2} \tag{1.4}$$

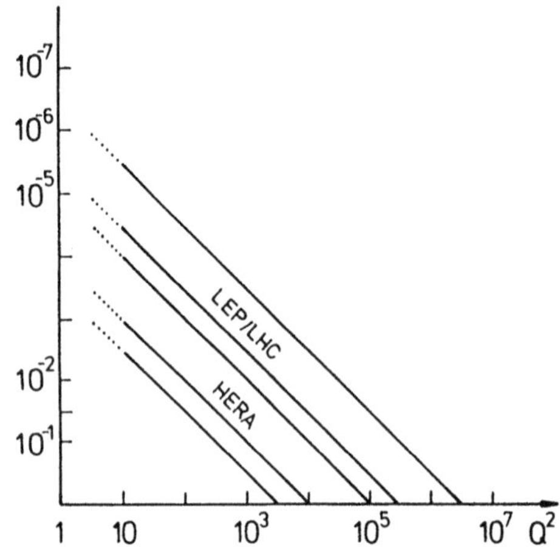

Figure 2. *Kinematic regions of HERA and LEP/LHC (only for $Q^2 > 3GeV^2$)*

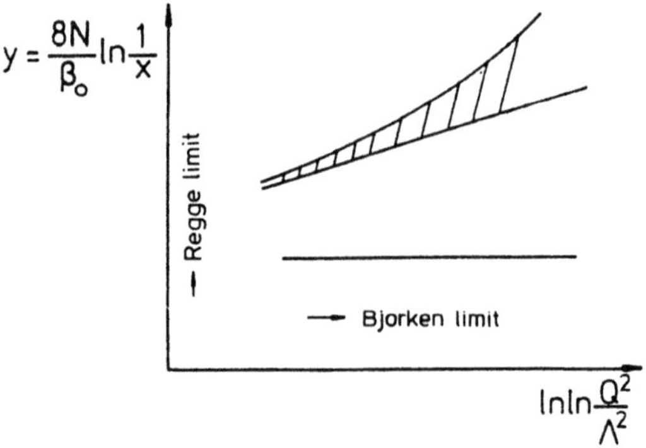

Figure 3. *Different High Energy Limits*

and

$$y = \frac{8N_c}{\beta_0} \ln \frac{1}{x}. \tag{1.5}$$

As usual, $q(x, Q^2)$ and $g(x, Q^2))$ define the probabilities of finding a quark and a gluon, resp., inside the nucleon. In the naive parton model, the nucleon structure functions F_1 and F_2 are given by:

$$F_1(x, Q^2) = \frac{1}{2} \sum_i Q_i^2 q_i(x, Q^2) \tag{1.6}$$

$$F_2(x, Q^2) = 2x F_1(x, Q^2). \tag{1.7}$$

In QCD, eq.(1.7) no longer holds, since the the longitudinal structure function is nonzero. The differential cross section is:

$$\frac{d^2\sigma}{dx dy} = \frac{4\pi\alpha^2}{Q^2} \left(y F_1(x, Q^2) + \frac{1-y}{y} F_2(x, Q^2) \right). \tag{1.8}$$

Throughout this talk, we shall use F_2, i.e. the momentum-weighted distribution functions. For simplicity, it will be denoted by $F(x, Q^2)$.

The theoretical situation of the various high energy limits shown in Fig.3 can be summarized as follows. The Bjorken limit (in the horizontal direction of Fig.3) is dominated by the light cone behavior of operator products and can therefore reliably be described by perturbation theory [2]. In particular, the renormalization group equation predicts [3,4,5] the Q^2 dependence of the moments of the structure functions. The Regge limit (in the vertical direction of Fig.3), on the other hand, has a strong nonperturbative component: for small momentum transfer scattering of hadrons is dominated by large transverse distances (impact parameter) and, hence, sensitive to the behavior of QCD at large distances. Instead of renormalization group equation and operator expansion, the most important "tool" in the Regge limit is unitarity: all attempts to study the Regge limit of QCD [6,7,8,9,10,11,12] start from perturbation theory and try to fulfill unitarity on both the s and t-channel. Solutions to this set of highly self-constrained equations will necessarily bring in nonperturbative contributions. All this, however, has not yet been worked out, and at present we have no QCD-based prediction for this region. In between these two regions lies the small-x limit of deep inelastic scattering: it probes the transition from the perturbative Bjorken limit to nonperturbative QCD and thus provides the opportunity to test theoretical ideas about the onset of nonperturbative dynamics in QCD, the change from the language of quarks and gluons to that of composite hadrons.

As it will be descibed below in more detail, this transition does not seem to be an abrupt one but has some "fine structure". Namely, according to the analysis of Gribov, Levin, and Ryskin [13] and Mueller and Qiu [14], there is a strip between the two regions in which, although the standard QCD evolution program is no longer valid, perturbative QCD can still be used. The linear evolution equations of Gribov, Lipatov, Altarelli and Parisi have to be replaced by another set of nonlinear equations, which in the following, I will name GLR-equations. In the derivation of these equations one uses elements of both the Regge limit and the Bjorken limit. When analysing these equations, it turns out that the beginning of the nonperturbative region is marked by a boundary ("critical") line which has the approximate form [15]:

$$y = \frac{1}{4} \exp 2\xi + \exp \xi + 2 \ln \ln \xi + const \tag{1.9}$$

The transition across this line appears to be smooth [16]. Due to our lack of a full understanding of the Regge limit, this GLR equation for the intermediate region seems, at present, to be the most useful tool at hand. Further work on its theoretical basis as well as an extended analytic and numerical analysis of its solutions seems most desirable.

This talk will be organized as follows. In the first part (section II) I will try to illustrate, within the parton picture, which modifications have to be included when one moves from the Bjorken limit towards smaller x-values. Then (section III) I turn to a more theoretical

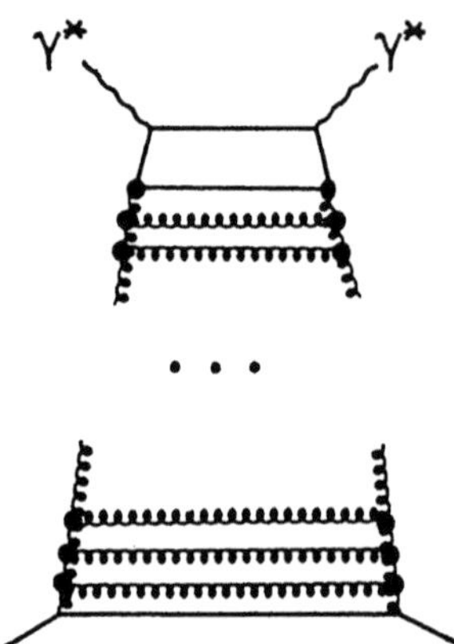

Figure 4. *QCD ladder diagrams describing the standard linear QCD evolution equation of the gluon structure function (quark contributions are left out)*

description of the small-x region, in particular to the Gribov-Levin-Ryskin equations and the origin of the critical line of Fig.3. Then I will turn to the Regge limit and describe some recent work in this field. In the final part (section V) I discuss phenomenological aspects of the small-x region: numerical solutions to the GLR- equations and the important question, how HERA can see the beginning of the new facets inside the nucleon.

2 The Parton Picture in the Small-x Region

Let me first recapitulate what goes wrong with the standard QCD evolution program when x is taken to be very small. The small-x behaviour of F_2 is dominated by gluon production and is of the form:

$$F(x, Q^2) \approx \frac{exp\left(\sqrt{2(\xi - \xi_0)y}\right)}{\sqrt{2\pi\sqrt{2(\xi - \xi_0)}}} \tag{2.1}$$

From this it follows that the total cross section for the scattering of a virtual photon off the proton would rise faster than any power of $\ln \frac{1}{x} \approx \ln \frac{s}{Q^2}$. Unitarity, on the other hand, requires that the growth cannot be faster than the square of the hadron radius:

$$\sigma_{tot}^{\gamma^{\bullet}} = \frac{4\pi^2 \alpha}{Q^2} F(x, Q^2) \leq 2\pi R(s)^2, \tag{2.2}$$

and the hadron radius grows as

$$R(s) \approx const \cdot \ln s. \tag{2.3}$$

Somewhere in the small-x region, therefore, the standard QCD description has to become invalid.

That something new has to come in at small values of x can be seen rather easily within the parton picture [13]. The standard QCD-evolution framework can be viewed as a cascade of partonic decay processes inside the nucleon: the photon picks a quark with momentum fraction x and virtuality Q^2 (which is approximately equal to the transverse momentum squared or the inverse square of the transverse radius). Such a quark represents the final result of a chain of subsequent decay processes, in course of which the partons become slower and, at the same time, gain a larger virtuality. This picture is nothing but a space-time interpretation of the QCD-ladder diagrams shown in Fig.4. In the transverse (impact parameter) plane the same situation is illustrated in Fig.5: partons at a low momentum scale Q^2 (or at the lower end of the ladder) have a rather larger transverse size and are drawn as being "fat" in Fig.5. Partons at the upper end have larger virtuality and are much "thinner" (Fig.5b and c). The number

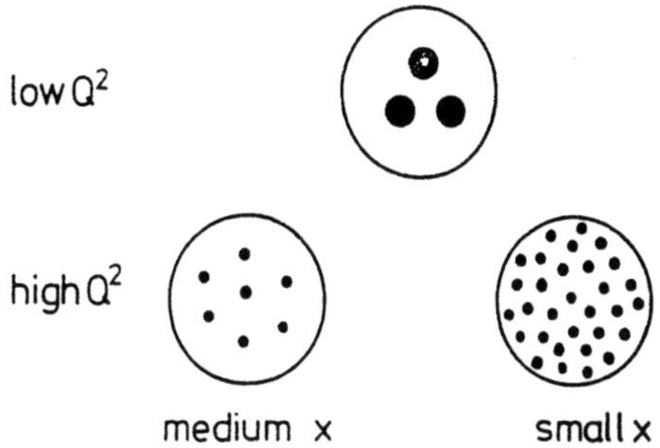

Figure 5. *Partons in the transverse plane (impact parameter plane)*

(or density) of the partons is determined by the structure functions: according to eq.(2.1) which represents the "standard QCD evolution", the number of small-x partons increases very fast. Hence, in the transverse plane their density becomes high and it becomes more and more probable that they will interact among each other. In the standard QCD framework, only one type of interaction has been taken into account, namely decay processes which cause the parton density to increase. It therefore appears to be extremely suggestive to expect also other interaction processes, e.g. recombination or annihilation processes which might balance

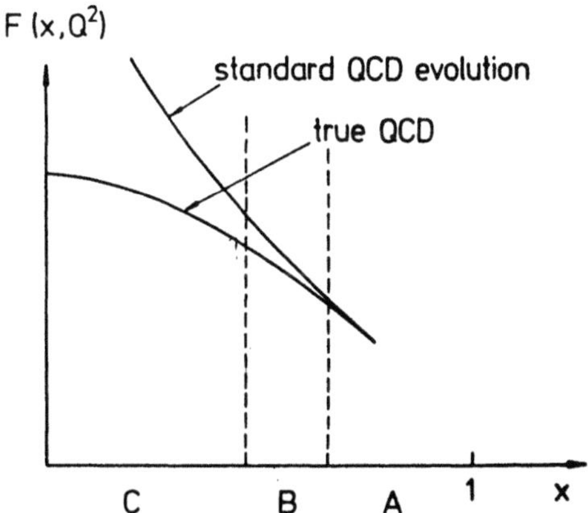

Figure 6. *Small-x behavior of the structure function: standard QCD evolution versus"true QCD" evolution: region A is the perturbative region, B the transition region, and C the non-perturbative region.*

the number of partons or even decrease their density. A possible result of these creation and annihilation processes could be a "saturation" as shown in Fig.6: the "true" QCD structure function no longer increases at small x, it approaches a constant value (up to powers of $\ln \frac{1}{x}$). Such a behaviour would also match the observed slow increase of the hadronic total cross section at low Q_0^2 in the Regge limit. Strong-field calculations within QED and QCD [17] indeed point into this direction.

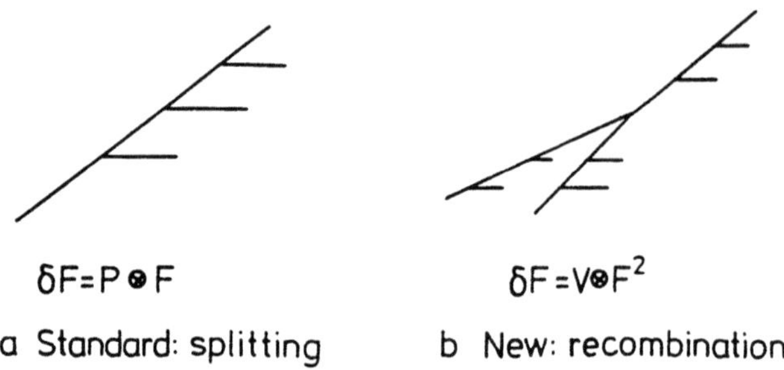

Figure 7. *Partonic subprocesses: (a) splitting of partons (b) partonic recombination process.*

A simple argument (Fig.7) may illustrate how such partonic annihilation processes will modify the QCD evolution equations. In the standard scenario (Fig.7a) a change of parton densities is obtained by a splitting of the incoming parton into two outgoing partons. Such a change δF is proportional to the probability of finding the initial parton, i.e. we have a linear evolution equation. Recombination processes (Fig.7b), on the other hand, must be proportional to the probability of finding two incoming partons. Most naively, this probability could be assumed to be proportional to the square of the probability of having one parton: one then obtains the nonlinear term indicated in Fig.7b. The strenght of this nonlinear term also depends upon the descent of the two incoming partons: if they originate from one common "parent" parton (e.g. one valence quark), it seems more likely that they have a chance to recombine again than for the case where they come from different valence quarks which are

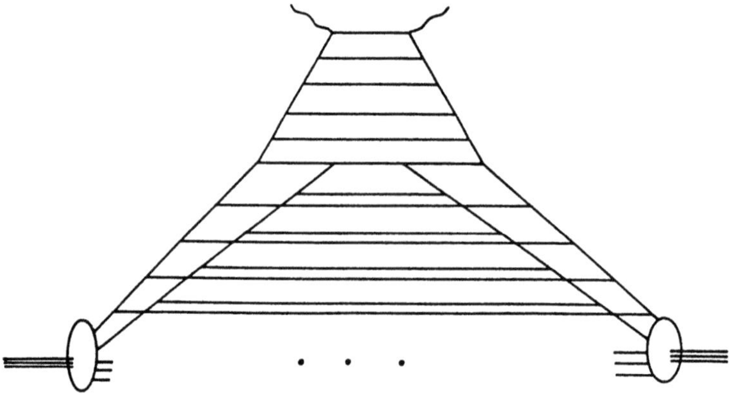

Figure 8. *Schematic view of QCD diagrams which contain the recombination process of Fig.7b (fan digram)*

spatially seperated. In the first case, the strenght would be proportional to the inverse square of the quark radius, in the latter to that of the hadron. Switching back to the language of Feynman diagrams, such a recombination process would be descibed by the (nonplanar) fan diagram shown in Fig.8: taking the energy discontinuity through the center, one obtains the recombination subprocess of Fig.7b. One should, however, note that there are other ways of cutting Fig.8: this indicates that the recombination process of Fig.7b is necessarily accompanied by other processes (rescattering in the initial and final states and diffractive dissociation of the photon).

This very qualitative description of small-x physics allows, in fact, a first estimate of the boundary between perturbative and nonperturbative physics in the ξ, y plane (Fig.2). Consider the quantity

$$W(x, Q^2) = \frac{\alpha_s(Q^2)}{Q^2} \frac{F(x, Q^2)}{R^2} \approx \frac{const}{Q^2 \ln Q^2} \frac{\exp \sqrt{2\xi y}}{y^2} \tag{2.4}$$

which measures the probability of partons to interact with each other. In order that the standard QCD casacde picture works, this probality should be less than one: neglecting all nonleading terms, this condition can be approximated by

$$y \leq const \cdot \exp 2\xi \tag{2.5}$$

which is a very crude approximation of the critical line in Fig.3.

To conclude this section, several remarks should be made. First, it has to be emphazised that the use of the parton language for the interpretation of QCD-Feynman diagrams becomes ambiguous if one goes beyond the leading-log Q^2 approximation. It is, therefore, by no means obvious how one could discuss this "new physics" consistently in the parton picture (i.e. without resorting to perturbation theory and Feynman diagrams). For example, it would be very useful to have a more intuitive derivation of the Gribov-Levin-Ryskin equation (which will be discussed below) in terms of partonic subprocesses. This equation is the result of a careful analysis of large classes of Feynman diagrams and takes into account nonleading twist contributions. The simple translation from QCD perturbation theory to the parton picture therefore no longer holds; as an example, one of the difficulties that one encounters are the so-called Abramovsky-Gribov-Kancheli cutting rules [18] which relate the different energy discontinuities of the fan diagrams to each other.

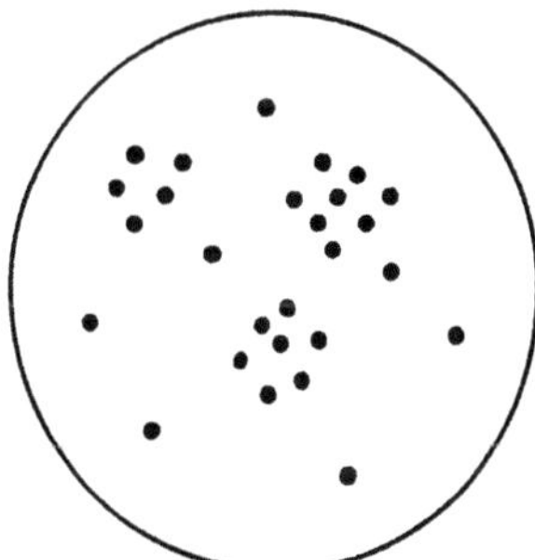

Figure 9. A "Hot spot" in the transverse plane

Second, the inclusion of the new partonic interactions such as annihilations or recombinations can describe only the beginning of new physics ("saturation") in the small-x region: near and beyond the critical line nonperturbative physics sets in, and theoretical progress will necessarily be linked to a better understanding of the nearby Regge limit.

Finally, for practical purposes to be disussed further below it may be necessary to develop a somewhat refined picture of the "saturation" [19]. Whereas the structure functions measure the parton density, averaged over the full transverse size of the nucleon, the saturation of partons may actually begin in a nonuniform fashion (Fig.9). In the simplest scenario, high density of partons may start to develop first close to the valence quarks, whereas other regions inside the nucleon may still remain empty for some range in x. The structure function would then still continue to grow with $\frac{1}{x}$, although locally saturation may have already been reached. Following A.Mueller, one could name these regions "Hot spots". To test this possibility, one has to look for special final states which are designed to probe small areas inside the nucleon. This will be discussed further in the third part of this review.

3 Quantitative Description: QCD and the Gribov-Levin-Ryskin Equation

3.1 General Remarks

After this more qualitative discussion of the small-x region it is necessary to adress the question: how can we make all this more quantitative, e.g. how can we compute these effects and how can we find out for what values of x and Q^2 these new effects start to become important?

It is instructive to first start from the operator expansion and discuss what happens when we approach the small-x limit. It may be useful to remember that, on very general grounds, moments of the structure function are expected to become singular near n=1: since hadronic total cross sections are slowly growing with energy (presently an increase proportional to $(\ln s)^2$ is observed), a similar behavior is expected for the total cross section for the scattering of a virtual photon off a nucleon. It then follows that the moment

$$\int_0^1 dx\, x^{n-1} W(x, q^2) = \frac{1}{2\pi} \int_0^1 dx\, x^{n-1} Im T_{\gamma^* p}(x, q^2) \tag{3.1}$$

should be singular near n=1. Taking the Mellin transform of (3.1), one immediately sees that the small-x behavior of $W(x, Q^2)$ will be dominated by this singular behavior. In perturbation theory this singularity is well-known: the anomalous dimension, e.g., γ_{VV} of the two-gluon operator

$$F^{\alpha \mu_1} D^{\mu_2} ... D^{\mu_{n-1}} F_\alpha^{\mu_n} \tag{3.2}$$

behaves as

$$\gamma_{VV} = \frac{4N}{\beta_0} \frac{1}{n-1} \tag{3.3}$$

near $n = 1$.

Now the important point to note is that this singularity will show up also in gluon operators of higher twists, e.g. in the anomalous dimension of the four-gluon operator:

$$(\partial)^{n_1} A_\mu (\partial)^{n_2} A^\mu (\partial)^{n_3} A_\nu (\partial)^{n_4} A^\nu \tag{3.4}$$

(evolution equations for operators of higher twist have been discussed, for example, in [20]). The usual argument of neglecting, for large Q^2, all operators of twist higher than two, therefore has to be modified in the small-x limit: depending upon the strength of the singularity near $n = 1$ and the smallness of x, operators of higher twist may become as important as or even more important than the twist-two contribution. In general, terms of all twists need to be investigated (i.e. the expansion in inverse powers of Q^2 is no longer useful) , and other arguments have to be used in order to decide which terms have to be kept. I believe that this should be done by starting from the Regge limit where unitarity is the powerful guiding principle, and then approaching the small-x part of the Bjorken limit in Fig.3 from above, i.e.from the opposite side compared to the perturbative Bjorken limit. I will say more on the Regge limit in the following section.

3.2 The GLR Equation

The first - and so far only - serious attempt for selecting those classes of QCD Feynman diagrams which become important when going from the Bjorken limit towards smaller x-values has been made about ten years ago by Gribov, Levin, and Ryskin [13] and, somewhat later, by Mueller and Qiu [14]. The first of these corrections to the standard QCD evolution is illustrated in Fig.10. The first term is a repeat of Fig.4, the sum of (gluonic) QCD ladder diagrams which leads to the evolution equations of Gribov, Lipatov, Altarelli, and Parisi. The second part illustrates the first correction (so-called fan diagram): the upper

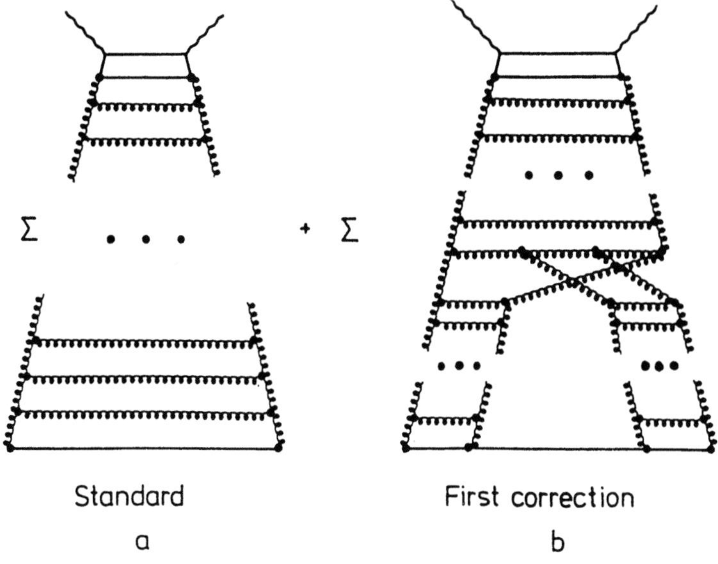

Standard   First correction

a   b

Figure 10. *QCD ladders (a) and the first correction, the so-called fan diagram (b).*

ladder branches into two ladders which at the lower end couple to the nucleon (in our picture only one quark line is shown; other possibilities include the coupling to two different quarks). The branching vertex (triple-ladder vertex) consists of the sum of several (nonplanar) diagrams, only one of which is shown. Taking the energy discontinuity of this set of diagrams in all possible ways, one obtains, among other contributions, the recombination process shown in Fig.7. However, since this set of diagrams allows for other energy cuts, the recombination process in Fig.7 must necessarily be accompanied by other contributions, e.g. the diffractive cut shown in Fig.11. A careful analysis of all these terms leads to the well-known Abramowski-Gribov-Kancheli rules (AGK-rules) [18]. In particular, the sum of all terms carries a minus sign, relative to the ladder diagrams which correspond to the standard evolution equations. Clearly, this minus sign is highly desirable to fight the strong increase of the structure function at small x.

In the approximation where only leading powers in both lnQ^2 and $ln\frac{1}{x}$ are kept (DLA approximation), the behavior of Fig.10 is rather simple. If we denote by $F_0(y,\xi)$ the expression for the right part of Fig.10, then the first correction is given by:

$$- const \int^y dy_1 \int^\xi d\xi_1 F_0(y - y_1, \xi - \xi_1)V(\xi_1)F_0^2(y_1, \xi_1) \tag{3.5}$$

where

$$V(\xi) = \frac{3\pi^2}{4\beta_0}\frac{Q_0^2}{\Lambda^2} \exp -e^\xi - \xi \tag{3.6}$$

stands for the triple ladder vertex.

Following Gribov, Levin, and Ryskin, the next corrections beyond the fan diagram of Fig.10 are generalizations indicated in Fig.12: moving

from the top to the bottom, QCD ladders branch into two ladders, thus increasing the number of ladders up to infinity. Diagrams with decreasing number of ladders or diagrams with number-conserving ladder-interaction vertices do not contribute in this transition region. Such contributions are essential in the Regge limit (where it is more appropriate to organize the diagrams in the form of reggeon diagrams), so one might expect that they become important for the small-x limit of DIS as soon as one approaches the critical line. From the point of view of the Regge limit, the fan diagrams of Gribov, Levin, and Ryskin thus represent a subset of

97

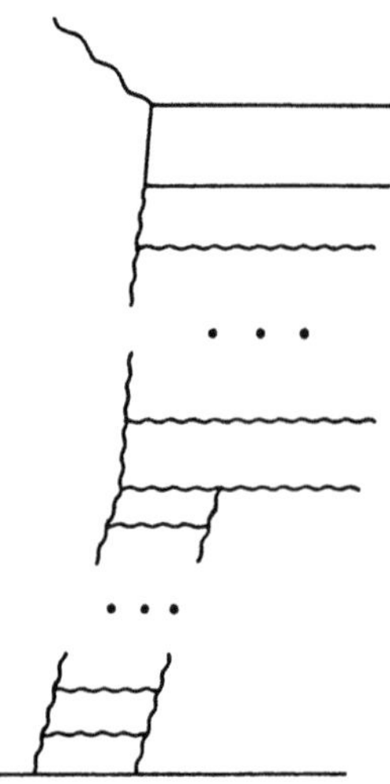

Figure 11. *Diffractive cut through the first fan diagram: it leads to the photon diffractive dissociation process discussed in Section IV.*

the reggeon diagrams which emerge in the Regge limit (to be precise: the ladders appearing in DIS are, for small x, equal to the large-Q^2 limit of the ladder diagrams which appear in the Regge limit). As it was said before, this close connection suggests, as an independent derivation of the GLR equation (or some alternative thereof), to investigate the large Q^2 limit of the reggeon diagrams. This would be the generalization of the well known result [5,21,22] that the large-Q^2 limit of the Lipatov-ladders leads to the singular pieces of the anomalous dimension of the twist-two gluon operator.

In order to proceed further, Gribov, Levin, and Ryskin made an assumption concerning the coupling of $n \geq 2$ ladders to the hadron: this coupling is assumed to be proportional to the n-th power of the single-ladder coupling. As a consequence, the probability of finding two gluons (at low momentum Q_0^2) with momentum fractions x_1 and x_2 is proportional to $g(x_1, Q_0^2) \cdot g(x_2, Q_0^2)$. This assumption then allows to find an equation for the sum of all fan diagrams, the GLR equation. It is a nonlinear integro-differential equation. In its simplest form, the DLA approximation, it is obtained by generalizing eq.(3.5) and performing the sum over all ladders:

$$F(y,\xi) = F_G(y,\xi) - \int_0^y dy' \int_{\xi_0}^\xi d\xi' F_0(y-y', \xi-\xi')C \exp\left(-e^{\xi'} - \xi'\right) F^2(y',\xi'). \qquad (3.7)$$

where C stands for the constant factors of (3.6), and F_G denotes the sum of QCD ladder diagrams with some initial distribution $G(y)$. The more accurate form of the equation is given as:

$$\frac{\partial \phi(x,q^2)}{\partial \ln\left(\frac{1}{x}\right)} = \int \bar{K}(q^2, q'^2)\phi(x,q'^2)\frac{4N\alpha_s(q'^2)}{4\pi}dq'^2 - \frac{1}{4\pi R^2}\left(\frac{\alpha_s(q^2)}{4\pi}\right)^2 V\phi^2(x,q^2). \qquad (3.8)$$

Here $\phi = \frac{\partial F}{\partial Q^2}$, R denotes the transverse radius of the hadron (or quark), and V stands for the triple ladder vertex, evaluated more accurately than in (3.7), namely in the semiclassical approximation.

As far as I know, there is no real theoretical justification of this assumption: further studies of this point seem to be important.

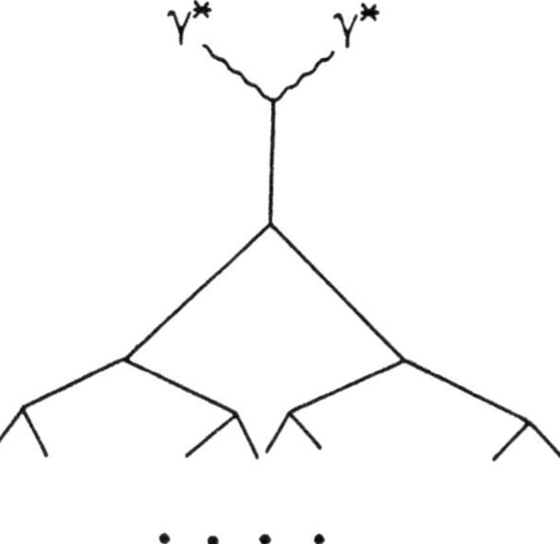

Figure 12. *The sum of fan diagrams which form the basis of the Gribov-Levin-Ryskin equation.*

3.3 Solutions in the Semiclassical Approximation

The first suggestion to solve these nonlinear equations in the semiclassical approximation is again due to Gribov, Levin, and Ryskin [13]. The idea is illustrated in Fig.13. Let me first consider the standard case, the linear evolution equation of Altarelli and Parisi (Fig.13a). According to the evolution equation

$$\frac{\partial G(x,Q^2)}{\partial \ln Q^2} = \frac{\alpha_s(Q^2)}{2\pi} \int_x^1 \frac{dz}{z} P_{gg}(z) G(\frac{x}{z}, Q^2) \tag{3.9}$$

we need to know the function $G(x,Q^2)$ at fixed Q^2 for the whole interval $(x,1)$, in order to calculate the change in the Q^2 direction. However, for small x and large Q^2 we really only need knowledge of a smaller and smaller x-interval (for small x, the integral (3.9) is dominated by the lower end of the integration region). This means that the evolution in Q^2 eventually only "propagates" along certain paths in the (Q^2, y) plane which become narrow if we move towards the upper right corner; conversely, in the lower left corner they become more and more diffuse and eventually are "washed out". For the case of the linear evolution equation (3.9) these paths can be shown to be straight lines ("rays"). Their precise intercepts and slopes depend upon the initial conditions of (3.9), i.e.the initial distribution at low Q^2. In any case, each point in the plane can be reached by exactly one ray which starts somewhere in the lower left corner. If we now switch from the linear case (3.9) to the nonlinear case (3.7) or (3.8), this pattern of paths changes (Fig.13b) [16]. In the left upper corner, a region appears which is unaccessible to trajectories that start in the lower left corner: the boundary line between this "empty" region and its complement is identified with the critical line of Fig.3 and has the characteristics of a caustic line, i.e. it is the envelope to the rays which now are slightly bent. It is the line where the validity of the GLR equation stops: above the line we are in the nonperturbative region which represents the large-Q^2 extension of the Regge limit. Far below this line, the rays are not affected by the nonlinearity, and the linear evolution equations can be used. It is only slightly below the line where the nonlinear term is felt: this is the "intermediate" region alluded to in the introduction. One also finds that at low Q^2 this intermediate region is a narrow strip just below the critical line, and it widens when Q^2 increases. It follows that measurements of the x-distribution of the structure function at rather low Q^2 values will quickly cross the intermediate x-region and enter the nonperturbative domain. As to the position of the critical line, it obviously will depend upon the starting values of the trajectories in the lower left corner, i.e. the initial y-distribution at the low momentum scale Q_0^2. The existence of this critical line is an important feature of the GLR equation: it means that the GLR equation predicts the limit of its validity. The fact that trajectories from the lower left corner never enter the "nonperturbative" part of the plane means that for the

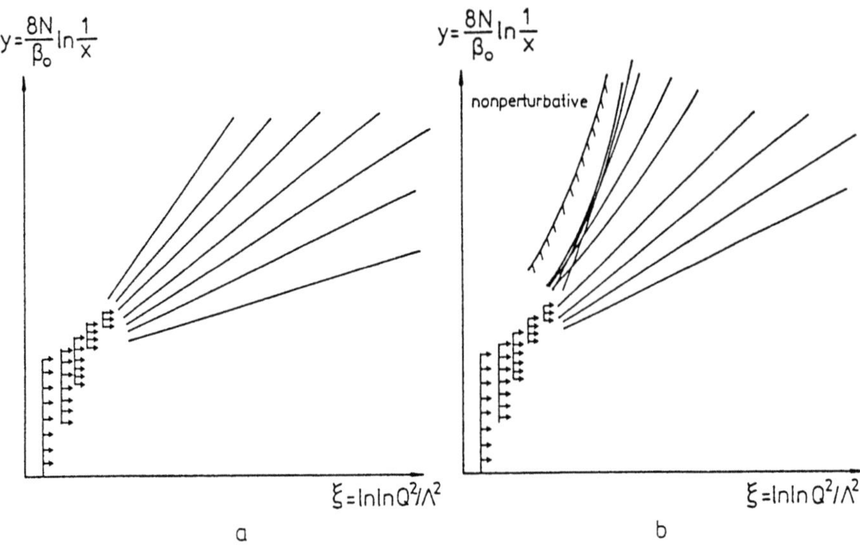

Figure 13. *Semiclassical paths of evolution in the x-Q^2 plane: (a) the linear case, (b) the nonlinear case.*

description of the "perturbative" region we only need to know the initial y-distribution inside a finite y-interval. All this discussion is based upon a detailed analysis [16] of eq.(3.8) with a particular set of initial conditions, but it is expected that the qualitative picture remains the same for a rather general class of initial distributions. Fig.14 shows some results of this analysis. There may, however, exist also initial distributions for which the character of the critical line of being a caustic is somewhat different [24].

4 The Regge Limit in QCD

In this section I wish to describe some recent work which I have done in the Regge limit. It is the third part of a series of papers which has been started several years ago [11]. As I have said already in the introduction, the best we seem to be able to do is to calculate perturbatively the high energy limit of QCD and incorporate as much unitarity as we can. Results of such an analysis can be put into a set of (an infinite number of) equations or, equivalently, be formulated as an effective two-dimensional (possibly conformal) field theory. Solutions will then bring in nonperturbative contributions. As to the reliability of such an approach, the following remark seems to be in place. In contrast to the "true" Regge limit in Fig.3, where it is difficult to justify even a start from perturbation theory, the $x \to 0$-limit at large Q^2 offers a more attractive possibility: when moving, in Fig.3, at some lage Q^2 upwards, the scale of the QCD coupling constant is large, at least initially. So we are entering the nonperturbative region with a small α_s, and we can hope to find a reliable answer, although we have started from perturbation theory. Whether the result of such an analysis will, eventually, also allow to reach the "true" Regge limit further to the right in Fig.3, is much harder to decide.

In my approach I start from a spontaneously broken gauge theory (for simplicity, I chose the SU(2) gauge group, and do not yet include fermions) and use analyticity properties of multiparticle production amplitudes in the Regge limit as a tool of computation. The mass of the gluon serves as an infrared cutoff and will be removed at the end of the perturbative part of the analysis. In the first two papers of [11] I constructed the leading-log approximation and the first step beyond; the third (and recent) paper computes unitarity corrections. I cannot hope to find *all* nonleading logs this way, but it seems that there exists a *subset* of terms which satisfies unitarity equations in both the s and t-channel and thus may provide the

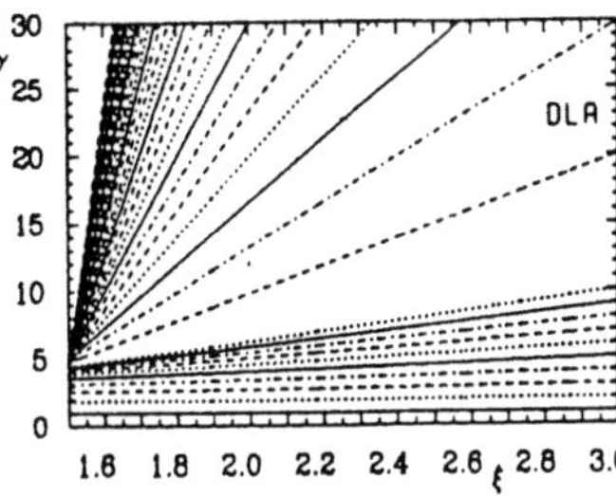
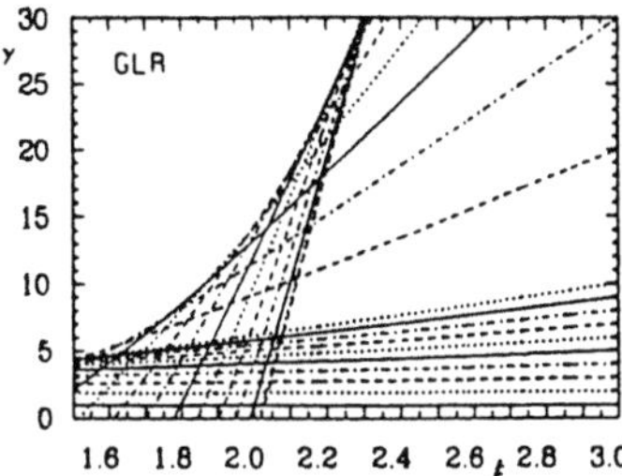
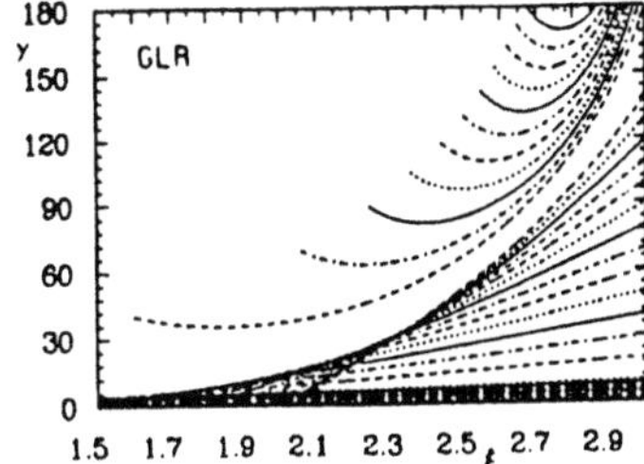

Figure 14. *The same as in Fig.13: results from a computer analysis (from [16]). In the third figure we show trajectories in the "nonperturbative" region: they do not cross the "critical line" between the two regions.*

"first" approximation to a completely unitary high energy theory. Since the gluon reggeizes, this high energy theory comes in the form of a reggeon field theory (with an infinite number of nonlocal interaction vertices). In this "first approximation" one then obtains the elements of the reggeon field theory (vertices, trajectory functions) in the leading power in g^2; taking into account *all* nonleading logs would give these elements to all orders g^2. It turns out, however, that there is another form of this unitary high energy approximation, "dual" to the reggeon field theory, which looks more suitable for formulating a two dimensional field theory. This version is also convenient for performing the transition to the small-x region of deep inelastic scattering.

Let me briefly describe the results of the calculations. For details I have to refer to the third paper of [11]. Starting point is the leading-log approximation ([6,7,8,9,10] and first paper of [11]) of the vacuum quantum number exchange amplitude (Pomeron). As illustrated in Fig.15a, it is obtained from the square of multi-gluon production amplitudes, all evaluated in the multiperipheral kinematic regime $Q^2 \ll s_{ij} \ll s$. Although this derivation, strictly speaking, is valid only for all internal momenta being of the order Q^2, the internal momentum integrals are convergent both in the infrared and the ultraviolet region. As a limitation of this leading-log calculation, the coupling constant remains fixed ($\alpha_s(Q^2)$), i.e. asymptotic freedom has to be included by hand. The leading angular momentum plane singularity of these QCD ladders is a fixed-cut singularity to the right of j=1; the internal region of phase space which is responsible for this unpleasant feature is that of very small and very large momenta, i.e. outside of the region of validity of this approximaion.

This makes it inevitable to look for unitarity corrections to this approximation. Generally speaking, what is missing are diagrams of the type shown in Fig.15b.: in order to satisfy reggeon unitarity in the t-channel, we need intermediate states with any number of reggeons. What is not so obvious, is the s-channel unitarity content of these diagrams. It can be shown (at least for subsets of the diagrams in Fig.15b) that inelastic production amplitudes which are built from reggeon diagrams with two, three,... reggeon intermediate states satisfy s-channel unitarity equations in all subchannels; this goes back to certain "bootstrap" relations

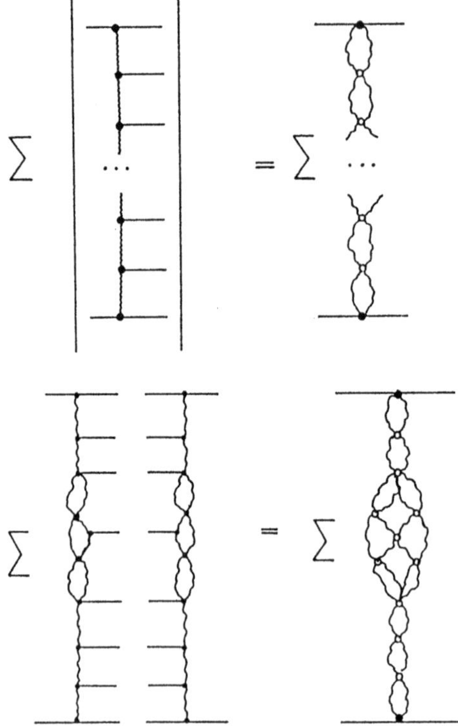

Figure 15. *(a) Diagrammatic illustration of the leading-log approximation of the Pomeron. The ladder diagrams of Lipatov et al. arise from the square of multi-gluon production amplitudes, all calculated in the multiperipheral kinematic region. Wavy lines denote reggeized gluons. (b) examples of diagrams which appear when unitarizing the leading-log approximation.*

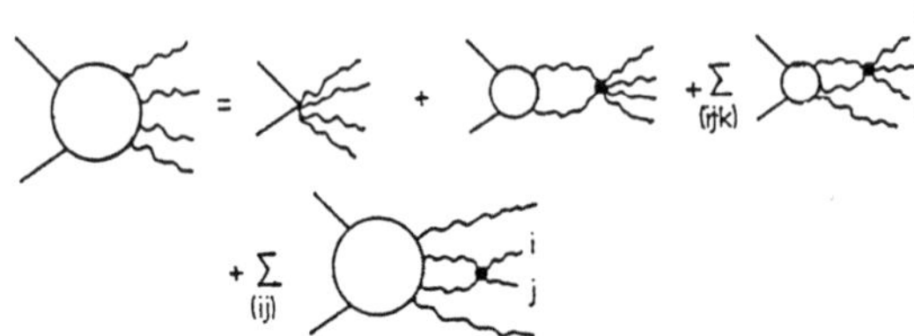

Figure 16. *The integral equation for the two-particle $\to$ four-reggeon amplitude.*

which generalize the one first found in [6,7,8,9,11] and are nothing but manifestations of the reggeization of the gluon.

As an illustration, we present the integral equation for the two particle $\to$ four reggeon vertex function (in leading order) in the color zero channel (Fig.16). The construction goes as follows: in the first step we calculate, from s-channel unitarity, multiple discontinuities. It is easy to write down integral equations for these objects, but they are not yet to be identified as partial waves to which Regge theory can be applied. This can be done only after we have introduced signature: one has to form combinations which are symmetric or antisymmetric under the simultaneous exchange of momenta and group indices. This set of partial wave amplitudes then satisfies coupled integral equations, an example of which is given here:

$$\omega D_4^{(000;+--)} = +\tfrac{4}{\sqrt{3}} D_2^{(0;+)} \otimes K_{2\to4}([k_1k_2][k_3k_4])$$

$$-2\sqrt{\tfrac{2}{3}} \left(D_3^{(01;+-)}(k_a\{k_bk_4\}) \otimes K_{2\to3}([k_1k_2]k_3) - D_3^{(01;+-)}(k_a\{k_bk_3\}) \otimes K_{2\to3}([k_1k_2]k_4) \right.$$

$$\left. + D_3^{(01;+-)}(\{k_1k_a\}k_b) \otimes K_{2\to3}(k_2[k_3k_4]) - D_3^{(01;+-)}(\{k_2k_a\}k_b) \otimes K_{2\to3}(k_1[k_3k_4]) \right)$$

$$+ D_4^{(000;+--)} \otimes \left(-2K_{2\to2}(k_1k_2) - 2K_{2\to2}(k_3k_4) + \sum \alpha(k_i) - 4 \right)$$

$$+ D_4^{(011;+--)} \otimes \tfrac{2}{\sqrt{3}} \left(K_{2\to2}(k_2k_3) - K_{2\to2}(k_1k_3) - K_{2\to2}(k_2k_4) + K_{2\to2}(k_1k_4) \right),$$

$$(4.10)$$

$$\omega D_4^{(011;+--)} = g^4 - D_2^{(0;+)} \otimes K_{2\to4}(\{k_1k_2\}\{k_3k_4\})$$

$$+ \tfrac{1}{\sqrt{2}} \left(D_3^{(01;+-)}(k_a\{k_bk_4\}) \otimes K_{2\to3}(\{k_1k_2\}k_3) + D_3^{(01;+-)}(k_a\{k_bk_3\}) \otimes K_{2\to3}(\{k_1k_2\}k_4) \right.$$

$$\left. + D_3^{(01;+-)}(\{k_1k_a\}k_b) \otimes K_{2\to3}(k_2\{k_3k_4\}) + D_3^{(01;+-)}(\{k_2k_a\}k_b) \otimes K_{2\to3}(k_2\{k_3k_4\}) \right)$$

$$+ D_4^{(000;+--)} \otimes \tfrac{2}{\sqrt{3}} \left(K_{2\to2}(k_2k_3) - K_{2\to2}(k_1k_3) - K_{2\to2}(k_2k_4) + K_{2\to2}(k_1k_4) \right)$$

$$+ D_4^{(011;+--)} \otimes \left(-K_{2\to2}(k_1k_2) - K_{2\to2}(k_3k_4) + \sum \alpha(k_i) - 4 \right.$$

$$\left. - \tfrac{1}{2} K_{2\to2}(k_2k_3) - \tfrac{1}{2} K_{2\to2}(k_1k_3) - \tfrac{1}{2} K_{2\to2}(k_2k_4) - \tfrac{1}{2} K_{2\to2}(k_1k_4) \right)$$

$$+ D_4^{(022;+--)} \otimes \tfrac{1}{2}\sqrt{\tfrac{5}{3}} \left(K_{2\to2}(k_2k_3) - K_{2\to2}(k_1k_3) - K_{2\to2}(k_2k_4) + K_{2\to2}(k_1k_4) \right),$$

$$(4.11)$$

$$\omega D_4^{(022;+--)} = \sqrt{\tfrac{5}{3}} D_2^{(0;+)} \otimes K_{2\to4}([k_1k_2][k_3k_4])$$

$$- \sqrt{\tfrac{5}{6}} \left(D_3^{(01;+-)}(k_a\{k_bk_4\}) \otimes K_{2\to3}([k_1k_2]k_3) - D_3^{(01;+-)}(k_a\{k_bk_3\}) \otimes K_{2\to3}([k_1k_2]k_4) \right.$$

$$\left. + D_3^{(01;+-)}(\{k_1k_a\}k_b) \otimes K_{2\to3}(k_2[k_3k_4]) - D_3^{(01;+-)}(\{k_2k_a\}k_b) \otimes K_{2\to3}(k_1[k_3k_4]) \right)$$

$$+ D_4^{(011;+--)} \otimes \tfrac{1}{2}\sqrt{\tfrac{5}{3}} \left(K_{2\to2}(k_2k_3) - K_{2\to2}(k_1k_3) - K_{2\to2}(k_2k_4) + K_{2\to2}(k_1k_4) \right)$$

$$+ D_4^{(022;+--)} \otimes \left(+K_{2\to2}(k_1k_2) + K_{2\to2}(k_3k_4) + \sum \alpha(k_i) - 4 \right.$$

$$\left. - \tfrac{3}{2}[K_{2\to2}(k_2k_3) + K_{2\to2}(k_1k_3) + K_{2\to2}(k_2k_4) + K_{2\to2}(k_1k_4)] \right).$$

$$(4.12)$$

(with expressions for the kernels $K_{2\to3,4}$ given in [11]). It is a remarkable property of these amplitudes that all infrared singularities cancel [23]. We also note that these equations could have been written in two different forms: originally they are equations for reggeon amplitudes and contain the usual reggeon energy denominators. After a certain amount of reshuffling during which the gluon trajectory functions become parts of the reggeon-reggeon interaction kernes, one arrives at the above equations which are conveniently interpreted as equations for "partonic amplitudes". This is the "dual" form alluded to in the beginning of this section. The solutions to the equations turn out to be very simple:

$$D_4^{(011;+--)}(k_1k_2k_3k_4;\omega) = \tfrac{1}{2} D_2^{(0;+)}(k_1+k_2, k_3+k_4;\omega) \cdot g^2$$

$$D_4^{(000;+--)}(k_1 k_2 k_3 k_4; \omega) = 0$$
$$D_4^{(022;+--)}(k_1 k_2 k_3 k_4; \omega) = 0. \tag{4.13}$$

The interpretation is easy: the pair of reggeized gluons (12) has the quantum numbers of the gluon itself, and the two external lines (12) "collapse" into a single line. The same holds for the pair (34). It is these consequences of the reggeization of the gluon which are crucial for s-channel unitarity [11].

As it was said before, the analysis of the diagrams a la Fig.15b is not yet complete: so far only particle $\rightarrow$ reggeon amplitudes have been defined and investigated, but not yet reggeon $\rightarrow$ reggeon amplitudes. Since the analytic properties of these objects are somewhat more complicated, their analysis is more involved and will be discussed in a forthcoming paper. It is, however, already obvious that this line of investigation will allow to find the elements (vertices and higher order contributions to the trajectory function) of the complete reggeon field theory. Alternatively, by writing integral equations in the same form as presented above, one obtains "partonic" amplitudes which may lead to the formulation of a two-dimensional field thoery.

There is an interesting question related to the small-x region in deep inelastic scattering: the diagrams of Fig.16, when coupled at the upper end to a deep-inelastic photon, should reduce, in the limit of large Q^2, to the first "fan diagram" (Fig.10b) of Gribov, Levin, and Ryskin. Obviously, some simplification must occur in order to agreement between the different approaches, and a study along these lines [23] will provide further insight into the GLR equations.

5   Phenomenology in the Small-x Region

In this final section I return to the small-x region of deep inelastic scattering and adress the question what the chances are to see the "new " effects at HERA. The first topic are numerical solutions of the GLR equation. So far there exist only a few calculations [16,25,26,27]. I shall comment on two of them, [16] and [25]. They both solve eq.(3.7), which can be written also in another form:

$$\frac{\partial F(y,\xi)}{\partial \xi} = \frac{1}{2} \int_0^y dy' F(y',\xi)(1 - 2C \exp\left(-e^\xi - \xi\right)F(y',\xi)). \tag{5.14}$$

This DLA approximation can be valid only for very small x-values. For a realistic numerical estimate it is necessary to improve this approximation. As a first step, one may replace, inside each ladder, the DLA approximation of the splitting function by the full Altarelli-Parisi kernel. Retaining only gluonic contributions one obtains:

$$\begin{aligned}
\frac{\partial F(y,\xi)}{\partial \xi} = {} & \frac{1}{2} \int_0^y dy' \left\{ \frac{z^2 G(y',\xi) - z G(y,\xi)}{1-z} + G(y',\xi)\left[-z + z^2(1-z)\right] \right. \\
& \left. + G(y',\xi)\left[1 - 2C \exp\left(-e^\xi - \xi\right)G(y',\xi)\right] \right\} \\
& + G(y,\xi)\left[\frac{4N}{\beta_0} \ln(1-x) + 1\right].
\end{aligned} \tag{5.15}$$

Here we defined $z = x/x' = \exp\left[-(y - y')\beta_0/(8N)\right]$. This is the improvement used in [16]. In [25] a similar improvement has been used; moreover, the nonlinear equation is applied only above a certain y-value, thus avoiding the problem of the large-x behavior.

As I have explained before, the numerical solution depends upon two input quantities, the initial y-distribution at some low Q_0^2 and the strenght C of the nonlinearity. [16] therefore uses two different input distributions, and both papers vary the parameter C by a factor of 10 or 2.5, resp. Predictions for the (momentum weighted) gluon distribution are shown in Fig.17 -19, for F_2 in Fig.20. In Fig.17, the full curves correspond to the nonlinear evolution: the upper ones to the (steeper) Morfin-Tung initial [28] distribution, the lower ones to the (almost flat) Eichten-Hinchclife-Lane-Quigg [29] distribution. The dashed and the dotted curves show

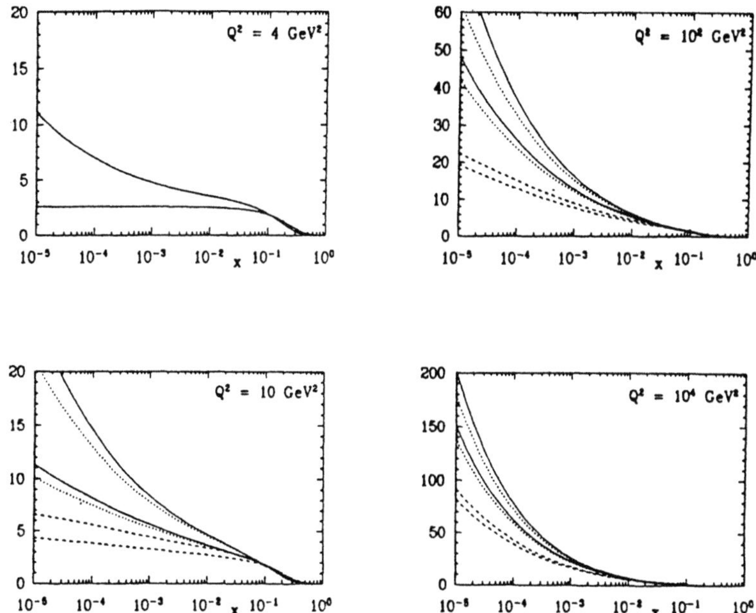

Figure 17. *x-distributions of $xg(x, Q^2)$ for the model of [16] for different values of Q^2. The full lines belong to the linear equation ($C = 0$), the dashed ones to the nonlinear case with C of eq.(2.3), and the dotted ones to the nonlinear case with $\frac{1}{10}C$. In each case, the upper curves belong to the steeper input distribution, the lower ones to the flat input.*

the corresponding results of the standard (linear) evolution programs for two different values of C: the lower curves belong to the larger value (3.9), for the upper ones C is divided by ten. The main point to be emphasized is the fact that e.g. for $Q^2 = 10 GeV^2$, deviations are visible already at x close to 10^{-2}. For the larger of the two C-values this effect is, of course, more pronounced than for the smaller one. In Fig.18 I show various estimates of the critical line. For illustration I have marked the kinematic regions of HERA and LEP/LHC; the 200 TeV Eloisatron machine would extend the kinematic region slightly above that of LEP/LHC. The more precise definition depends upon the method of extracting structure functions from a $p\bar{p}$ collider. Line "1" belongs to the parametrization of Levin and Ryskin [30]:

$$Q_c^2(x) = Q_0^2 + \Lambda^2 \exp\left(3.56\sqrt{\ln\frac{1}{x}}\right) \tag{5.16}$$

with

$$\Lambda = 52 MeV \,, Q_0^2 = 2 GeV^2. \tag{5.17}$$

whereas line "4" denotes the analytic form of eq.(1.9) with the constant being adjusted such that for large Q^2 it agrees with "1". Again, the point to be stressed is the fact that deviations are sizable at not so small x, and the nonperturbative region will be reached both at HERA and LEP/LHC. An analysis analogous to Fig.14 has not yet been done for any of the two model calculations. Figs.19 and 20 show results of [25] for $xg(x, Q^2)$ and $F_2(x, Q^2)$, resp. In this case the input distribution is again a rather steeply rising function in y; its analytic form has been motivated by the analysis of [31]. Deviations between the linear and the nonlinear evolution start at x between 10^{-2} and 10^{-3}, quite in agreement with the findings of [16]. [25] also presents results for F_2: it is, however, not clear whether this calculation takes into account certain difficulties which have been pointed out in [32]: once the gluon structure function has been calculated from the GLR-equation, the quark loop which couples the gluons to the deep inelastic photon, can no longer be used in the leading-Q^2 approximation.

The important message of these two papers seems to be that, provided the x-distribution at low Q^2 is as steeply rising as it was assumed in most of these calculations, HERA will enter the region where the "new" physics is at work. To be more precise, since at these relatively small values of $Q^2 \approx 10 GeV^2$ the "intermediate" region between perturbative and nonperturbative QCD is very narrow, HERA not only enters this transition region but also reaches into the domain of nonperturbative QCD. Here, strictly speaking, we have no reliable theory. As a first estimate, one might try to use the solution of the GLR equation: it predicts saturation [31] and thus avoids the obvious conflict with s-channel unitarity. It should, however, be repeated that this equation does not claim to be applicable in the nonperturbative region and, hence, cannot be taken as describing the "true QCD behavior". Again, what is lacking is the analysis of the Regge limit of QCD.

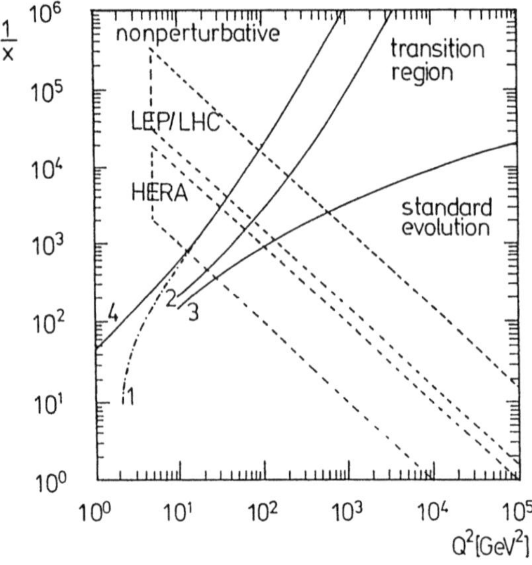

Figure 18. *Numerical estimates of the "critical line": Line "1" corresponds to the parametrization of Levin and Ryskin (eq. (5.12)), lines "2" and "3" are estimates from [16], and "4" belongs to eq.(1.9), with the constant being adjusted in such a way that for large Q^2 it agrees with the line of Levin and Ryskin.*

The other remark that has to be made concerns the question whether it is possible to conclude from HERA measurements beyond any doubt, whether one sees this new physics or not. Due to our ignorance of the correct x-distribution at low Q_0^2 which always enters the calculation as an input, it is very well possible that the measurement of some flatter structure function can be descibed by the standard evolution, using some modified input function. Conversely, a continuing rise of the structure function at low x is not necessarily a reliable indication that the "new physics" is not yet at work.The best way to distinguish is the comparison of evolution in Q^2: linear and nonlinear evolution equations predict different Q^2-dependence. In order to extract this from data we need: (i) a range in Q^2 as large as possible and (ii) computer algorithms for both the linear and the nonlinear evolution equations. Whether HERA will satisfy the first requirement, is not yet clear to me. As to the second demand, more theoretical work is necessary, both for the analytic analysis of the GLR equation and for numerical solutions.

Because of these difficulties it is inevitable to look for experiments which are specially designed to probe the "new physics". So far, two such experiments have been proposed. The one is due to A.Mueller [17] and refers to the "hot spots" mentioned before, the other

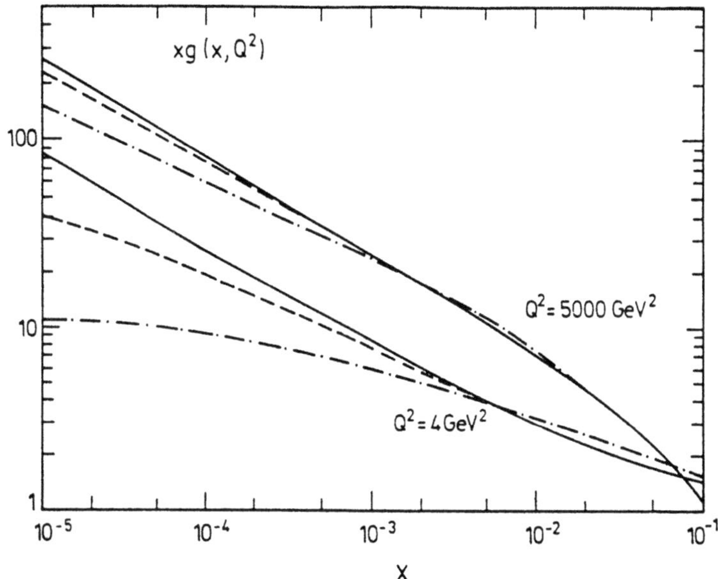

Figure 19. x-distributions of $xg(x, Q^2)$ for different values of Q^2 (from [25]): the solid line belong to the linear evolution. The dashed and the dot-dashed lines are the results of the nonlinear evolution; in the second case the nonlinearity parameter is increased by a factor of 2.5, compared to the first one.

uses the diffractive dissociation of the photon and has been proposed by Ryskin [33]. At the end of section II I mentioned the possibility that "saturation" of soft gluons may set in nonuniformly, i.e. there may be regions inside the nucleon (Fig.5) which are more densely populated by soft gluons than others. Since the structure function F_2 ("total photon hadron cross section") measures the parton density, averaged over the full size of the hadron, one needs to look at exclusive final states which measure the distribution of partons inside limited subregions of the hadron (Fig.5b). Such an experiment is illustrated in Fig.21: measure the cross section of jets which carry the momentum fraction x (not to be confused with the usual Bjorken x_B which belongs to the photon at the upper end) and the momentum scale (transverse momentum squared) $k_\perp^2$. The latter should be not much smaller than Q^2, the mass of the photon: $R_{parton}^2 \approx \frac{1}{Q^2}$ is the transverse radius of the parton struck by the photon, and $R_{jet}^2 \approx \frac{1}{k_\perp^2}$ is the square of the average transverse distance, by which the parent-gluon of the jet is seperated from this parton. If $k_\perp^2 \leq Q^2 \ll Q_0^2$, then $R_{parton}^2 \leq R_{jet}^2 \ll R_{hadron}^2$, and this process explores the parton cloud of diameter R_{jet} around the struck parton. The cross section should be measured as a function of the ratio $\frac{x_B}{x}$, and the region of interest is where $x_B \ll x$. The cross section for this process is, in the QCD-parton model, given by:

$$\frac{k_\perp^2 x d(\nu W_2)}{dx dk_\perp^2} = C\alpha(Q^2) \left(xg(x, k_\perp^2) + \frac{4}{9}(xq(x, k_\perp^2) + x\bar{q}(x, k_\perp^2)) \right) F(\frac{x_B}{x}, Q^2, k_\perp^2) \qquad (5.18)$$

where C is a normalization factor (see [34,35]), $q(x, q^2)$ and $g(x, q^2)$ are the quark and gluon distribution functions, resp., and $F(\frac{x_B}{x}, Q^2, k_\perp^2)$ stands for the sum of QCD ladders above the jet, evaluated in the limit $\frac{x_B}{x} \to 0$ (this is one of the places where the so-called "Lipatov Pomeron" [10] described in Section IV could be tested). This formula predicts a steep rise as $\frac{x_B}{x}$ becomes small:

$$F(\frac{x_B}{x}, Q^2, k_\perp^2) \approx C \frac{\exp\left(\frac{12\alpha(Q^2)}{\pi} \ln 2 \ln \frac{x}{x_B}\right)}{\sqrt{\ln \frac{x}{x_B}}} \qquad (5.19)$$

and a deviation from this would be indicative of "local saturation", i.e. the existence of a "hot spot". Corrections to this formula can (and should) be calculated: these are not fan-diagrams

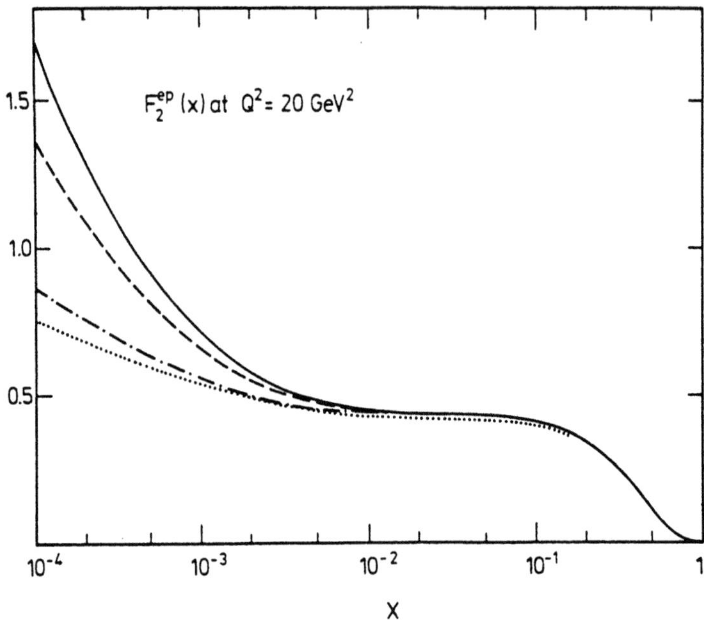

Figure 20. *x-distributionsfor $F_2(x,Q^2)$ (from [25]): for the upper three curves the notation is the same as for Fig.21. The dotted curve corresponds to another input distribution (which is flat for $x \to 0$).*

but reggeon diagrams, since both Q^2 and $k_\perp^2$ are of the same order. Eq.(4.14) has not been evaluated numerically: this will clearly be necessary in order to estimate the number of events and the event structure.

Another strong prediction based upon the existence of such "hot spots" has been made by Ryskin [33]. Instead of the "one-jet inclusive cross section" of Fig.21 one considers the photon diffractive dissociation processes shown in Fig.22 (its connection with the Pomeron structure function [37,38,39] has been discussed in [36]): there is a rapidity gap between the proton at the lower end and the missing mass cluster above. As an example, this missing mass final state could consist of three jets, originating from $q\bar{q}g$. The final states are further restricted by the requirement that the jet in the direction of the Pomeron should have a controllable $k_\perp$. It is then the dependence upon this $k_\perp$ which distinguishes between "saturation" and "standard" QCD behavior: with saturation the cross section should be substantially smaller, and for the integrated cross section there could be a difference up to a factor of one hundred!.

There are clearly more measurements that might be suitable to test the presence of the "new physics" described in this talk: this will not be discussed here and should be a topic of further theoretical work.

6 Summary

This talk tried to describe our present understanding of the small-x behavior of deep inelastic structure functions and of the Regge limit in QCD, partially as preparation for HERA experiments which are expected to start in 1991, partially in order to stimulate further theoretical efforts towards studying QCD at this interface between perturbative and nonperturbative high energy limits.

From the theoretical viewpoint, the low-x region in deep inelastic scattering describes the transition from perturbative to nonperturbative QCD. In the former region we have the well-tested evolution equations of Gribov, Lipatov, Altarelli, and Parisi, whereas for the non-perturbative small-x limit which is the large-Q^2 continuation of the Regge limit we do not yet

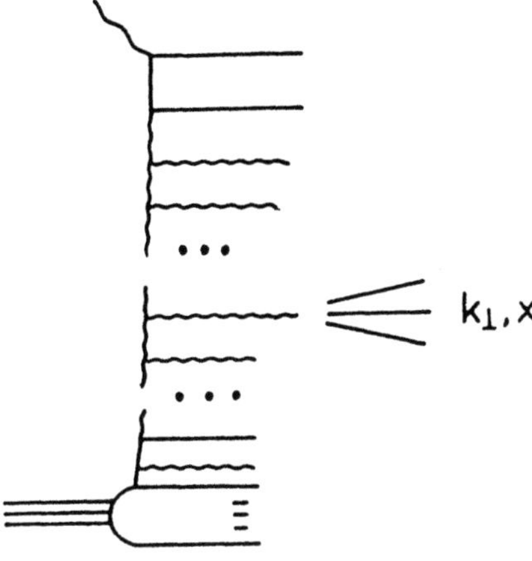

Figure 21. *A final state configuration which probes the "Hot spot"*

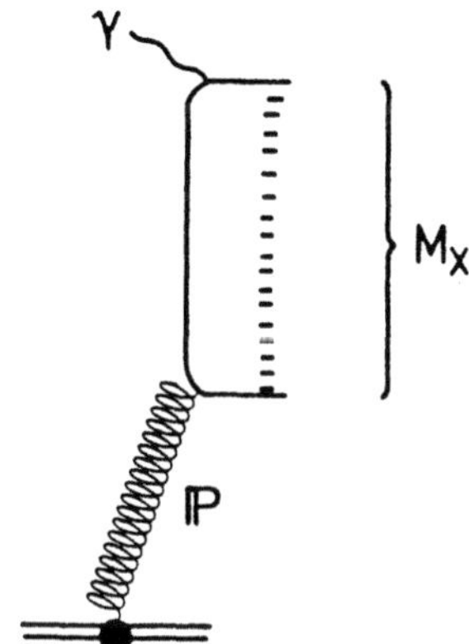

Figure 22. *Final state configuration for photon diffractive dissociation*

know what QCD predicts. In between these two regions, however, there is an intermediate regime, for which Gribov, Levin, and Ryskin suggested a new, nonlinear evolution equation which is still based upon perturbative QCD. Although its validity in the x-Q^2 plane is limited and does not extend into the nonperturbative region, its predictions for this region are consistent with unitarity and, in particular, the idea of "saturation". The theoretical basis of this equation lies in the analysis of classes of Feynman diagrams: since it is such an important equation (and also relies on a few assumptions which need further justification!), it seems inevitable to look further into the derivation of this equation (or to find alternatives). For the Regge limit, although it can have only a nonperturative solution, extensive investigations of classes of Feynman diagrams are under the way and will, hopefully, lead to a unitary high energy theory. But clearly, we still have a long way to go.

More general, the theoretical investigation of the transition from perturbative to nonperturbative QCD seems to offer an excellent opportunity for testing ideas on confinement dynamics in QCD: when approaching, at some large Q^2, smaller and smaller values of x, interactions between the (initially free) partons become more and more frequent. Beyond a certain point in x, one seems to reach a state in which a description in terms of partonic degrees of freedom is no longer useful, and one has to switch to hadronic degrees of freedom. To follow this transition in more detail, appears to be the theoretical challenge.

The analysis of the GLR equation has just been started: apart from attempts to use the semiclassical approximation for an analytic solution, only two numerical computer calculations

are available. In contrast to the standard linear evolution equations, the nonlinearities make
the calculations much more difficult. The numerical results mark the strong dependence of the
results on the (unknown) input distribution and on the (unknown) strenght of the nonlinearity.
Correspondingly, there is a principal uncertainty where (in x and Q^2) these new effects become
relevant. Moreover, even if we are in the right regime, this uncertainty is also likely to make
the interpretation of experimental data difficult: if deviations from the small-x prediction of
some parametrization of structure functions are observed, it may happen that these effects can
be compensated by a change of the input distribution and/or the strenght of the nonlinearity.
Measuring not only the x-distribution at some fixed Q^2 but also the Q^2-evolution of the
structure function in the small-x region will certainly be of help.

This makes it mandatory to look for other measurements which test the onset of this "new
physiscs" more directly. Such experiments have been proposed, but they need more theoretical
preparation.

One of the most important questions of practical relevance is the location in the x-Q^2 plane
of the transition from perturbative to nonperturbative QCD. At present the most reliable
estimates come from the two numerical evaluations of the GLR equation, but as it was said
before, there is the uncertainty due to the input parameters. Both calculations indicate that
HERA will enter this transition region, and, at low Q^2, even reach into the nonperturbative
domain where the parton picture no longer holds.

Acknowledgements

I have profitted very much from discussions with J.Bluemlein, G.Ingelman, E.Levin, L.Lipatov,
A.Mueller, O.Nachtmann, M.Ryskin, and G.Schuler.

References

[1] *Proceedings of the Small-x Workshop*, held at DESY, May 1990 (ed.A.Ali and J.Bartels),
Nucl.Phys.B (Proc.Suppl.), **18C**, 1991.

[2] For general reviews on perturbative QCD see, e.g.: H.D.Politzer, *Phys.Rep.* **14C**,
129(1974); C.H.Llewellyn Smith, Schladming Lectures 1987, *Acta Phys.Austr. Suppl.XIX*,
331(1978); Yu.Dokshitzer, D.I.Dyakonov, S.I.Troyan, *Phys.Rep.* **58**, 269(1980); E.Reya,
Phys.Rep. **69**, 195(1981); A.H.Mueller, *Phys.Rep.* **73**, 237(1981); G.Altarelli, *Phys.Rep.*
81,1 (1982).

[3] V.N.Gribov and L.N.Lipatov, *Sov.Journ.Nucl.Phys.* **15**, 438 and 675 (1972).

[4] G.Altarelli and G.Parisi, *Nucl.Phys.* **126**, 297(1977).

[5] Yu.L..Dokshitser, *Sov.Phys.JETP* **46**, 641(1977).

[6] E.A.Kuraev, L.N.Lipatov, V.S.Fadin, *Sov.Phys.JETP* **44**, 443(1976).

[7] E.A.Kuraev, L.N.Lipatov, V.S.Fadin, *Sov.Phys.JETP* **45**, 199(1977).

[8] Ya.Ya.Balitzky, L.N.Lipatov *Sov.Jour.Nucl.Phys.* **28**, 822(1978).

[9] Ya.Ya.Balitzky, L.N.Lipatov *JETP Letters* **30**, 355(1979).

[10] L.N.Lipatov, *Sov.Phys.JETP* **63**, 904(1986) and references therein.

[11] J.Bartels, *Nucl.Phys.* **B151**, 293(1979); *Nucl.Phys.* **175**, 365(1980); *Acta Phys.Pol.* **B11**, 281(1980); DESY-91-074.

[12] A.R.White, ANL-HEP-PR-90-28 and references therein.

[13] L.V.Gribov, E.M.Levin, and M.G.Ryskin, *Phys. Rep.* **100**, 1 (1982)

[14] A.H.Mueller and J.Qiu, *Nucl.Phys.* **B268**, 427 (1986).

[15] J.Bartels, I.Hamann, DESY-preprint in preparation.

[16] J.Bartels, J.Blümlein, G.Schuler, *Zeitschr. f.Phys.* **C 50**, 91 (1991).

[17] A.H.Mueller, *Nucl.Phys.* **B307**, 34 (1988); *Nucl.Phys.* **B317**, 573(1989); *Nucl.Phys.***B335**, 115(1991).

[18] V.A.Abramovsky, V.N.Gribov, O.V.Kancheli,*Sov.Journ. Nucl.Phys.* **18**, 595(1973).

[19] A.H.Mueller, in [1],p.125.

[20] A.P.Bukhvostov, G.V.Frolov, L.N..Lipatov, *Nucl.Phys.* **B258**, 601(1986).

[21] T.Jaroscewicz, *Phys.Lett.* **116B**, 291(1982).

[22] J.C.Collins, J.Kwiecinsky, *Nucl.Phys.* **B316**, 307(1988).

[23] J.Bartels, DESY-preprint in preparation.

[24] E.M.Levin and M.G.Ryskin, private communication.

[25] J.Kwiecinski, A.D.Martin, W.J.Stirling, and R.G. Roberts, *Phys.Rev.* **D42**, 3645 (1990).

[26] V.T.Kim, M.G.Ryskin, DESY-91-064.

[27] J.Kwiecinsky, A.D.Martin, P.J.Sutton, *Phys.Let.* **B264**, 199 (1991).

[28] J.Morfin and Wu-Ki Tung, preprint Fermilab–Pub 90/74 (1990).

[29] E. Eichten et al., Rev. Mod. Phys. **56** (1984) 579 and Erratum **58** (1986) 1065.

[30] M.G.Ryskin, *Sov.Journ.Nucl.Physics* **47**, 230 (1988).

[31] J.C.Collins, J.Kwiecinsky, *Nucl.Phys.* **B335**, 89 (1990).

[32] E.M.Levin and M.G.Ryskin, Frascati Preprint April 1990.

[33] M.G.Ryskin, in [1], p.162.

[34] J.Bartels, A.DeRoeck, M.Loewe, DESY-preprint in preparation.

[35] W.-K.Tang, Columbia-preprint in preparation.

[36] J.Bartels, G.Ingelman, *Phys.Lett.* **235B**, 175 (1990).

[37] G.Ingelman and P.Schlein, *Phys.Lett.* **152B**, 256(1985).

[38] H.Fritzsch and H.H.Streng, *Phys.Lett.* **154B**, 391(1985).

[39] E.L.Berger, J.C.Collins, D.E.Soper, G.Sterman, *Nucl.Physics* **B286**, 704 (1987).

EXPLORING HIGGS BOSONS/ELECTROWEAK SYMMMETRY BREAKING PHYSICS AT 200 TEV[*]

J.F. Gunion

Department of Physics
University of California, Davis, CA 95616

ABSTRACT

The role of a 200 TeV hadron collider in exploring Higgs/EWSB physics that cannot be easily accessed at the LHC or SSC is illustrated using a number of Standard Model and Minimal Supersymmetric Model scenario examples.

INTRODUCTION

It is widely appreciated that a major goal of high energy physics is to understand the source of electroweak symmetry breaking (EWSB), *i.e.* to discover the origin of mass. Many possible mechanisms for EWSB have been discussed, but certainly theories which lead to elementary Higgs boson(s) remain a prime candidate. Among such theories the minimal Standard Model (SM) is certainly the simplest, having just one doublet Higgs field resulting in a single physical Higgs boson, the ϕ^0. But the SM is only one of many possibilities. Any extension of the Higgs sector that contains only doublet (and singlet) fields still avoids difficulty with the W to Z mass ratio at tree-level. Supersymmetric models, which are well-motivated theoretically in that they provide a solution to the hierarchy and naturalness problems, actually require at least two doublet fields. The simplest supersymmetric model is termed the minimal supersymmetric model (MSSM) and contains exactly two doublets. The physical Higgs bosons of this model are: two neutral CP-even Higgs bosons (h^0 and H^0, with $m_{H^0} \geq m_{h^0}$); one neutral CP-odd Higgs (A^0); and a charged Higgs pair ($H^\pm$).

Necessarily, the experimental accessibility of Higgs boson(s) is model and accelerator dependent. The LHC and SSC hadron colliders provide substantial opportunity for discovering elementary Higgs bosons. For instance, in the case of the SSC, it is now clear that the SM ϕ^0 can be found if it lies anywhere in the mass range $m_Z \lesssim m_{\phi^0} \lesssim 1$ TeV, although detection in the Intermediate Mass Region $m_Z \lesssim m_{\phi^0} \lesssim 135$ GeV and Obese Mass Region $m_{\phi^0} \gtrsim 1$ TeV may require two to three times the canonical integrated luminosity of $L = 10$ fb^{-1}. Techniques for detecting the Higgs bosons of the MSSM over much of parameter space have also been developed. (For a general review, see, for example, Ref. [1]. Recently developed detection techniques for an intermediate

---

[*] Work supported in part by the U.S. Department of Energy.

mass ϕ^0, a $\gtrsim 1$ TeV ϕ^0, and for the MSSM Higgs bosons are reviewed in Ref. [2].) Nonetheless, there are some MSSM scenarios for which detection of some (if not all) of the Higgs bosons may prove extremely difficult, and certainly for many SM and MSSM scenarios event rates at the LHC and SSC may prove to be only just adequate for discovery; detailed confirmation and studies of the properties of a new resonance discovered in an appropriate channel could be difficult or impossible.

Thus, two useful litmus tests for a successor accelerator are: a) its potential for discovering those Higgs bosons of the MSSM that would not be found by the LHC and SSC in certain regions of model parameter space; and b) the ability to confirm and study in reasonable detail the properties of any SM or MSSM Higgs boson that is discovered at the LHC and SSC, but that is produced with low event rate and/or is detected in a mode with severe background problems. An example of a) is detection of the heavy CP-even H^0 of the MSSM. Possibly the only viable technique for its detection at a hadron collider will turn out to be via its $H^0 \to ZZ \to l^+l^-l^+l^-$ decay mode. But there are regions of MSSM parameter space where this mode is severely suppressed, and event rates at the SSC will be too small relative to the $ZZ \to l^+l^-l^+l^-$ continuum background for a meaningful signal. Can the next accelerator detect such a H^0? As useful examples of b), suppose that an intermediate mass or very heavy SM ϕ^0 is detected (with just adequate event rate) at the LHC or SSC. Will the next accelerator allow us to verify that its couplings are as predicted?

An important issue is whether a high energy e^+e^- linear collider or a very high energy hadron collider is best suited to the type of follow-up tasks that might prove necessary in order to study in detail the properties of the Higgs boson/EWSB sector. For a collider to be an appropriate successor to the LHC and SSC will require that production of the Higgs boson(s) in a number of modes/channels have high event rate and that backgrounds be controllable. In this review, I will use the scenarios mentioned above to illustrate the ability of a 200 TeV hadron collider to fill gaps in the experimental information that would be provided by the LHC and SSC.

TWO STANDARD MODEL SCENARIOS

At the SSC energy of $\sqrt{s} = 40$ TeV and with an integrated luminosity of 10 fb^{-1}, detection of the SM ϕ^0 in the channel $\phi^0 \to ZZ$ or $ZZ^* \to l^+l^-l^+l^-$ is clearly possible in the mass range 130 GeV $\lesssim m_{\phi^0} \lesssim 800 - 900$ GeV. For lower m_{ϕ^0}, $BR(\phi^0 \to l^+l^-l^+l^-)$ drops rapidly and the event rate in this $4l$ channel becomes tiny. The upper reach of the $4l$ channel depends in detail upon m_t (which determines the magnitude of the gg fusion cross section) and upon the available integrated luminosity. For $m_t = 150$ GeV and $L = 10$ fb^{-1}, it is difficult to recognize the broad Higgs resonance against the $q\bar{q} \to ZZ \to 4l$ continuum background once $m_{\phi^0} \gtrsim 800$ GeV. The $\phi^0 \to ZZ \to l^+l^-\nu\bar{\nu}$ channel may allow ϕ^0 detection at $L = 10$ fb^{-1} for m_{ϕ^0} above 800 GeV (perhaps up to ~ 1 TeV). (For recent Monte Carlo studies of ϕ^0 detection in these final states at the SSC, including experimental detector simulation, see, for example, Ref. [3].) Of course, the LHC discovery range is quite restricted at $L = 10$ fb^{-1}; to even be competitive with the SSC an integrated luminosity of order $L = 100$ fb^{-1} would be required. Thus, for $m_{\phi^0} \lesssim 130$ GeV and $m_{\phi^0} \gtrsim 1$ TeV, specialized techniques for detecting the presence of the ϕ^0 have had to be developed. These techniques, which shall be reviewed below, lead to event rates (in the relevant mass regions) for $L = 10$ fb^{-1} that are marginal at the SSC (and impossibly small at the LHC). Consequently, it is the very low and very high ϕ^0 masses for which a hadron collider with substantially larger energy than that of the SSC could have the greatest impact. We shall focus our discussion on such masses.

However, before proceeding, it is useful to remark that even in the central 130 GeV $\lesssim m_{\phi^0} \lesssim 800$ GeV mass region a 200 TeV machine might well provide crucial information regarding the couplings of the ϕ^0 that could not be extracted at the LHC or SSC. In particular, it is possible that the only means for extracting the relative magnitude of the $t\bar{t}\phi^0$ and $W^+W^-\phi^0$, $ZZ\phi^0$ couplings at a hadron collider will be by determining the relative size of the $gg \to \phi^0$ and $W^+W^- \to \phi^0$ fusion production cross sections, respectively. Separation of the two production mechanisms is, in principle, possible because of the different final state topologies expected in the two cases. The W^+W^- fusion mechanism subprocess is of the type $qq \to qqW^+W^-$, implying that there are two spectator jets (with p_T of order m_W and with large energy). By either anti-tagging against or explicitly tagging these spectator jets one can hope to distinguish events arising from gg fusion from those coming from W^+W^- fusion. However, both anti-tagging and tagging can only be done with limited efficiency for retaining the events of interest. This implies that a determination of the relative size of the gg and W^+W^- fusion cross sections will require a large number of ϕ^0 production events, and is not likely to be possible at the SSC for $L = 10$ fb^{-1} in the $l^+l^-l^+l^-$ final state mode over more than a portion of the central mass region. In contrast, at a 200 TeV hadron collider very large numbers of events will be available in the $130 \lesssim m_{\phi^0} \lesssim 1$ TeV region and the required separation of the the gg and W^+W^- fusion events on a statistical basis should be relatively straightforward. Of course, detailed Monte Carlo simulations will be required in order to specify with precision the relative abilities of the LHC, the SSC, and a 200 TeV collider to perform this task. Thus, for now we leave aside the central mass region and return to the $m_{\phi^0} \lesssim 130$ GeV and $m_{\phi^0} \gtrsim 1$ TeV mass regions where event rates for ϕ^0 *discovery* will be marginal at the SSC.

$m_Z \lesssim m_{\phi^0} \lesssim 130$ GeV

It is now widely appreciated that the best means for detecting a SM Higgs boson in this mass region at a hadron collider is in the $l\gamma\gamma X$ final state obtained via associated $W + \phi^0 + X$ production, followed by $W \to l\nu$ and $\phi^0 \to \gamma\gamma$. The largest contribution to the event rate comes from the process $gg \to t\bar{t}\phi^0$ with one of the top quarks emitting the leptonically decaying W.[4,5] The reaction $q\bar{q} \to W^* \to W\phi^0$ can also lead to the $l\gamma\gamma X$ final state, but with a substantially smaller cross section.[6-8] A number of studies have shown that after imposing appropriate kinematic cuts, after requiring isolation for the lepton and the two photons, and after implementing photon-jet discrimination at the level of $R_{\gamma j} = 5 \times 10^{-4}$, backgrounds to the $l\gamma\gamma X$ final state are relatively small, provided $\gamma\gamma$ mass resolution of order $\Delta M_{\gamma\gamma}/M_{\gamma\gamma} \sim 3\%$ is possible. The surviving backgrounds derive primarily from $W\gamma\gamma$, $W\gamma j$ and $t\bar{t}\gamma\gamma$ production.[6,7,9] These $R_{\gamma j}$ and $\Delta M_{\gamma\gamma}$ resolution requirements appear to be within the capabilities of a general purpose detector such as the SDC detector designed for operation at the SSC.[3] The result is that at the SSC one achieves at $m_{\phi^0} = 110$ GeV a signal to background event ratio of order $S/B = 24/9$ for a canonical yearly luminosity of 10 fb^{-1}. Of the S events 20 derive from the $t\bar{t}\phi^0$ process and 4 from the $W^* \to W\phi^0$ reaction. The B events comprise about 3 $t\bar{t}\gamma\gamma$ events, 4 $W\gamma\gamma$ events and 3 $W\gamma j$ events. (Lepton coverage out to rapidities $|y| \leq 2.5$ is necessary to achieve such substantial signal event rates.) Assuming that the background can be normalized outside the Higgs peak region, this implies a statistical significance for the signal of order 8σ. (At the LHC, $L_{year} \sim 50$ fb^{-1} is required for a 8σ level signal.) The result quoted assumes a top quark mass of order $m_t = 150$ GeV, but is rather insensitive to m_t. The statistical significance of the signal is smaller for significantly higher or lower m_{ϕ^0}, but does not fall below 4σ until $m_{\phi^0} \gtrsim 160$ GeV.

Thus, it is clear that the SSC is rather likely to discover the ϕ^0 if its mass lies in the $m_Z \lesssim m_{\phi^0} \lesssim 130$ GeV region. However, the small number of expected signal events precludes several theoretically crucial studies. First, with more events the angular distribution of the $\gamma\gamma$ pairs in the ϕ^0 rest frame could be used to verify that the observed resonance is indeed spin zero. Second, we note that the $W^* \to W\phi^0$ and $gg \to t\bar{t}\phi^0$ production mechanisms rely on the two most interesting couplings of the ϕ^0: the $W^+W^-\phi^0$ and $t\bar{t}\phi^0$ couplings, respectively. Thus, it would be highly desirable to be able to separate the two classes of events. Since the $t\bar{t}\phi^0$ events have a much higher level of jet activity than do the W^* events, the W^* events could be isolated by vetoing against jet activity, while the $t\bar{t}\phi^0$ events can be isolated by requiring a high level of jet activity. The ratio of the number of events in each class would then yield a direct determination of the ratio of the WW and $t\bar{t}$ couplings of the ϕ^0. In order to determine the coupling magnitudes separately, $BR(\phi^0 \to \gamma\gamma)$ must be known. Since, in the present scenario, this could not be experimentally determined, the theoretical calculation of the $\phi^0 \to \gamma\gamma$ coupling, which is mainly determined by W and t loops involving these very same couplings, would be employed. At the SSC, the main difficulty in carrying out this program is the small number of W^*-induced events expected for $L = 10$ fb^{-1}. Even assuming that the $t\bar{t}$-related S and B events could be completely eliminated by jet vetoing, at $m_{\phi^0} = 110$ GeV we would be left with ~ 4 W^* signal events compared to a $W\gamma\gamma + W\gamma j$ background of about 7 events. Roughly $L = 100$ fb^{-1} (ten times the canonical luminosity) would be required to achieve a reasonably good W^* signal rate determination.

While enhanced luminosity at the SSC is under consideration, it is by no means a

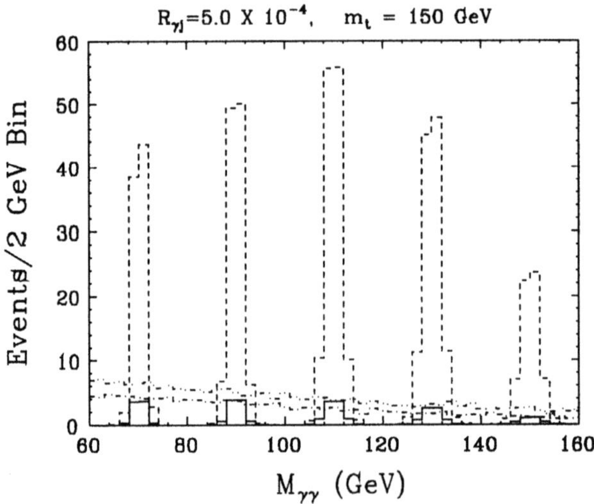

Figure 1. $W^* \to W\phi^0$ (solid), $t\bar{t}\phi^0 \to W\phi^0 X$ (dashes), $W\gamma\gamma$ (dotdash), and $W\gamma j$ (dotdotdash) (assuming $R_{\gamma j} = 5 \times 10^{-4}$) event rates are plotted for pp collisions with $L = 10$ fb^{-1} at $\sqrt{s} = 200$ TeV.

certainty. Thus, it is appropriate to ask how much a 200 TeV collider with $L = 10$ fb^{-1} would improve our ability to study a ϕ^0 in this mass range. The situation is illustrated in Fig. 1. At $m_{\phi^0} = 110$ GeV we obtain 7.5 $W^* \to W\phi^0$ events and about 110 $t\bar{t}\phi^0$ events in the $l\gamma\gamma X$ channel after cuts. The $W\gamma\gamma$ and $W\gamma j$ backgrounds to the W^* mechanism sum to a total of about 12 events under the Higgs peak.[10] The $t\bar{t}\gamma\gamma$ background to the $t\bar{t}\phi^0$ events is not plotted, but is of order 12 events also. The conclusions are apparent. A 200 TeV collider would certainly yield enough total signal events that analysis of the $\phi^0 \to \gamma\gamma$ angular decay distribution could be used to determine the spin of the resonance, but isolation of the W^* portion of the signal remains difficult, especially in light of the increased dominance of the $t\bar{t}\phi^0$ production mechanism that would have to be vetoed in order to uncover the W^* signal. Assuming complete vetoing is possible, $L = 50$ fb^{-1} would be required for a reliable calibration of the W^* process rate. The relatively small gain in our access to the W^* signal is due to the fact that it is induced by $q\bar{q}$ collisions, the luminosity for which is not significantly enhanced in going to higher beam energy at low ϕ^0 masses.

This situation should be considered in conjunction with that which might arise at a e^+e^- collider with $\sqrt{s} \sim 500$ GeV. There, the primary $b\bar{b}$, $t\bar{t}$, and W^+W^- couplings of a ϕ^0 in this mass range could almost certainly be determined, but one would be unable to see the $\phi^0 \to \gamma\gamma$ decays. This implies an important complementarity between the e^+e^- collider and a 200 TeV hadron collider; the latter would allow an excellent determination of $BR(\phi^0 \to \gamma\gamma)$ once the $t\bar{t}\phi^0$ coupling was determined from the e^+e^- results. The $\phi^0 \to \gamma\gamma$ branching ratio is, of course, of great interest because of its potential sensitivity to new particles of arbitrarily high mass that would yield new loop diagrams in the $\phi^0\gamma\gamma$ coupling calculation. For instance, a new generation of very heavy fermions would reduce $BR(\phi^0 \to \gamma\gamma)$ by a factor of order 6 or 7 compared to the three-family prediction. Despite this reduction, the event rate would be such that the $l\gamma\gamma$ signal could still be easily detected at the 200 TeV hadron collider, and the smaller-than-expected rate would constitute incontrovertible evidence for new physics at a mass scale that might be impossible to access directly.*

Finally, we should note that over a large region of parameter space in the minimal supersymmetric model, namely for $m_{H^0} \gtrsim 2m_Z$, the lighter CP-even h^0 has SM-like couplings to both VV ($V = W, Z$) and $q\bar{q}$ channels. Further, even though the MSSM predicts that $m_{h^0} \leq m_Z$ at tree level, after including one-loop radiative corrections[11] m_{h^0} is rather likely to lie in the region between m_Z and 130 GeV if the top quark mass is $\gtrsim 150$ GeV and $\tan\beta$ is not small. (The precise result depends, as well, on other supersymmetric model parameters, such as the squark masses.) Minimal extensions of the MSSM predict similar results for the h^0. Thus, it is quite possible that the search and study techniques discussed above for the SM ϕ^0 would be exactly those appropriate for the h^0 if supersymmetry turns out to be nature's choice. In determining the relative virtues of different machines for exploring Higgs boson physics, this has two important consequences. First, considerable weight should be attached to the type of scenario we have discussed in this section. But, second, it should be noted that if only the ϕ^0 or h^0 were to be detected in this general mass range and determined to have SM-like couplings, it would be impossible to know if the SM were correct or if the light Higgs boson found was part of a larger spectrum such as that predicted by the MSSM. Thus,

---

$\star$ Of course, the 200 TeV collider would be able to directly produce new quarks at a detectable level unless their mass were very large indeed.

the ability of a particular machine to detect the other Higgs bosons of an extended Higgs sector will also be of great importance. Our later discussion of the H^0 will provide an illustrative example.

$\underline{m_{\phi^0} \gtrsim 1 \text{ TeV}}$

In this section, I shall argue that one of the strongest motivations for a hadron collider at 200 TeV is that it would allow us to study the strong interactions between the longitudinally polarized states of W (and Z) bosons that would arise at WW center of mass energies in the TeV range in the case that there is no light Higgs boson. Most theorists agree that, although the SM model with $m_{\phi^0} \gtrsim 1$ TeV may not be a completely consistent theory, it does provide one reasonable model for the strength of interactions of longitudinally polarized W bosons in the TeV energy range. Equally important, it provides our only guide to the kinematic distributions of longitudinally polarized W bosons (and their decay products) produced via WW scattering. I shall employ this model to illustrate the potential of a 200 TeV collider to explore such interactions.

In order to focus the discussion, let me consider a specific channel of interest — the like-sign dilepton signal, $W^+W^+ \to l^+l^+\nu\nu$. This particular leptonic final state channel has the dual advantages of *both* a relatively small irreducible background (in particular, there is no analogue to the $q\bar{q} \to W^+W^- \to l^+l^-$ continuum background to the unlike-sign channel) *and* a relatively high event rate (compared, for instance, to the $ZZ \to l^+l^-l^+l^-$ channel). Consequently, the like-sign dilepton spectrum as a means of detecting strong scattering of longitudinally polarized W^+'s has been frequently discussed.[12–14] Most importantly, considerable effort has gone into developing techniques for overcoming the background from production of transversely polarized W^+'s deriving from $qq \to qqW^+W^+$ subprocesses. Efficient techniques for suppressing the transverse polarization background were established in Refs. [15] and [16], and complementary work has recently appeared in Ref. [17]. In particular, in Refs. [15] and [16] it was found that anti-tagging against energetic jets produced in association with the like-sign leptons was very effective in discriminating against events containing transversely polarized W^+'s in the final state, while at the same time allowing retention of most events in which both W^+'s are produced with longitudinal polarization. Cuts favoring back-to-back leptons were developed in Ref. [16] that also significantly suppress the transverse polarization backgrounds. Altogether, the techniques developed allow one to suppress the 'irreducible' background from events in which one or both of the W^+'s are transversely polarized to a level below that coming from the longitudinal polarization signal of interest, provided the signal is larger than or comparable in magnitude to that predicted in the SM computation in the case of $m_{\phi^0} = 1$ TeV. However, as we shall see, overall event rates at the SSC are very small.

The only important 'reducible' background was first explored in Ref. [16]; there it was found that additional cuts are needed in order to adequately suppress the background from $t\bar{t}$ production events in which one obtains like-sign leptons from the chain $t \to bW^+$, $W^+ \to l^+\nu$, and $\bar{t} \to \bar{b}\,jets$, $\bar{b} \to \bar{c}\,l^+\nu$. One possible procedure involving the tagging of a single jet was developed in Ref. [16]. This technique was explored in greater depth in Ref. [18], including $t\bar{t}g$ final states. In this latter reference, a possible alternative employing a requirement of lepton isolation was also examined.[†] We found that implementation of either procedure will almost certainly enable one to severely suppress the $t\bar{t}$-induced backgrounds while retaining the bulk of the events in which longitudinally polarized W^+'s are produced.

In order to discuss quantitatively our ability to see the excess events that would arise from scattering of longitudinally polarized W^+'s when $m_{\phi^0} \sim 1$ TeV, as compared to the number of events expected when m_{ϕ^0} is small, we define the signal event rate S as the *difference* between the $m_{\phi^0} = 1$ TeV rate and the $m_{\phi^0} = 50$ GeV event rate as computed in the SM. The most important irreducible background event rate is denoted by B and is computed as the event rate predicted from *purely electroweak* SM processes for $m_{\phi^0} = 50$ GeV. There is an additional background deriving from mixed electroweak/QCD processes involving gluon exchange diagrams that we do not discuss in detail here since it is always much smaller than the B event rate once the cuts we have discussed are imposed.

The results of Ref. [18] as a function of machine energy appear in Table 1. There, we compare the LHC and SSC to a machine with 200 TeV center of mass energy. We quote event rates for an integrated luminosity of 10 fb^{-1}. For clarity, we omit the above-mentioned 'gluon-exchange' and $t\bar{t}g$ backgrounds. (See Ref. [18] for further details.) Event rates are quoted in the order of increasing machine energy: LHC, SSC, 200 TeV. For details of the cut definitions please refer to Ref. [18]. Here we simply note that the lepton cut of the first row of the table is also imposed for the second row event rates, and that both the lepton cuts of the first and second row are imposed for all the following rows. The $p_T^{\max,5}$ cut is a jet vetoing cut in which one looks for the most energetic jet with $|y| < 5$ and demands that it have p_T below some maximum amount. For the $M_{jl}^{\min}$ cut, we tag one or more jets in the final state and construct the invariant masses of all jet-lepton pairs. We then demand that the minimum such invariant mass ($M_{jl}^{\min}$) be larger than some critical value. The isolation cut on $E_{\Delta R=0.25}^{\max}$ is a requirement that both the final l^+'s have very small energy in a surrounding cone of appropriate size (chosen to be $\Delta R = 0.25$). In the table, event rates quoted in parentheses have been obtained after imposing additional cuts on the energetic jets: namely there must be no more than two jets with $|y_j| < 5$ and $p_T^j \geq 30$ GeV, and any such energetic jet must be separated from the nearest l^+ by at least $\Delta R = 0.4$. These latter cuts are designed to reduce the $t\bar{t}$-induced background.

The results appearing in Table 1 indicate that the LHC will have great difficulty with this type of signal. For instance, after the $E_{\Delta R=0.25}^{\max} < 8$ GeV cut, the LHC yields only about 1 signal event per year vs. a background of about 0.6 events. Indeed, the LHC at $L = 100$ fb^{-1} yields only slightly greater statistical significance for the signal than does the SSC with $L = 10$ fb^{-1}. We estimate that at least $L = 20$ fb^{-1} is required at the SSC for detection of the like-sign dilepton Standard Model 1 TeV Higgs signal, while at the LHC about 190 fb^{-1} would be required (where in both cases we assume that the $E_{\Delta R=0.25}^{\max}$ procedure is employed and we sum l^+l^+ and l^-l^- channels). The like-sign channel event rates at 200 TeV for $L = 10$ fb^{-1} are much larger than in the case of the SSC, *e.g.* by a factor of about 13 for both the signal and background after the $E_{\Delta R=0.25}^{\max} < 8$ GeV cut. Thus, the 200 TeV energy choice yields an increase of

---

[†] The extent to which lepton isolation can be implemented is currently being studied by E. Wang and others, the main question being the ability to separate leptonic from hadronic energy in a detector with limited segmentation and shower development information.

Table 1. We give the event rates for integrated luminosity of $L = 10$ fb^{-1} for the electroweak signal, S, the $\mathcal{O}(\alpha_W^2)$ background, B, and the $t\bar{t}$ background, after imposing various cuts. The format and cut definitions for this table are those discussed briefly in the text and in more detail in Ref. [18]. Only the l^+l^+ event rates are given here. The triplets of numbers correspond to the LHC, SSC and 200 TeV machine energy choices, respectively. The W^-W^- final state yields about 1/3 as many S and B events in the l^-l^- channel as the number of l^+l^+ events listed in this table. The $t\bar{t}$ background event rate is the same in the l^-l^- channel as in the l^+l^+ channel.

Cut	S	B	$t\bar{t}$
$M_{ll} > 300,\ \|y_l\| < 3.5,\ p_T^l > 75$	1.6,12,180	5.6,46,550	1000,10000,1.6×10^5
	(1.6,12,170)	(5.2,43,540)	(83,660,8300)
$z_{ll} \leq -0.8,\ \delta p_T^{ll} \geq 200$	1.5,11,150	2.8,21,250	870,8900,1.4×10^5
	(1.4,11,150)	(2.6,19,240)	(58,470,5800)
$p_T^{\mathrm{max},5} \leq 125$	1.1,8.2,110	0.65,3.5,44	410,3300,42000
	(1.1,8.2,110)	(0.63,3.4,44)	(42,330,3500)
$p_T^{\mathrm{max},5} \leq 125,\ M_{jl}^{\mathrm{min}} \geq 200$	(0.83,6.5,73)	(0.29,1.9,26)	(0.062,0.42,6.3)
$p_T^{\mathrm{max},5} \leq 125,\ E_{\Delta R=0.25}^{\mathrm{max}} < 8$	1.1,8.2,110	0.65,3.5,44	0.01,0.016,0.19

event rate that would be appropriate for a worthwhile successor to the SSC; it would certainly be capable of accumulating enough events after several years of running to allow an actual measurement of the M_{ll} spectrum.

To illustrate this in more detail, we plot in Fig. 2 the M_{ll} spectrum for a 200 TeV machine as compared to that for the SSC. For each machine two curves appear: the upper curve is that found for $m_{\phi^0} = 1$ TeV; the lower curve is that obtained for $m_{\phi^0} = 50$ GeV. The excess of the upper curve over the lower curve is the signal arising from the scattering of longitudinally polarized W^+'s. One immediately sees that at the SSC an actual measurement of the M_{ll} spectrum is impossible for $L = 10$ fb^{-1}, regardless of the mass of the ϕ^0. In contrast, at the 200 TeV machine there are a reasonable number of events in each of the first ten 50 GeV bins when $m_{\phi^0} = 1$ TeV, and even if $m_{\phi^0} = 50$ GeV a rough idea of the M_{ll} spectrum would emerge. The ability to measure the M_{ll} spectrum when the ϕ^0 is light would enable us confirm that the discovered light Higgs boson does indeed cure all potential unitarity/high energy behavior problems in the W^+W^+ channel. This important cross check of the SM with a light ϕ^0 is only possible with $L = 10$ fb^{-1} at a ~ 200 GeV machine.

As discussed in Ref. [18], the lepton and M_{jl}^{min} cut techniques can be extended to other purely leptonic final state channels as well, with similar results. Thus, a 200 TeV collider would offer the opportunity to study the scattering of all types of longitudinally polarized vector boson pairs in the clean purely leptonic final states. A thorough test of the predictions of the SM (or any other model with similar or larger event rates) for these interactions would thus be feasible. In order to be competitive, a e^+e^- collider would need to have a center of mass energy of order $4 - 5$ TeV.

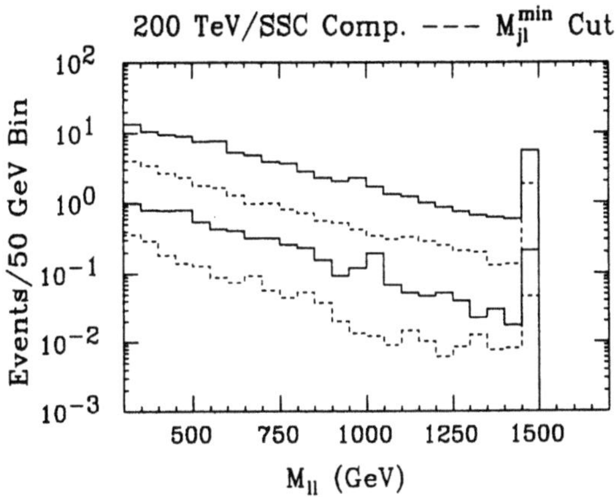

Figure 2. The M_{ll} spectrum for four W^+W^+ production cases. In order of decreasing magnitude: 200 TeV, $m_{\phi^0} = 1$ TeV; 200 TeV, $m_{\phi^0} = 50$ GeV; SSC, $m_{\phi^0} = 1$ TeV; SSC, $m_{\phi^0} = 50$ GeV. Results are those obtained after imposing lepton cuts, the $p_T^{max,5}$ jet-vetoing cut and the M_{jl}^{min} cut. $L = 10$ fb^{-1} is assumed. The bin size is 50 GeV. The final bin accumulates all events with $M_{ll} \geq 1450$ GeV.

THE H^0 OF THE MINIMAL SUPERSYMMETRIC MODEL

The final comparison that I shall discuss is the detectability of the MSSM H^0 in its $H^0 \to ZZ \to l^+l^-l^+l^-$ (4l, for short) decay mode.[2,19] (Recall that the H^0 is the heavier of the CP-even Higgs bosons of the MSSM.) As reviewed earlier, the effectiveness of searching for a CP-even Higgs boson in the 4l final state has been thoroughly documented in the case of the the SM ϕ^0.[1,3] In particular, the only significant background derives from the $q\bar{q}, gg \to ZZ \to 4l$ irreducible subprocesses. As we noted earlier, if a detector is built with reasonable rapidity coverage and momentum resolutions for the four leptons, then detection of the ϕ^0 is possible in this mode from roughly $m_{\phi^0} \gtrsim 130$ GeV all the way up to $\gtrsim 800$ GeV, depending upon the top quark mass and available integrated luminosity. It would be highly desirable to be able to detect the H^0 (which over much of model parameter space is heavy enough to decay to two Z's) in this same way. However, in the region of parameter space where $m_{H^0} \geq 2m_Z$ the H^0ZZ coupling is generally rather severely suppressed in comparison to the SM ϕ^0ZZ coupling.[1] Consequently, it might be supposed that detection in this mode would be extremely difficult. Fortunately, this turns out to not necessarily be the case.

Indeed, there are a number of crucial facts which result in the possibility of H^0 detection in the 4l mode over a significant, but machine dependent, portion of parameter space. First, even though the $H^0 \to ZZ$ decay *width* is suppressed, $BR(H^0 \to ZZ)$ remains close to the corresponding value for the ϕ^0 so long as $\Gamma(H^0 \to ZZ) \gtrsim \Gamma(H^0 \to Q\bar{Q})$, where Q is the heaviest quark kinematically allowed in H^0 decays. In practice, this is always the case for $m_{H^0} \lesssim 2m_t$, so long as $\tan\beta$ (the ratio of the vacuum expectation values of the two doublet fields)* is not very large, *i.e.* so long as the $H^0 \to b\bar{b}$ decays are not greatly enhanced. Second, the generally much smaller values of $\Gamma(H^0 \to ZZ, W^+W^-,)$ compared to $\Gamma(\phi^0 \to ZZ, W^+W^-)$ (with $m_{\phi^0} = m_{H^0}$) implies that the H^0 is always a very narrow resonance. This has the consequence that the $q\bar{q}, gg \to ZZ \to 4l$ background is much smaller than in the ϕ^0 case, and is entirely determined by the resolution in the $4l$ channel. For our computations in this paper we will consider the $4l$ mass resolution (which is a slowly varying function of M_{ZZ}) found for the design parameters for the SDC detector.[3] Finally, the production cross section for the H^0 at a hadron supercollider is generally quite large.

The main sources for H^0 production are gluon-gluon fusion into H^0 or $H^0 Q\bar{Q}$ (where $Q=t$ or b quark); they provide an ample event sample at the SSC (and LHC). (W^+W^- fusion, which makes a significant contribution to the ϕ^0 cross section at larger m_{ϕ^0}, is never important for the H^0 because of the suppression of the $H^0W^+W^-$ coupling mentioned earlier.) The relative importance of the gg fusion mechanisms depends crucially upon both m_t and $\tan\beta$. $gg \to H^0$ is dominant over most of the relevant m_{H^0} range, for moderate or small $\tan\beta$, provided m_t is not small (a good bet given the CDF limit of $m_t \gtrsim 90$ GeV). This is because, for large m_t, the $gg \to H^0$ fusion mechanism is dominated by the t loop if $\tan\beta$ is not too large — if $\tan\beta$ is large the $H^0 t\bar{t}$ coupling is suppressed. In contrast, the $gg \to H^0 b\bar{b}$ mechanism tends to be dominant for large values of $\tan\beta$ where the $H^0 b\bar{b}$ coupling is enhanced. Indeed, at large $\tan\beta$ the production cross section for the H^0 can even be larger than that for a ϕ^0 of similar mass. As a result, even though $BR(H^0 \to 4l)$ declines rapidly with increasing $\tan\beta$ due to the rapid increase of $\Gamma(H^0 \to b\bar{b})$, the associated rise of the $gg \to H^0 b\bar{b}$ cross section means that only for large $\tan\beta$ values will H^0 detection in the $4l$ mode be completely impossible.

The relative abilities of the SSC and a 200 TeV collider to detect the H^0 in its $4l$ decay mode can be seen in Fig. 3. There we plot the $4l$ event rates for $L = 10$ fb^{-1} and $m_t = 200$ GeV for a number of $\tan\beta$ values, in comparison to the event rate required in order to achieve a 4σ signal over the $ZZ \to 4l$ continuum background. We note that the gg-induced signal event rate is typically a factor of 10 higher at 200 TeV as compared to the SSC, whereas the ZZ continuum background only increases by about a factor of 4, implying that the 4σ event level requirement, proportional to the square root of the background, increases by only a factor of 2. Not surprisingly, therefore, the range of $m_{H^0} - \tan\beta$ parameter space over which $H^0 \to 4l$ detection is possible is much greater at a 200 TeV machine than at the SSC.

Regarding the role of a e^+e^- collider in H^0 detection we note that since the only viable production mode is $Z^* \to H^0 + A^0$ and since $m_{H^0} \sim m_{A^0}$ (once $m_{H^0} \gtrsim 2m_Z$), the region of accessible m_{H^0} is restricted to $m_{H^0} \lesssim \sqrt{s}/2$. Thus, from the figure, we see that, depending upon $\tan\beta$ and m_t, crudely speaking a 1 TeV e^+e^- collider would be required to be competitive with a 200 TeV hadron collider. Of course, detection of the H^0 at a e^+e^- collider would be in its (dominant) $b\bar{b}$ or $t\bar{t}$ decay mode, and it would be unlikely that the (small) branching ratio for $H^0 \to ZZ$ could be measured. So, if the H^0 were to be detected at both machines, the information on the H^0 couplings would be nicely complementary. To extract the $H^0 ZZ$ coupling from the combined

---

$\star$ See Ref. 1 for details regarding this MSSM parameter and a general theoretical background to the Higgs sector of the MSSM.

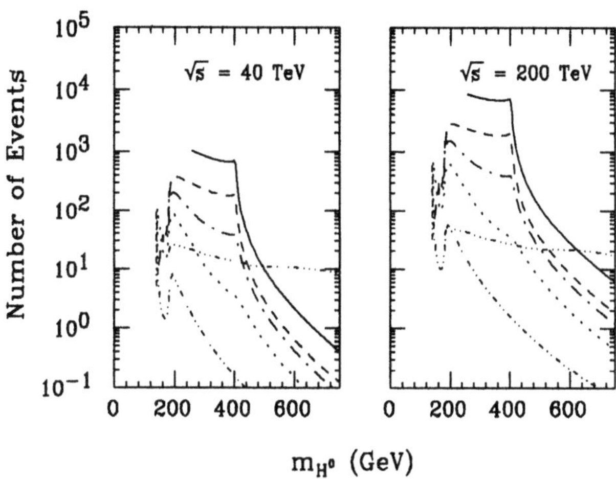

Figure 3. $H^0 \to 4l$ event rates at the SSC and at 200 TeV are compared for various $\tan\beta$ values: $\tan\beta = 0.5$, 1, 2, 5, and 20 in order of decreasing event rate. Also shown as the dashdotdotdot curve is the 4σ event level requirement in order to see the narrow H^0 peak over the $q\bar{q}, gg \to ZZ \to 4l$ background. We assume $L = 10$ fb^{-1} and have taken $m_t = 200$ GeV.

data would, however, require a number of steps. First, we presume that the lighter CP-even state of the MSSM, the h^0, would already have been discovered. Given m_t, m_{h^0}, and m_{H^0}, and squark masses (presumably observed, if not at the SSC, then certainly at the 200 TeV machine), all Higgs sector parameters are determined. (Squark masses and the value of m_t are needed to fix the radiative corrections to tree-level predictions for the Higgs sector — the latter depend only on the two parameters m_{h^0} and m_{H^0}.) One could then use the e^+e^- results to check that at least the $H^0 Q\bar{Q}$ coupling for the dominant $H^0 \to Q\bar{Q}$ decay channel is as predicted. Meanwhile, the predicted and/or measured $H^0 Q\bar{Q}$ couplings could be used to predict the H^0 production rate at the hadron collider, which could then be combined with the observed $H^0 \to 4l$ event rate to determine $BR(H^0 \to ZZ)$ and, thence, the $H^0 ZZ$ coupling.

CONCLUSIONS

Thus, we see that a 200 TeV collider would, indeed, expand considerably our ability to uncover the nature of the Higgs boson/electroweak symmetry breaking sector. As I have discussed, its greatest advantage over any e^+e^- collider other than one with super high energy (> 4 TeV) would emerge if there is no light Higgs boson (or equivalent), implying that there would be strong interactions between W and Z bosons at energy scales at and above a TeV. However, even if one or more light Higgs bosons do exist, a 200 TeV collider can perform many tasks that the SSC cannot and Higgs studies performed at it could end up nicely complementing the information available from a 0.5 to 2 TeV e^+e^- collider.

Acknowledgements

I would like to thank the Ettore Majorana Centre for its hospitality and organization of the 'QCD at 200 TeV' conference where the comparisons presented here were initiated.

REFERENCES

1. J.F. Gunion, H.E. Haber, G.L. Kane, and S. Dawson, *The Higgs Hunter's Guide*, Addison Wesley, Frontier in Physics Series.

2. J.F. Gunion, 'An Overview of, and Selected Signatures for, Higgs Physics at Colliders', to appear in *Proceedings of the 'Workshop in High Energy Physics Phenomenology — II'*, Calcutta (January, 1991).

3. SDC Collaboration, Letter of Intent, publication SDC-90-00151.

4. J.F. Gunion, *Phys. Lett.* **261B** (1991) 510.

5. W. Marciano and F. Paige, *Phys. Rev. Lett.* **66** (1991) 2433.

6. M. Mangano, SDC Collaboration Note SSC-SDC-90-00113.

7. R. Kleiss, Z. Kunszt, and W.J. Stirling, *Phys. Lett.* **253B** (1991) 269.

8. J.F. Gunion, G.L. Kane *et al.*, 'Overview and Recent Progress in Higgs Boson Physics at the SSC', to appear in *Proceedings of the 1990 Snowmass Summer Study on "Research Directions for the Decade"*, preprint PRINT-91-0010 (1990).

9. Z. Kunszt, Z. Trocsanyi, and W.J. Stirling, preprint ETH-TH-91-17 (1991); A. Ballestrero and E. Maina, preprint DFTT 29/91 (1991).

10. These backgrounds rates were provided by M. Mangano.

11. J. Ellis, G. Ridolfi, F. Zwirner, *Phys. Lett.* **257B** (1991) 83; H.E. Haber, R. Hempfling, *Phys. Rev. Lett.* **66** (1991) 1815; Y. Okada, M. Yamagushi, T. Yanagida, *Prog. Theor. Phys.* **85** (1991) 1.

12. M.S. Chanowitz and M.K. Gaillard, *Nucl. Phys.* **B261** (1985) 379.

13. M.S. Chanowitz and M. Golden, *Phys. Rev. Lett.* **61** (1988) 1053, and Erratum, *Phys. Rev. Lett.* **63** (1989) 446.

14. R. Vega and D.A. Dicus, *Nucl. Phys.* **B329** (1990) 533.

15. V. Barger, K. Cheung, T. Han, and R.J.N. Phillips, *Phys. Rev.* **D42** (1990) 3052.

16. D. Dicus, J.F. Gunion and R. Vega, *Phys. Lett.* **258B** (1991) 475.

17. M. Berger and M. Chanowitz, *Phys. Lett.* **263B** (1991) 509.

18. D.A. Dicus, J.F. Gunion, L. Orr, and R. Vega, preprint UCD-91-10 (1991).

19. R. Bork, J.F. Gunion, H.E. Haber, A. Seiden, preprint UCD-91-23 (September, 1991).

BARYON NUMBER VIOLATION AND INSTANTONS IN THE STANDARD MODEL*

V. V. Khoze

Center for Theoretical Physics
Laboratory for Nuclear Science and Department of Physics
Massachusetts Institute of Technology
Cambridge, Massachusetts 02139 U.S.A.

ABSTRACT

We report on the recent developments in the theory of high energy collisions mediated by an instanton. Theoretical methods of studying the non-perturbative dynamics of the Standard Model at high energies in association with multi-particle-production and (B+L)-violation are reviewed. Particularly, we concentrate on the optical theorem approach based on the valley method. A pre-history and a current status of the theory of electroweak (B+L)-violation in high energy collisions are discussed.

I. INTRODUCTION

Recently there has been growing interest in non-perturbative effects in the Standard Model. For example, it has been suggested that electroweak interactions become strong at high energies, showing completely new physical phenomena like high-multiplicity events involving "weakly" interacting particles and baryon (B) plus lepton (L) number violation.[1]

Baryon number violation is a problem which has been of much interest in theoretical physics. It was pointed out by Sakharov[2] that the baryon asymmetry may be generated in the early stages of the Universe if the hot plasma particle interactions violate C, CP and B. With a proper understanding of the dynamics of baryon number violation, one may hope that it should be possible to compute from first principles the observed baryon asymmetry of the Universe. Until recently, it was believed that the only baryon number changing processes relevant for cosmology were those whose natural physical scale is of order of $E \geq 10^{15}$ GeV. Such processes are widely discussed beyond the Standard Model.

---

* This work is supported in part by funds provided by the U. S. Department of Energy (D.O.E.) under contract #DE-AC02-76ER03069.

However, it has been well known[3] that there is another class of baryon number changing processes associated with electroweak theory within the framework of the Standard Model*. The Standard Model of elementary particle physics is based on the dynamics of gauge theories. In particular, weak interactions are described by a non-Abelian SU(2) gauge theory: quarks and leptons interact through the exchange of massive vector bosons, $W^\pm$, Z^0, and the Higgs boson, H^0. A specific feature of *non-Abelian* gauge theories is a non-trivial topological structure[7] of their vacuum-states. Let us consider the fundamental SU(2)-Higgs model, defined by the following action (in Minkowski space),

$$S[W,\Phi] \equiv S_g[W]+S_h[W,\Phi] \equiv \int d^4x \left\{ -\frac{1}{2}\mathrm{tr}(F_{\mu\nu}F_{\mu\nu})+ \mid D_\mu\Phi \mid^2 -\lambda\left(\mid \Phi \mid^2 -\frac{v^2}{2}\right)^2\right\}.$$
$$(1.1)$$

This model corresponds to the bosonic part of the electroweak theory in the limit of vanishing Weinberg angle.

Classical vacuum configurations, *i.e.* static fields with vanishing action, are pure gauges. Let us choose for convenience the temporal gauge, $W_0 \equiv 0$. In this case the classical vacua can be written in general as

$$\vec{W}_{\mathrm{vac}}(\vec{x}) = -\frac{i}{g}\left[\vec{\nabla}\Omega(\vec{x})\right]\Omega^{-1}(\vec{x}) \ , \ \Phi_{\mathrm{vac}}(\vec{x}) = \Omega(\vec{x})\Phi_0, \qquad (1.2)$$

where $\Omega(\vec{x})$ is a static SU(2) gauge transformation and $\Phi_0 = (v/\sqrt{2})(0\ 1)^T$ denotes the vacuum expectation value. The allowed spatial gauge transformations Ω are continuous (in order to have a finite action) mappings from $\mathbf{R}^3 \cup \{\infty\} \sim S^3$ onto SU(2) $\sim S^3$. Such mappings fall into homotopy classes, which may be classified by integer winding-number ν. The winding-number is the number of times the image $\Omega(\vec{x})$ wraps around the group manifold SU(2) $\sim S^3$ as $\vec{x}$ runs through the spatial 3-sphere. Gauge transformations (and the corresponding classical vacua) with different winding-numbers are topologically inequivalent, *i.e.* they cannot be continuously deformed into each other. Physically this means[7] that the potential in the Yang-Mills theory in the temporal gauge is a periodic function of the winding number and has local minima for integer values of ν (see Fig. 1). These minima represent a set of topologically inequivalent vacua. Indeed, the potential in these minima is equal to zero and any two minima can not be related by a non-singular gauge transformation.

In quantum theory there is a probability to penetrate the barrier between two different vacua. If this probability is small, it can be evaluated by semiclassical methods. To this end one should find the classical under-barrier-trajectory in imaginary time, *i.e.* the Euclidian classical solution – *instanton* (I).[8] The probability of quantum tunneling is proportional to $\exp(-2S_I)$, where S_I is the instanton action. The

---

* We remind that C, CP and B violations are not conserved in the Standard Model of electroweak interactions.[4] Charge conjugation, C, is violated because of the (V-A)-structure of weak interactions, CP is non-conserved in the three-family Standard Model due to a CP-violating phase in the weak mixing matrix[5] and, finally, B and L are violated because of the chiral anomaly,[6] which leads to a non-conservation of the baryon-number (and lepton-number) current[3] due to the (V-A)-structure of weak interactions. These features of the Standard Model may be crucial for the formation of the baryon asymmetry of the Universe. However, the purpose of the present lecture is to discuss (B+L)-violation in high energy collisions rather than its cosmological consequences.

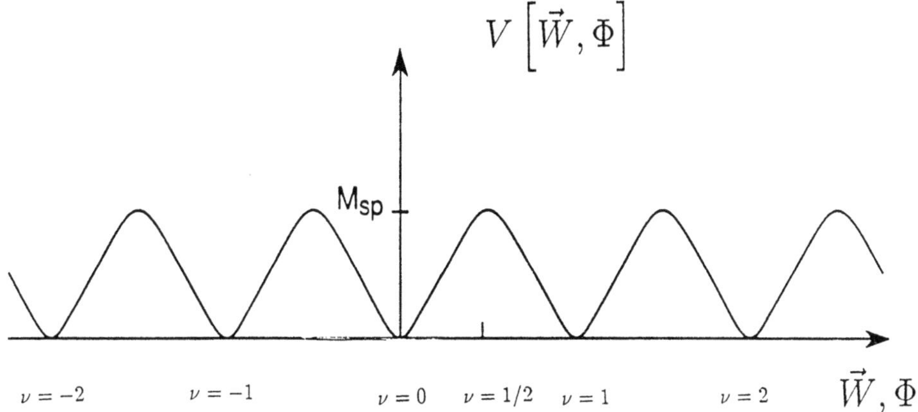

Fig. 1. The electroweak potential V as the functional of gauge and Higgs fields. The minima are labeled by integer winding numbers ν. Only one particular direction is shown on this picture so the sphaleron looks here as the maximum of V. In fact the sphaleron is rather a saddle point, than a maximum of V since V increases in other directions.

transitions between inequivalent vacua have drastic effects due to an anomaly in the axial-vector current.[6] Namely, the non-trivial vacuum structure is intimately connected with a (B+L)-violation, which is, in contrast to similar processes in grand unified theories of strong and electroweak interactions, entirely of non-perturbative nature. For notational convenience we shall call such processes *anomalous*. Due to the (V-A)-structure of weak interactions, the non-Abelian SU(2) gauge fields couple only to the left-handed fermion doublets, $\psi^{(i)}, i = e, \mu, \ldots$. The fermion-number current of each doublet, $J_\mu^{(i)} \equiv \overline{\psi}^{(i)} \gamma_\mu \psi^{(i)}$, is conserved at the classical level, at the quantum level, however, it is not conserved due to the chiral anomaly,[6]

$$\partial_\mu J_{(i)}^\mu = -\frac{g^2}{16\pi^2} \text{tr}\left(F_{\mu\nu}\tilde{F}^{\mu\nu}\right). \tag{1.3}$$

Integrating the anomaly (1.3) we get

$$N^{(i)}(t = \infty) - N^{(i)}(t = -\infty) = -\frac{g^2}{16\pi^2} \int_{-\infty}^{+\infty} dt \int d^3x \ \text{tr}\left(F_{\mu\nu}\tilde{F}^{\mu\nu}\right) \equiv -Q. \tag{1.4}$$

Q is an integer if the gauge fields at the times $t = \pm\infty$ are vacua, and it coincides in this case exactly with the winding-number difference between those vacua, *i.e.* $Q = \triangle\nu$. Q is therefore called a topological charge.

In summary, transitions between topologically inequivalent vacua are associated with a change in the fermion-numbers, $\triangle N^{(i)} = -\triangle\nu$. Since the change is equal for every fermion doublet, we obtain the following selection rule for the anomaly-induced (B+L)-violation in the three-family Standard Model,

$$\triangle N_e = \triangle N_\mu = \triangle N_\tau = \frac{1}{9}\triangle N_q = -\triangle\nu, \tag{1.5}$$

where $N_e(N_\mu, N_\tau)$ denotes the electron-(muon-,tau-)-lepton-number and N_q stands for the quark-fermion-number. Electrical charge, SU(3)-color and B-L are conserved. An example of a ($Q = \triangle\nu = 1$)-process is shown in Fig. 2. The amplitudes for

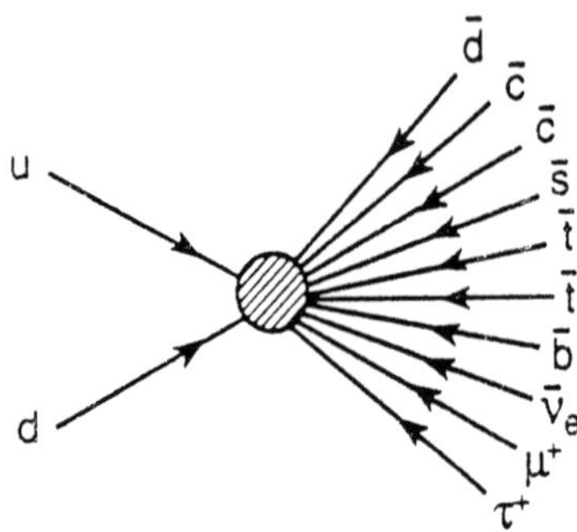

Fig. 2.   Example of (B + L)-violating processes in the Standard Model.

such instanton-induced processes were estimated to be of the order $e^{-2\pi/\alpha_W}$, where α_W is the SU(2) coupling constant. This factor is of order 10^{-80}, and is so small that if there is no large factor to compensate this exponential suppression, then such processes are of no practical phenomenological significance.

It was shown by Manton and Klinkhammer[9] that the minimum barrier height between inequivalent vacua is finite in electroweak theory and is given by the mass (static energy) $M_{\rm sp}$ of *the sphaleron* – an unstable static solution of the field equations which represents a height of a barrier in Fig. 1, $(\vec{W}_{\rm sp}, \Phi_{\rm sp})$. This unstable solution has been found in Ref. [10] and studied in detail in Ref. [9]. The mass and radius of the sphaleron are of the order

$$M_{\rm sp} \simeq \pi \frac{m_w}{\alpha_w} = 7.2 \text{ TeV} , \ R_{\rm sp} \simeq m_w^{-1}. \tag{1.6}$$

Taking the potential barrier picture seriously, it was suggested[11−13] that these anomalous processes may be unsuppressed if high temperatures or high energies are involved: instead of tunneling through the barrier the field can pass over the barrier. It is believed that there is by now a consensus that $(B + L)$-violating effects are important[11,13,14] in a hot plasma. Putting aside some minor questions with the thermal case, we shall consider in this paper only the case of high energies.

The sphaleron mass sets the scale above which the transitions between topologically inequivalent vacua may become important. The enormous height of the barrier explains why the usual perturbation theory works so well for the description of electroweak processes at presently available energies. The physical difference between the low and high energy pictures is roughly similar to the difference between *spontaneous* and *induced* decay of heavy nucleus. The probability of the decay induced by the nucleon with the energy of order of the height of the potential barrier may be higher by several orders of magnitude then the probability of spontaneous decay.

The new input to this problem was done by Ringwald,[1] see also Ref. [15], who advocated in the framework of the so-called leading-order instanton approach that the 't Hooft suppression $e^{-2\pi/\alpha_W}$ of an amplitude for the fermion number violating processes can be compensated by some energy-dependent factor for sufficiently high energy of colliding particles $E \sim M_{\rm sp}$. In the leading-order approximation the total cross section for anomalous processes grows exponentially with energy. The compensating factor appears due to multi-particle production in the instanton-induced high-energy collision. The problem, however, is that far before the instanton-induced amplitudes become large, the leading-order instanton approach used to compute their effects breaks down.[16-21] That means that even the first corrections to the naive instanton calculation must be significant at high energies and may drastically change the final result.

The organization of the rest of the paper is as follows. In sect. 2 we give the idea of Euclidean functional methods (instanton approach) and review the leading-order calculation[1,15] of the cross-section for (B+L)-violating processes. The problem of classification and taking into account in some self-consistent way the corrections to the naive instanton calculation became one of the most interesting and important problems of the Standard Model. This problem is discussed in sects. 3-5. Namely, the valley method as a tool to investigate instanton-induced total cross-sections beyond the naive instanton approximation is discussed in sect. 3. In sect. 4 we push the valley method to its limits and present the latest speculations on the behavior of instanton-induced cross-sections above 10 TeV. Quantum (due to hard particles) and classical (due to multi-instantons) corrections are discussed in sect. 5. Our summary is presented in sect. 6.

II. $(B + L)$-VIOLATION IN HIGH ENERGY COLLISIONS

We are interested in a computation of the amplitude of a (B+L)-violating quark-quark collision. For sufficiently high energy of incoming quarks where we must have in the final state in addition to the minimum number of anti-fermions also a great number of W-, Z- and Higgs-bosons (see Fig. 3). It is easy to see that this amplitude gets no contribution from usual perturbation theory about the trivial vacuum, since there are no any connected Feynman diagram with only incoming fermion lines. This means that we have to use another (more general) calculation scheme in order to compute the amplitude. For this end it is convenient to use the functional integral representation of quantum field theory. In order to get the complete physical answer in such a formalism one should perform the integrations in the functional space over all the fields with the statistical factor $\exp(-S(\phi))$, where S is the Euclidean action of the field ϕ. This is the reminiscence of the classical principle of the minimal action. Indeed, the action is huge or even infinite for almost all the fields in the functional space and one can hope that only such fields which have the minimal action

$$\phi_{\text{cl}} \; : \; \frac{\delta S}{\delta \phi}\Big|_{\phi_{\text{cl}}} = 0, \tag{2.1}$$

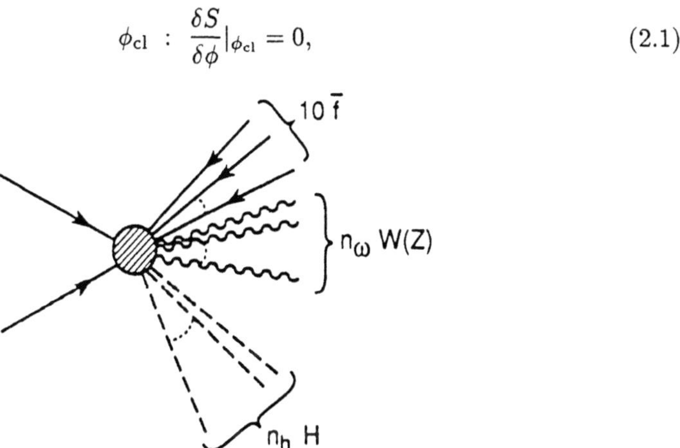

Fig. 3. $(B + L)$-violating process with many gauge and Higgs bosons.

are significant. Thus, it seems to be a good approximation if one integrates only over the small fluctuations of the fields in the vicinity of classical solutions. The perturbation theory which is, probably, the most elaborated field theoretical approach deals with the fluctuations around *trivial* classical solution $\phi_{\rm cl} = 0$. This means that integrations over fluctuations about such a solution reproduce the perturbative contribution (*p.t.*). In general, a more complete answer is, of course, the sum of *p.t.* contribution and integrations about *non-trivial* classical solutions (I and $\overline{I}$), however, in our case *p.t.* contribution is absent. Up to analytic continuation and LSZ reduction,[22] the amplitude of the process is given by the Euclidean functional integral over the fields corresponding to the external particles in Fig. 3,

$$
\mathcal{T}_{n_w,n_h}^{\Delta(\mathrm{B+L})} \sim \int \mathcal{D}W\mathcal{D}\Phi\mathcal{D}\psi\mathcal{D}\overline{\psi}\; e^{-S[W,\Phi,\overline{\psi},\psi]} \prod_{i=1}^{12} \overline{\psi}^{(i)}(x_i) \prod_{j=1}^{n_w} W_{\mu_j}(y_j) \prod_{k=1}^{n_h} \Phi(z_k), \quad (2.2)
$$

where,

$$
S[W,\Phi,\overline{\psi},\psi] \equiv \int d^4x \left\{ \frac{1}{2}\mathrm{tr}(F_{\mu\nu}F_{\mu\nu}) + |\, D_\mu\Phi\,|^2 + \lambda\left(|\,\Phi\,|^2 - \frac{v^2}{2}\right)^2 \right.
$$
$$
\left. + \sum_{i=1}^{12} \overline{\psi}^{(i)}(i\sigma_\mu D_\mu)\psi^{(i)} \right\} , \tag{2.3}
$$

is the Euclidean action of the theory of weak interactions with 12 massless fermion left-handed doublets (12 fermion flavors correspond to 3 generations of quarks and leptons and 3 color of quarks) in the limit of vanishing Weinberg angle and $\sigma_\mu \equiv (i,\sigma^a)$, where σ^a stands for usual Pauli matrices.

Because of the anomaly, Eq. (1.4), this functional integral receives[3, 23] only contributions from gauge fields which interpolate between vacua with winding-number difference $\Delta\nu = 1$, *i.e.* which have topological charge $Q = 1$.

Now we have to find a minimum of the Euclidean action, *i.e.* a classical solution, with topological charge $Q = 1$, and expand the integrand about that. These classical solutions are called *instantons*.[8, 3, 24] The instanton is actually a family of field configurations,

$$
\phi_{\tau_I}(x) \equiv \phi_I(x;\tau_I) \equiv \left\{ W_I(x;\tau_I)\,,\; \Phi_I(x;\tau_I)\,,\; \psi_I(x;\tau_I) \right\}, \tag{2.4}
$$

labelled by a set of (collective coordinates), $\tau_I = \{x_I, \rho_I, U_I\}$, where instanton position x_I, scale-size ρ_I and orientation U_I reflect invariance of the theory under translations, dilations and rotations correspondingly. The action attains its minimum for all values of these collective coordinates which means that we should integrate exactly over these parameters rather than in Gaussian approximation. Actually, some complication arises in the electroweak theory due to its mass scale introduced by symmetry breaking: by scaling arguments one may show that strictly speaking no exact solution exists.[24] Nevertheless, there exists a constrained minimum of the Euclidean action, which means that one may find a minimum of the action, under the constraint that the scale-size ρ_I is fixed.[24]

As we already mentioned, one should perform the integrations in Eq. (2.2) over the fluctuations $\delta\phi$,

$$
\delta\phi \; : \; \phi = \phi_{\tau_I} + \delta\phi,
$$

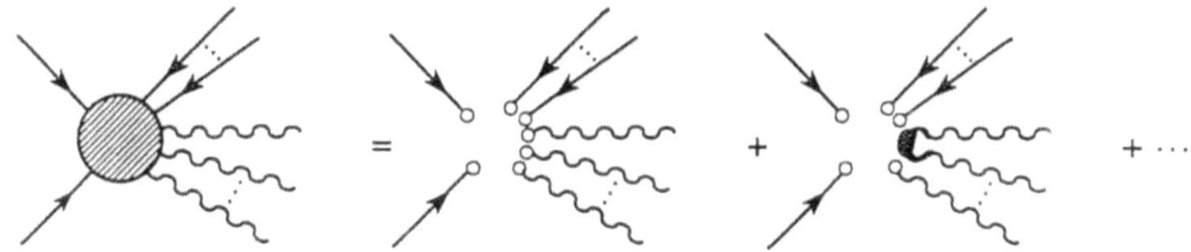

Fig. 4. The semiclassical expansion of (B + L)-violating amplitude. The first term on the right-hand side corresponds to the leading-order result, Eq. (2.5). The next term is an $O(\alpha_W)$ correction: two external lines are connected by a propagator in the instanton background.

in the vicinity of classical solutions (2.4). In the leading-order instanton approach[1] we neglect the fluctuations in the preexponent on the r.h.s. of Eq. (2.2) and leave in the exponent only the leading (quadratic) term in fluctuations. The integration over $\delta\phi$ is, thus, performed in Gaussian approximation. In this approach we get our amplitude (2.2) to be proportional to the product of the classical solutions (see Fig. 4), weighted by the exponential factor $\exp(-S_{cl})$, where $S_{cl} = 2\pi/\alpha_w + \pi^2 v^2 \rho_I^2$ is the classical action of the instanton. We see immediately[3] that amplitudes for (B+L)-violating processes have always to fight against the exponentially small factor $\exp(-2\pi/\alpha_w)$.

The fact that in leading-order approximation the amplitude (2.2) is just proportional to a product of classical solutions has as consequence that, on the mass-shell of the external particles, it behaves very similar to a point-like vertex. Explicitly we obtain[1,15]

$$T_{n_w,n_h}^{\Delta(B+L)} \sim \frac{(n_w + n_h)!}{m_w^{n_w+n_h}} g^{n_w+n_h} e^{-2\pi/\alpha_w} \prod_{j=1}^{n_w} \left(\frac{E_j}{m_w}\right), \qquad (2.5)$$

where E_j is the energy of the jth W-boson. The particles will be produced essentially only in the s-wave, *i.e.* they will come out of the vertex isotropically (in the CMS).

At high energies these point-like amplitudes violate s-wave unitarity, since according to Eq. (2.5) the corresponding cross-sections grow at least like phase-space, $\sigma_{n_w,n_h}^{\Delta(B+L)} \sim E^{4n_w+2n_h}$, where $E \equiv \sqrt{s}$ is the CM energy of the processes. At which energy does this happen? Let us consider the total cross-section for (B+L)-violation which, in our approximation Eq. (2.5), is given by[17]

$$\sigma_{\text{tot}}^{\Delta(B+L)} \equiv \sum_{n_w,n_h} \sigma_{n_w,n_h}^{\Delta(B+L)} \sim \exp\left[\frac{4\pi}{\alpha_w} F_0(\epsilon)\right], \qquad (2.6)$$

where $\epsilon \equiv E/M_{\text{sp}}$. The function F_0 grows with the energy like

$$F_0(\epsilon) \simeq -1 + \left(\frac{9}{16\sqrt{8}}\right)^{2/3} \epsilon^{4/3}. \qquad (2.7)$$

In the first term in Eq. (2.7) we recognize the 't Hooft suppression, whereas the second term, which leads to an exponential growth, originates from the associated W- and Z-production The mean multiplicity of gauge bosons is directly related to the exponent in Eq. (2.6) and reads[17]

$$\bar{n}_w \simeq \frac{4\pi}{3\alpha_w}\left[F_0(\epsilon) + 1\right] = \frac{4\pi}{3\alpha_w}\left(\frac{9}{16\sqrt{8}}\right)^{2/3} \epsilon^{4/3}. \qquad (2.8)$$

The leading-order approach is based on the assumption that one can neglect fluctuations, $\delta\phi \sim 1$, about an instanton field, $\phi_{\tau_I} \sim 1/g$, in the preexponent on the r.h.s. of Eq. (2.2). However, this is valid only at energies much lower than the sphaleron mass. At high energies, $\epsilon \sim 1$, the mean multiplicity of final particles (2.8) is parametrically large, $\overline{n} \sim 1/\alpha$, and the final state corrections (soft-soft quantum corrections) to the leading-order result are important,

$$(\phi)^{\overline{n}} = (\phi_{\tau_I} + \delta\phi)^{\overline{n}} \sim (\phi_{\tau_I})^{\overline{n}}(1 + \mathcal{O}(\overline{n}^2 \alpha_w) + ...). \tag{2.9}$$

It was shown[18,19,21] that soft-soft corrections exponentiate and contribute to the next-to-leading-order terms in the expansion of $\Gamma(\epsilon)$,

$$\Gamma(\epsilon) \equiv \frac{4\pi}{\alpha_w} F(\epsilon) \equiv \ln \sigma_{\text{tot}}^{\text{anom}}, \tag{2.10}$$

in terms of ϵ, while the Ringwald approach gives the leading-order result (this explains why the Ringwald method is called the leading order approach).

Thus, we proved that at low energies, $\epsilon \ll 1$, the anomalous cross-section (2.6) grows exponentially with energy. At high energies, $\epsilon \sim 1$, (which is the most interesting case) the leading-order approach breaks down. In the next section we demonstrate an effective and clever way[25,26] to go beyond the leading-order result and find the next-to-leading-order term in Eq. (2.10).

III. BEYOND LEADING-ORDER WITH THE VALLEY METHOD

The higher order corrections in α_w to the amplitudes with fixed multiplicities, as indicated in Fig. 4, lead to form factors by tying together external legs by propagators in the instanton background. One expects that these corrections soften the amplitudes. In the literature these corrections have been divided into two classes, namely into corrections connecting the large number of soft particles in the final state[18,19,21] (soft-soft corrections) which may be large due to combinatorics, and into corrections connecting the hard particles in the initial state[20] with each other (hard-hard corrections) or with soft final-state particles (hard-soft corrections), which may be large due to the high energy stored in the initial-state particles. We briefly discuss hard quantum corrections in sect.5.

What one would like to have is a systematic way to calculate directly the total cross-section for (B+L)-violation, without referring to the fixed-multiplicity amplitudes. Such a systematic procedure was proposed in Refs. [25,26]. (for another approach see Ref. [19]). It was argued (see also an earlier work of Zakharov[17]) that the total cross-section for (B+L)-violation could be inferred from a calculation of the imaginary part of the forward elastic scattering amplitude (see Fig. 5),

$$\sigma_{\text{tot}} \sim \frac{1}{s} \cdot \text{Im } \mathcal{T}(2 \to 2; s; t = 0), \tag{3.1}$$

if one takes into account in the functional integral expression for the $2 \to 2$ scattering amplitude, which may be written in a multi-component notation for the field variables as

$$\mathcal{T}(2 \to 2) \sim \int \mathcal{D}\phi \prod_{i=1}^{4} \phi(x_i) \exp(-S[\phi]), \tag{3.2}$$

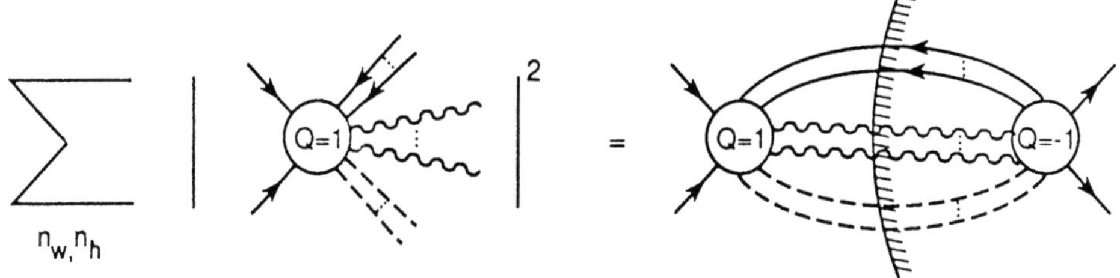

Fig. 5.   Optical theorem for an anomalous process.

only the contribution from the instanton(I)/anti-instanton($\bar{I}$) configuration.

The $I\bar{I}$ contribution to the functional integral can be calculated systematically by semiclassical methods.[27-29] Let $\phi_{\tau_I(\tau_{\bar{I}})}$ be the $I(\bar{I})$ solution, Eq. (2.1), which is characterized by a set of collective coordinates $\tau_{I(\bar{I})} = \{x_{I(\bar{I})}, \rho_{I(\bar{I})}, U_{I(\bar{I})}\}$. The sum of I and $\bar{I}$ is an approximate solution at large separation $R = x_I - x_{\bar{I}}$. At small separation, however, it ceases to be an approximate solution, because of the $I\bar{I}$ interactions. Therefore, to expand only around the sum of I and $\bar{I}$ is not reliable, since there are large linear terms in the exponent due to interactions.

The valley method[27-29] gives a means to circumvent these problems. The idea is the following. Suppose we want to expand the functional integral about a given set of field configurations ("trajectory"), $\phi_\tau \equiv \phi(x; \tau_I, \tau_{\bar{I}})$, labelled by the double set of I and $\bar{I}$ collective coordinates, $\tau = \{\tau_I, \tau_{\bar{I}}\}$. For this end we should introduce by means of the Faddeev-Popov procedure a unity into the functional integral measure*,

$$ 1 = \triangle_{\text{FP}}(\phi) \int d\tau \; \delta\left(\left\langle (\phi - \phi_\tau), \frac{\partial \phi_\tau}{\partial \tau_i} \right\rangle_w \right). \tag{3.3} $$

where the scalar product** reads $\langle f, g \rangle_w = \int d^4x \; w(x; \tau) g(x) f(x)$. The introduction of this unity means that we divide the original functional integral in two parts, namely into the integration along our trajectory and into the integration orthogonal to this trajectory.

So far this can be done for any given trajectory. But in order to be able to do a systematic perturbation expansion about the trajectory, we should find a constrained minimum of the action: we should take as background field such a trajectory ("valley trajectory"), ϕ_τ, which, for any fixed value of the collective coordinates, is a minimum of the action within the subspace of fields orthogonal to it (see Fig. 6),

$$ \left\{ \phi \; : \; \left\langle (\phi - \phi_\tau), \frac{\partial \phi_\tau}{\partial \tau} \right\rangle_w = 0 \right\}. \tag{3.4} $$

The action minimization gives the following equation for the valley trajectory ϕ_τ ("valley equation"),

$$ \frac{\delta S[\phi]}{\delta \phi}\bigg|_{\phi = \phi_\tau} = \sum_i l_i(\tau) w(x; \tau) \frac{\partial \phi_\tau}{\partial \tau_i}, \tag{3.5} $$

---

* For notational simplicity we omit the gauge fixing condition. For more details about the valley method see Ref. [29].

** Here we have used our freedom in the choice of the scalar product by introducing some positive weight function w.

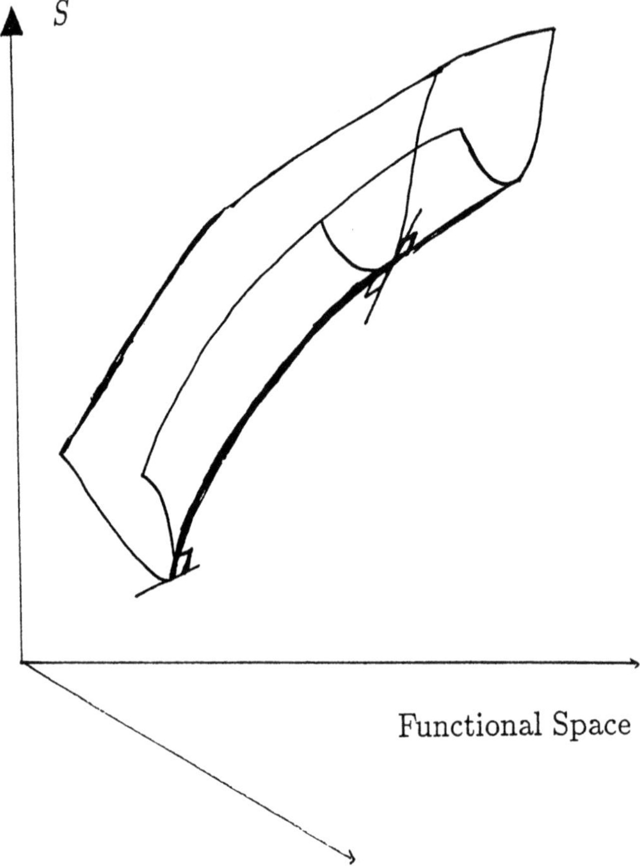

Fig. 6. Schematic picture of the valley.

where $l_i(\tau)$ are Lagrange multipliers. The $I\bar{I}$ valley trajectory ("streamline") satisfies the boundary condition

$$\phi(x;\tau_I,\tau_{\bar{I}})\,|_{R\to\infty}=\phi_I(x;\tau_I)+\phi_{\bar{I}}(x;\tau_{\bar{I}}). \tag{3.6}$$

Suppose, we have found the valley trajectory. Then the functional integral corresponding to the $2\to2$ amplitude,

$$T(2\to2)\sim\int d\tau\ \int\mathcal{D}\phi\ \Delta_{\mathrm{FP}}(\phi)\ \delta\!\left(\langle(\phi-\phi_\tau),\frac{\partial\phi_\tau}{\partial\tau}\rangle_w\right)\times \tag{3.7}$$

$$\times\prod_{i=1}^{4}\phi(x_i)\ \exp\left[-S[\phi_\tau]-\frac{1}{2}\langle(\phi-\phi_\tau),\frac{\delta^2 S}{\delta\phi^2}[\phi_\tau](\phi-\phi_\tau)\rangle-S_{\mathrm{int}}\right],$$

where S_{int} contains the cubic and quartic interactions, may be consistently expanded about it. In this context it is important to note that there is no linear term in the exponent due to the presence of the δ-function, provided ϕ_τ satisfies the valley equation (3.5). This makes the perturbative expansion of Eq. (3.7) in terms of $\delta\phi$ well defined (see Fig. 7).

We will see later that, on general grounds, the $I\bar{I}$ valley trajectory collapses for $R\equiv x_I-x_{\bar{I}}\to0$ to the perturbative vacuum, $i.e.$ $\phi_{\tau|R=0}=0$. This means that, in

Fig. 7. The perturbative expansion of the four-point function in the valley background.

the expansion around the valley trajectory, also the conventional perturbation theory (expansion around the perturbative vacuum) is contained as a special case ($R = 0$). Note that the conventional perturbative graphs do not appear in the first few terms in the expansion around the valley which are shown in Fig. 7, since these terms vanish for $R = 0$. They appear first in terms which are quartic in the fluctuation $\delta\phi = \phi - \phi_\tau$. Generally speaking, of course, the forward elastic scattering amplitude (total cross-section) cannot be divided into a perturbative and a non-perturbative piece. But in the saddle-point approximation one may consider the general answer as the sum of contributions from different saddle-points. We will find that for finite energies there is an $I\bar{I}$-induced saddle-point, giving rise to an imaginary part of the forward elastic scattering amplitude, which is well separated from the trivial one (perturbative vacuum).

Up to now we made no approximation; Eq. (3.7) should be generally valid (up to eventual problems with gauge copies[30,31]). However, for our purposes, we are interested only in the leading-order (in α_w) approximation around the $I\bar{I}$ valley trajectory. To this end we we keep only the first term on the r.h.s. of Fig. 7,

$$T^{(I\bar{I})}(2 \to 2) \sim \int d\tau\, n(\tau) \prod_{i=1}^{4} \phi_\tau(x_i)\, \exp\left(-S[\phi_\tau]\right), \qquad (3.8)$$

where the measure $n(\tau)$ is given by the integration over the Gaussian fluctuations orthogonal to the trajectory and may be expressed in terms of functional determinants and Jacobians.

By an analytic continuation to Minkowski space* and by application of the LSZ reduction formula[22] we obtain the instanton-induced contribution to the total cross-section, which in the CM frame reads generically,[25,26]

$$\sigma_{\text{tot}}^{(I)} \sim \text{Im}\, T^{(I\bar{I})}(2 \to 2; t = 0) \sim$$

$$\text{Im} \int d^4R\, d\rho_I\, d\rho_{\bar{I}}\, n(R, \rho_I, \rho_{\bar{I}})\, P(p_1, p_2; R, \rho_I, \rho_{\bar{I}})\, \exp\left[ET - S(R, \rho_I, \rho_{\bar{I}})\right], \qquad (3.9)$$

where $R \equiv (T, \vec{X})$. The pre-exponential factor, P, and the exponent ET arise due to the LSZ-reduced Fourier transform of the valley trajectory. Note that in general

---

* Strictly speaking we should first perform the collective coordinate integrals in Eq. (3.8) and continue afterwards to Minkowski space. In the steepest descent approximation we are allowed to reverse the order of these manipulations. The continuation of E to Minkowski space must be done simultaneously with the changing of the contours of integrations over collective coordinates (in order to have convergent integrals). The rotated contours should pass through the saddle-point which gives the main contribution to the integral.

the valley trajectory is a multi-component object similar to Eq. (2.4). Therefore, P depends on the types of particles for which we compute the total, instanton-induced cross-section. The exponential part, on the other hand, is universal. To find the saddle-point one has to study the exponent,

$$\Gamma(E; R, \rho_I, \rho_{\bar{I}}) \equiv \frac{4\pi}{\alpha_w} F(E; R, \rho_I, \rho_{\bar{I}}) \equiv ET - S(R, \rho_I, \rho_{\bar{I}}), \qquad (3.10)$$

$$\frac{\partial \Gamma}{\partial \tau}\Big|_{\tau_*} = 0. \qquad (3.11)$$

An imaginary part will arise if there is one negative mode of $\partial^2 \Gamma / \partial \tau_i \partial \tau_j \mid \tau_*$, *i.e.* if τ_* is a true saddle-point. The saddle-point value is at $\vec{X}_* = 0$ and the saddle-point for T can be found from Eq. (3.11). We are sure that we will get a non-perturbative, instanton-induced contribution to the total cross-section from Eq. (3.9) as long as R_* is non-zero.

To proceed, one has to know $S(\tau)$, the action on the valley trajectory. The $I\bar{I}$ valley in the Higgs model is known at large separation, $\rho/T \ll 1$, in terms of an expansion in ρ/T.[28] Using this expansion one may find[25] the perturbative expansion of $\tau_*(\epsilon)$ and the result is that the total cross-section has an exponential form similar to Eq. (2.6),

$$\sigma_{\text{tot}}^{(I)} \sim \exp\left[\frac{4\pi}{\alpha_w} F_*(\epsilon)\right] , \quad (\epsilon \equiv E/M_{\text{sp}}) , \qquad (3.12)$$

where now

$$F_*(\epsilon) = -1 + \left(\frac{9}{16\sqrt{8}}\right)^{2/3} \epsilon^{4/3} - \frac{3}{32}\epsilon^2 + \mathcal{O}\left((\frac{\lambda}{g^2} + 1)\epsilon^{8/3}\right). \qquad (3.13)$$

At low energies F_* coincides with the naive instanton result, Eq. (2.7). The third term on the r.h.s. of Eq. (3.13), which gives the first correction to the naive instanton result, Eq. (2.7), has been reproduced meanwhile directly by computing the higher order corrections to Eqs. (2.6),(2.7) in the one-instanton-sector.[32−34] Therefore, at low energies or large separation only (B+L)-violation contributes to the total, non-perturbative cross-section, inferred from the leading-order $I\bar{I}$ contribution to the 4-point function. Unfortunately, since the neglect of the higher-order terms in Eq. (3.13) is justified only at $\epsilon \ll 1$, we cannot draw any conclusions about the behavior of the total cross-section above the sphaleron scale.

To conclude, at low energies the direct calculation of higher order corrections in the I sector or the indirect calculation via the $I\bar{I}$ valley and the optical theorem give the same result (for a formal proof see Ref. [35]). The usefulness of the valley method lies in the fact that it helps us to approach the goal to sum up effectively the soft-soft quantum corrections for (B+L)-violation beyond perturbation theory in E/M_{sp} as we will demonstrate in the next section.

IV. BEYOND THE SOFT-SOFT-EXPANSION

In order to get the instanton-induced cross-section beyond a perturbative series in E/M_{sp} it is necessary to know the action on the valley trajectory for arbitrary separation, in particular beyond a perturbative expansion in ρ/T.

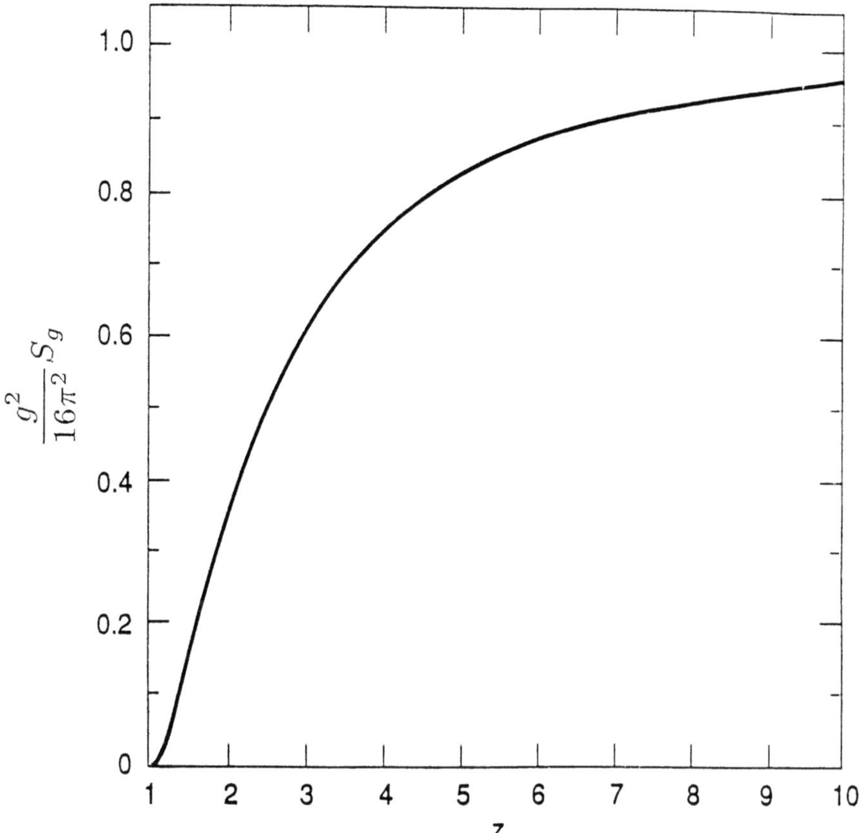

Fig. 8.   The action of the $I\bar{I}$ valley trajectory in pure $SU(2)$ gauge theory.

Unfortunately, the exact solution of the valley equation in the SU(2)-Higgs model is not known. However, it is known in pure SU(2) gauge theory[29] (see also Refs. [26,36]). Due to conformal invariance the action on the valley trajectory, for the case of maximal $I\bar{I}$ attraction, is a function of only one variable z, which may be expressed in terms of the collective coordinates as

$$z = \frac{R^2 + \rho_I^2 + \rho_{\bar{I}}^2 + \sqrt{(R^2 + \rho_I^2 + \rho_{\bar{I}}^2)^2 - 4\rho_I^2\rho_{\bar{I}}^2}}{2\rho_I\rho_{\bar{I}}} \;, \; 1 \leq z < \infty \;, \qquad (4.1)$$

and reads (see Fig. 8)

$$S_g(z) = \frac{48\pi^2}{g^2} \cdot \left[\frac{6z^2 - 14}{(z - 1/z)^2} - \frac{17}{3} - \ln(z)\left(\frac{(z - 5/z)(z + 1/z)^2}{(z - 1/z)^3} - 1 \right) \right]. \qquad (4.2)$$

For $z \to \infty$, *i.e.* for large separation, the action goes to two times the I action, whereas for $z \to 1$, *i.e.* for small separation and $\rho_I = \rho_{\bar{I}}$, it goes to zero, reflecting the $I\bar{I}$ collapse into the perturbative vacuum. These two limits reflect the general behavior of the action on the $I\bar{I}$ valley trajectory: the collapse into the perturbative vacuum is inevitable[28] since there is no non-trivial solution in the zero-topological-charge sector. If somewhere the decrease of the action were stopped it would follow from the valley equation that there exists an exact stable non-trivial solution of the Yang-Mills classical equations of motion with zero topological charge.

We may use above result directly[26] for the computation of the instanton-induced contribution to the total cross-section in pure gluo-dynamics. We should consider the scattering of off-shell gluons, since, due to confinement, the gluons are not the correct asymptotic states. We will consider here gluons with Minkowski-four-momenta $k_{1/2} = (E/2, \pm\vec{k}), k^2 \equiv k_{1/2}^2 < 0$. The I-induced contribution to the total cross-section for the process gluon (k_1) + gluon $(k_2) \to$ anything is given by[26]

$$\sigma_{\text{tot}}^{(I)} \sim \text{Im} \int dR d\rho_I d\rho_{\bar{I}} \ldots \exp\left[ET - S_{\text{g}}(z) - 2\sqrt{-k^2}(\rho_I + \rho_{\bar{I}})\right]. \qquad (4.3)$$

The third term in the exponent, which differs from Eq. (3.9), arises after Fourier transformation of the four-point Green function since the scattering gluons are off-shell.

Evaluating the collective coordinate integral in the saddle-point approximation we obtain $\sigma_{\text{tot}}^{(I)} \sim \exp \Gamma_*(\epsilon)$, where we define in this case $\epsilon \equiv E/\sqrt{-k^2}$ and where the function $\Gamma_*(\epsilon)$ denotes the exponent in Eq. (4.3) for $\rho_I = \rho_{\bar{I}} \equiv \rho$,

$$\Gamma(E, k^2; \chi, \rho) \equiv E\rho\chi - S_{\text{g}}(\chi) - 4\sqrt{-k^2}\rho \; ; \; (\chi \equiv T/\rho), \qquad (4.4)$$

taken at the saddle-point $R_* = (T*, \vec{X}_* \equiv 0)$. We find (see Fig. 9)

$$\chi_* = 4\epsilon^{-1} \; , \; \rho_* = \frac{1}{E} \frac{\partial S_{\text{g}}}{\partial \chi}\bigg|_{\chi_*} \; , \; \Gamma_* = -S_{\text{g}}(\chi_*). \qquad (4.5)$$

The total cross-section induced by the non-perturbative $I\bar{I}$ valley is exponentially suppressed for high virtualities, but this suppression seems to disappear for close-to-mass-shell gluon collisions. It should be noted that the instanton size is always bounded. In particular, we obtain, at large $E/\sqrt{-k^2} \gg 1$, $\rho_* \sim (16\pi^2/g^2)(\sqrt{-k^2}/E^2)$: at very high energies, but small virtualities, the dominant instantons are very small.

Let us consider now the SU(2)-Higgs model. According to our ideology we should find the complete valley trajectory in this model. Unfortunately, the solution of the valley equations in the Higgs model is not known beyond perturbation theory in $\rho/T \ll 1$.[28] However, let us assume that the dominant contribution to the I-induced total cross-section comes from the multiple production of W- and Z-bosons and that we can neglect Higgs production (as it is the case at low energies). In addition let us neglect W-mass effects. In this case it seems reasonable to work with the pure SU(2) gauge theory action, Eq. (4.2) and to omit the Higgs-induced part of the interaction, $i.e.$ to write[26]

$$\sigma_{\text{tot}}^{(I)} \sim \text{Im} \int dR d\rho_I d\rho_{\bar{I}} \ldots \exp\left[ET - S_{\text{g}}(z) - \pi^2 v^2(\rho_I^2 + \rho_{\bar{I}}^2)\right]. \qquad (4.6)$$

It should be noted that the action in the exponent (the last two terms in Eq. (4.6)), although not being the exact valley action, has the correct qualitative behavior: at large separation and $\rho_I = \rho_{\bar{I}}$ it goes to two times the instanton action whereas at small separation and small scale-sizes it goes to zero, representing the collapse into the perturbative vacuum.

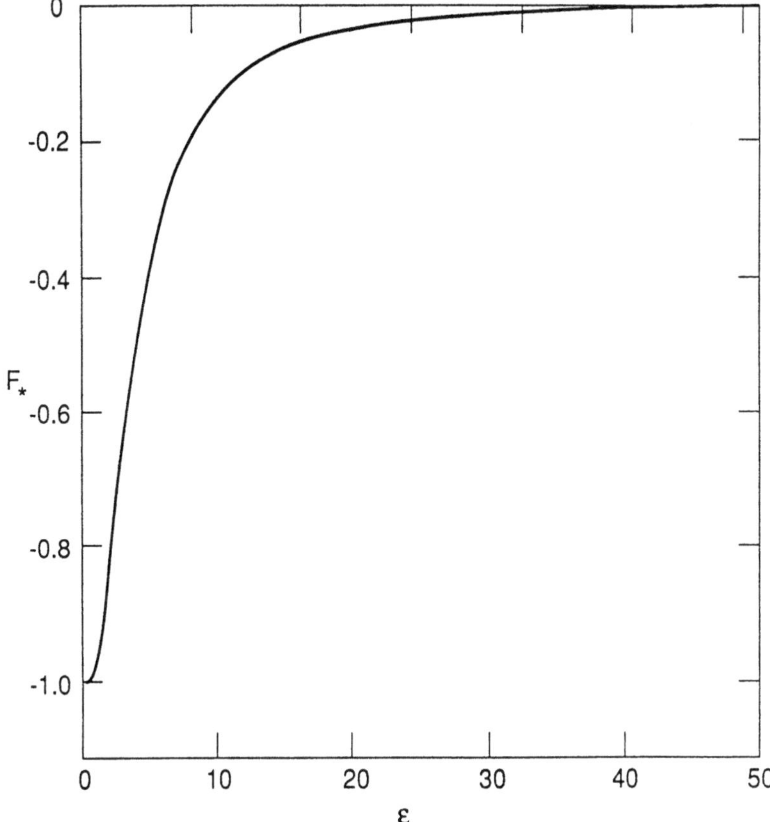

Fig. 9. The logarithm of the I-induced contribution of the cross section for off-shell gluon scattering, $F_* \sim (g^2/16\pi^2) \log \sigma_{\text{tot}}$, as a function of $\epsilon \equiv E/\sqrt{-k^2}$.

In the saddle-point approximation we have again $\sigma_{\text{tot}}^{(I)} \sim \exp[(4\pi/\alpha_w)F_*(\epsilon)]$, where now, as in previous sections, $\epsilon \equiv E/M_{\text{sp}}$. It is convenient to define the variables $\chi \equiv T/\rho$, $\xi \equiv m_w\rho$. In terms of these variables, the function $F_*(\epsilon)$ denotes

$$F(\epsilon; \chi, \xi) \equiv \frac{1}{4}\epsilon\xi\chi - \frac{\alpha_w}{4\pi}S_g(\chi) - \frac{1}{2}\xi^2, \tag{4.7}$$

taken at the saddle-point $R_* = (T_*, \vec{X}_* \equiv 0)$. The solid line in Fig. 10 shows the function $F_*(\epsilon)$. At low energies the total cross-section rises exponentially, in agreement with Eqs. (2.6),(2.7). At high energies F_* approaches zero - there seems to be no problem with unitarity. The dashed line in Fig. 10 shows the instanton scale-size at the saddle point. At low energies the instanton size grows, reaching a maximum size $\xi_*^{\text{max}} \simeq 0.67$ at $\epsilon \simeq 2.7$. For higher energies it decreases, asymptotically approaching zero. The parameter χ_* decreases monotonically with increasing ϵ ($\chi_* = 1$ at $\epsilon \simeq 2.7$) and goes to zero asymptotically: at asymptotic energies the saddle-point configuration approaches the perturbative vacuum* Strictly speaking, the asymptotic

---

* One may be worried that for $\chi_* < 1$, i.e. when I and $\bar{I}$ are overlapping, the valley-induced total cross-section is not guaranteed to be only (B+L)-violating. It can, however, be shown that only (B+L)-violation contributes to the leading-order valley-induced total cross-section.[35, 37]

region of F_* is outside the region of validity of our investigation, since the saddle-point configuration approaches the perturbative vacuum. However, it is important to note that for $\epsilon \leq 6$ the dominant configuration is still non-perturbative in the sense that $S_* \geq 1/g^2$, whereas nevertheless, at $\epsilon \sim 6$, the cross-section is essentially unsuppressed, $(16\pi^2/g^2)(F_*/S_*) \sim 10^{-2}$.

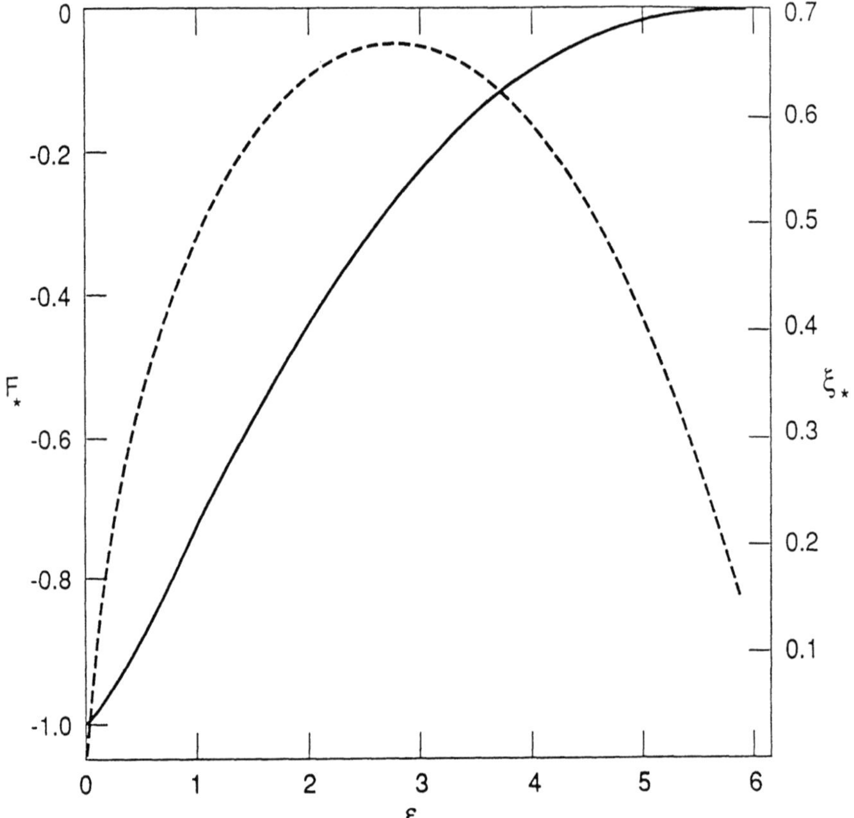

Fig. 10. *The logarithm of the I-induced contribution to the total cross-section for W-production in the SU(2)-Higgs model, F_* (solid line) and instanton scale size, ξ_* (dashed line), both as a function of $\epsilon \equiv E/M_{SP}$.*

Since the mean multiplicity of produced W-bosons is directly related[19,32 38] to the exponent of the instanton-induced total cross-section, $\bar{n}_w \simeq (4\pi/(3\alpha_w))[F_*(\epsilon)+1]$, we find from Fig. 10 that it behaves at low energies according to the naive instanton result, Eq. (2.8), whereas at high energies it goes to $\bar{n}_w \sim \pi/\alpha_w = \mathcal{O}(100)$.[38]

So far we neglected the higher-order quantum corrections, as indicated in Fig. 7. One may argue that, in the language of perturbation around the instanton (see Fig. 4), our semiclassical $I\bar{I}$ calculation takes into account all tree-graph corrections involving the final-state particles (soft-soft corrections).[18,19,32-35] Corrections involving the hard initial-state particles[20] are not taken into account. These initial-state corrections are contained, in the valley approach, in the higher-order corrections in Fig. 7.

140

V. QUANTUM AND CLASSICAL CORRECTIONS

The problem of influence of hard quantum corrections on the leading-order valley result (3.8) is the subject of intensive study[20,39−42] and attracts a lot of interest. It was argued that hard-hard corrections may exponentiate[20,39] and change (*e.g.* damp) the leading-order result (3.8). There are also some indirect signals about exponentiation of hard-soft corrections.[41] The existence and the magnitude of the latter effect seems to be related with the freedom of the choice of the weight function, w, in Eqs. (3.3)-(3.5). Indeed, it may happen that there exist different valleys for different choices of w which may lead to different leading-order results. The final result, however, cannot depend on the way of introduction of unity, Eq. (3.3). That means that some terms in the formal perturbative expansion of Eq. (3.7) must eliminate the discrepancy in the leading-order results and essentially change them.

We note that any method which deals with instantons cannot avoid a freedom of the w-choice in Eq. (3.3) in order to introduce collective coordinates in the functional integral. The formal correspondence between the weight functions in the R-term method[19] and the valley method is stated in a recent paper.[35]

Our strategy is, certainly, to work with the most convenient choice of w which enables us to solve the $I\bar{I}$ valley equation and then work with the systematic perturbative expansion of Eq. (3.7) in the valley background. The latter expansion, however, has no small parameter for $E \geq M_{\mathrm{sp}}$: the coupling constant α_w disappears if we express everything in terms of $\epsilon \equiv E/M_{\mathrm{sp}}$. Thus, until we are able to sum up the formal perturbative expansion in the $I\bar{I}$ valley background we are forced to work in the region $\epsilon \ll 1$. However, in such a situation we know for sure that quantum corrections do not change the first few terms of the leading-order result in the expansion of $\Gamma(\epsilon) \equiv \ln \sigma^{\mathrm{anom}}$ in terms of $\epsilon \ll 1$ and Eq. (3.13) is valid. Namely, hard-hard corrections may contribute to $\Gamma(\epsilon)$ starting from $\epsilon^{10/3}$ term and hard-soft corrections may start from $\epsilon^{8/3}$ term. For more details about new developments in the field of quantum corrections see the original papers.[20,39−42] Thus, one may conclude that at low energies the cross-section of anomalous processes grows with energy exponentially while its behavior at high energies, $E \geq M_{\mathrm{sp}}$, and observability of such processes remains unclear.

However, it was suggested by Zakharov[43] and later by Maggiore and Shifman[44] that multi-I/multi-$\bar{I}$ contributions to the pure single-I result become crucial at some energy scale, $E < M_{\mathrm{sp}}$, and lead to suppressed (unobservable) final result for the anomalous cross-section, $\sigma^{\mathrm{anom}} < \sim \exp(-S_I)$.

In the rest of this section we shall argue that these classical corrections, being unimportant at low energies, $E \ll M_{\mathrm{sp}}$, where their effect is suppressed by the factor $\exp(-2S_I)$, may compete with the quantum corrections only at energy $E \sim M_{\mathrm{sp}}$. More concrete,

$E \ll M_{\mathrm{sp}}$: multi$-I$/multi$-\bar{I}$ corrs. $\ll$ hard quantum corrs. $\ll$ single$-I$ result,

$E \sim M_{\mathrm{sp}}$: multi$-I$/multi$-\bar{I}$ corrs. $\sim$ hard quantum corrs. $\sim$ single$-I$ result,

which indicates rather the breakdown of the semiclassical Euclidean methods at $E \sim M_{\mathrm{sp}}$ than unobservability of anomalous processes at $E > M_{\mathrm{sp}}$.[45]

To proceed we note that multi-I/multi-$\bar{I}$ contributions to the anomalous processes may be taken into account on the same ground as the single-I contribution in the previous section. Indeed, the optical theorem (3.1) relates the anomalous cross-section and the forward elastic scattering amplitude which is given by the functional integral in the zero topological charge sector, Eq. (3.2). In sect. 3 we took into account only the single-I contribution to the anomalous cross-section expanding the functional integral (3.2) in the vicinity of the $I\bar{I}$ valley trajectory. Note, that multi-I/multi-$\bar{I}$ ($I^M\bar{I}^M$) configuration may and must be defined also as a valley. Indeed, if all the separations $\{R_{i,j}\}_{i,j=1}^M$ between the I_i and $\bar{I}_j$ space-time positions are infinite, the $I^M\bar{I}^M$ configuration is an exact classical solution with $2pM$ zero modes*. At finite separations the $I^M\bar{I}^M$ configuration is not a classical solution and we define it as the solution of *constrained* equations of motion (3.5) with the constraint (3.4), where $\tau \equiv \{\tau_i\}_{i=1}^{2pM}$ and $R \equiv \{R_{i,j}\}_{i,j=1}^M$. Thus, the $I^M\bar{I}^M$ configuration is the solution of the valley equations (3.5) with the general boundary conditions of the type (3.6). The $I^M\bar{I}^M$-valley-induced contribution to the $2 \to 2$ forward elastic scattering amplitude is given by Eq. (3.7). Note, that the $I^M\bar{I}^M$-valley-induced contribution is more general than the $I\bar{I}$ one and contains it as a special case. Indeed, $I^M\bar{I}^M|_{R_{i,j}=0} = I^{M-1}\bar{I}^{M-1}$, which reflects the $I\bar{I}$-collapse at zero separations. Thus, the r.h.s. of Eq. (3.7) which contains integrations over all $R_{i,j}$ ($\int d\tau$) takes into account contributions from $I\bar{I}$, $I^2\bar{I}^2$, ..., $I^{M-1}\bar{I}^{M-1}$, $I^M\bar{I}^M$.

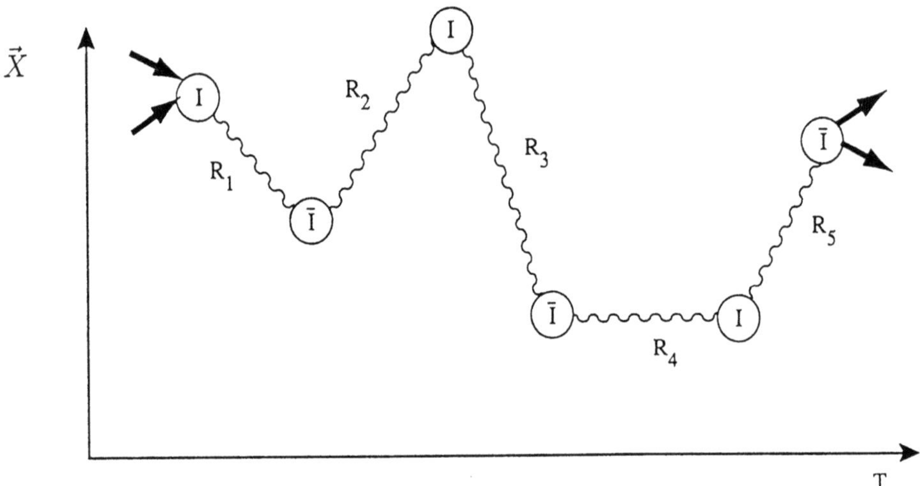

Fig. 11. Example of multi-I/multi-$\bar{I}$ contribution to the $2 \to 2$ amplitude. Both incoming (outgoing) hard-particle-lines are attached to the same I ($\bar{I}$). Wavy lines denote $I/\bar{I}$ "chemical bonds" or "springs" due to $I/\bar{I}$ interaction (attraction).

To proceed let us assume[43,44] that the classical (multi-I/multi-$\bar{I}$) corrections become more important than the quantum ones at some energy scale, $E < M_{\text{sp}}$ and we can neglect the latter corrections. With all this in mind we arrive at Eq. (3.8). In analogy with Eq. (3.9) the optical theorem for anomalous (multi-I/multi-$\bar{I}$) processes reads, see Fig. 11,

$$\sigma_{\text{tot}}^{\text{anom}} \sim \text{Im} \int \prod_{i=2}^{2M} d^4R_i \; n(R) \; P(p_1, p_2; R) \; \exp[E \sum_{i=2}^{2M} T_i - S(R)], \qquad (5.1)$$

---

* p stands here for a number of zero modes of single-I.

where in order to follow the arguments of Refs. [43,44] we consider the case where both incoming (outgoing) lines are attached to the same $I(\bar{I})$. The main contributions to the r.h.s. of Eq. (5.1) should come in the steepest descent method from the saddle points and the limits of integration. The limits of integrations, $R_i = 0$ for $2 \leq i \leq 2M$, describes the collapse of the corresponding $I\bar{I}$ pair which reduces the $I^M\bar{I}^M$ configuration to the $I^N\bar{I}^N$ one ($N < M$). For every such a field, $I^N\bar{I}^N$, there is the only saddle-point, $R_i^* = (T_i^*, X_i^*)$,

$$X_i^* \ : \ X_i^* = 0,$$

$$T_i^* \ : \ E = \frac{\partial S}{\partial T_i}\big|_{T_i^*} \ , \ i = 2, ..., 2N. \tag{5.2}$$

Equations (5.2) mean that the main contributions to the integral (5.1) come from the configurations of the chain-type shown in the Fig. 12. It is easy to see that this chain-valley indeed corresponds to the saddle-point configuration. The left-right deviations of one of the intermediate $I(\bar{I})$ in the horizontal (time) direction lead to monotonously decreasing action on the chain. At the point of the "maximal" time-deviation $I(\bar{I})$ collapses with one of the neighboring $\bar{I}(I)$. The deviations in the vertical (space) direction lead to monotonously increasing action.

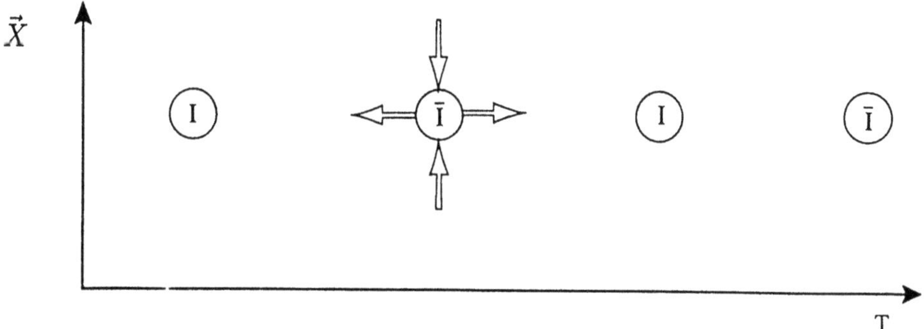

Fig. 12. *Multi-I/multi-$\bar{I}$ configuration of the T-chain type. Arrows denote here "streamlines directions." Horizontal deviations of $\bar{I}$ lead to decreasing action, while vertical deviations lead to increasing action o the chain.*

Thus, we conclude in agreement with Refs. [43,44] that the main contribution to the forward elastic scattering amplitude (and correspondingly to the anomalous cross-section) is the sum of the chain-valley contributions shown on Fig. 13.

In order to follow the ideas of Refs. [43,44] we assume that all the $I\bar{I}$ separations are large and we can use the binary-forces approximation,

$$S^{I^N\bar{I}^N} = 2NS_I - \sum_{i,j=1}^{N} U_{ij}^{\text{int}}. \tag{5.3}$$

Moreover, we assume that $I(\bar{I})$ interacts only with its closest neighbors. Both these assumptions may be understood only in the region $E < M_{\text{sp}}$. First, Refs. [43,44] suggest that the classical corrections become crucial at some energy scale, $E < M_{\text{sp}}$, and only in such a situation we we may skip the quantum corrections and come from

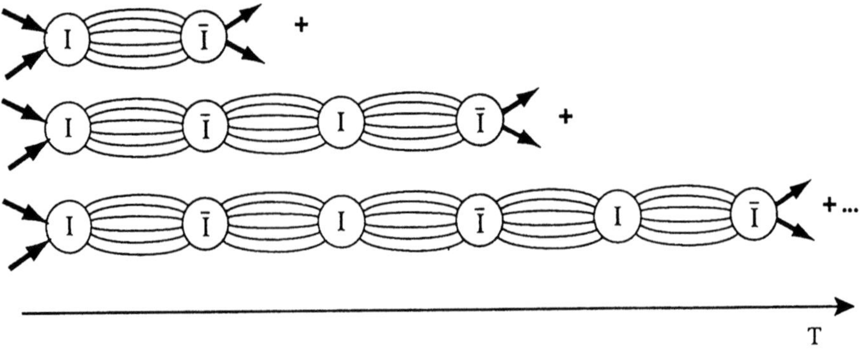

Fig. 13. Main classical contribution to the forward elastic scattering amplitude. Binary $I/\bar{I}$ interactions (wavy lines of Fig. 11) are interpreted here as the exchange of intermediate particles.

Eq. (3.8) to Eq. (5.1) (it is clear that at $E \sim M_{\rm sp}$ one cannot distinguish quantum and classical result and treat the corresponding corrections simultaneously). It is also easy to see that the low energy limit corresponds to the well-separated multi-I/multi-$\bar{I}$ medium. Note, that the action on the 1-chain, $S^{I\bar{I}}(T)$ is much smaller then the action on the M-chain ($M \to \infty$), $S^{I^N\bar{I}^N}(T)$ for the above mentioned assumptions,

$$S^{I\bar{I}}(T) = 2S_I + U^{\rm int}(T) \ll S^{I^N\bar{I}^N}(T) \simeq 2NS_I + (2N-1)U^{\rm int}(T), \qquad (5.4)$$

which means that

$$U^{\rm int}(T) \ll S_I. \qquad (5.5)$$

Let us now estimate the contribution of the process shown on Fig. 13 in the binary-forces approximation,

$$\sigma_{tot}^{\rm anom} \sim {\rm Im} \sum_{N=1}^{M} \int [\prod_{i=2}^{2N} dT_i dX_i]\, \exp[E \sum_{i=2}^{2N} T_i - 2NS_I + (2N-1)U^{\rm int}(T)] \sim$$

$$ {\rm Im}\, e^{-S_i} \sum_{N=1}^{M} \Big(\int dT \exp[ET - S_I + U^{\rm int}(T)] \Big)^{2N-1} \sim, \qquad (5.6)$$

$$ e^{-S_i} \sum_{N=1}^{M} (-1)^{N-1} \Big(\int dT \exp[\Gamma(E) + S_I] \Big)^{2N-1}, $$

where,

$$ i\, \exp[\Gamma(E)] \sim \int dT\, \exp[ET - 2S_I + U^{\rm int}(T)]^{2N-1}, \qquad (5.7)$$

denotes the single-I/single-$\bar{I}$ contribution to the forward elastic scattering amplitude. At low energies, $E : \Gamma(E) + S_I < 0$, the r.h.s. of Eq. (5.6) may be formally summed up and gives[43,44]

$$ \sigma_{\rm tot}^{\rm anom}\Big|_{E\,:\,\Gamma(E)+S_I<0} \sim \frac{e^{\Gamma(E)}}{1 + \big(e^{\Gamma(E)+S_I}\big)^2}, \qquad (5.8)$$

At low energies, $E : \Gamma(E) + S_I < 0$, we have that $\sigma_{\rm tot}^{\rm anom} \sim e^{\Gamma(E)} < \sim e^{-S_I}$ which coincides with the single-I result and the multi-I (classical) corrections are not important, however at high energies, E : $\Gamma(E)+S_I > 0$, the r.h.s. of Eq. (5.8) rapidly decreases showing a resonance-like behavior. The latter observation was treated by the Zakharov and later by Maggiore and Shifman as the crucial damping of the anomalous processes by the classical corrections, $\sigma_{\text{tot}}^{\text{anom}} < \sim e^{-S_I}$, which should lead to unobservability of such processes.

It seems, however, that this conclusion is "too optimistic". The expression on the r.h.s. of Eq. (5.8) has sense only in the region where it was derived, E : $\Gamma(E)+S_I < 0$, and in this region the classical corrections are simply not of great importance. However, in the region, E : $\Gamma(E)+S_I \geq 0$, the formal sum in Eq. (5.6) is divergent. More physically, this means that at some energy,

$$E' \; : \; \Gamma(E') = S_I, \tag{5.9}$$

all the chains of the Fig. 13 give the same contributions. Therefore any insertions of many→many-particle transitions mediated by an $I(\bar{I})$, see Fig. 14, do not change the corresponding amplitude. This means that such transitions reach unity at the energy level (5.9) and this energy corresponds to the sphaleron mass.* This suggestion may be checked in Abelian Higgs model.[45]

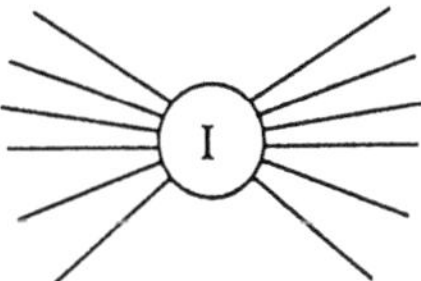

Fig. 14. Many→many-particle transition mediated by I.

We would like to mention that in the pure gauge theory with linear cutoff (sect. 4) the energy E' corresponds to $U^{\text{int}}(T') = S_I$ which invalidates, see Eq. (5.5), the binary-forces approximation.

VI. SUMMARY

In this lecture we have discussed the possibility that anomalous electroweak (B+L)-violating processes with the associated production of many W- and Z-bosons may become important in qq collisions with CM energies above the sphaleron scale, $M_{\text{sp}} \sim m_w/\alpha_w \sim 10$ TeV.

It is by now well established that the total cross-section for electroweak (B+L)-violation grows exponentially at energies below the sphaleron scale and becomes much larger than the instanton suppression factor, $\exp(-4\pi/\alpha_w)$, would suggest. If it becomes of observable size is, however, unknown. The valley method turned out to be extremely useful to sum up all the higher-order (tree graph) corrections involving only final-state particles. With the help of this method we could establish that these corrections will unitarize the amplitudes in the sense that there is, at high energies, no exponential blow-up of the instanton-induced total cross-section. Taking our result at face value we find that the exponential suppression of instanton-induced processes disappears at energies above the sphaleron scale (see Fig. 10) and that the mean

---

* Note that 2 →many-paticle anomalous transition is still exponentially suppressed for the energy (5.9).

multiplicity of W- and Z-bosons goes to $\bar{n}_w \sim \pi/\alpha_w = \mathcal{O}(100)$. However, we have not shown that including initial-state corrections and after that classical corrections will not change our answer.

We think that the new step – some more powerful method which would generalize and unify all known ideas should be found in order to deal with the Standard Model at energies above the sphaleron. We hope to return to this question in a future work.

VII. ACKNOWLEDGEMENTS

The author is grateful to Ian Balitsky, Jochen Kripfganz, Andreas Ringwald and V.I. Zakharov for stimulating discussions. Andreas Ringwald is especially acknowledged for his kind permission to use some pedagogical ideas and explanations from the lecture[46] in the present paper. It is a pleasure to thank CERN Theory Division where this work was started for hospitality and support. This work was supported in part by INFN Eloisatron Project.

REFERENCES

1. A. Ringwald, *Nucl. Phys.* **B330** (1990) 1.

2. A. Sakharov, *JETP Lett.* **5** (1967) 24.

3. G. 't Hooft, *Phys. Rev. Lett.* **37** (1976) 8; *Phys. Rev.* **D14** (1976) 3432; (E) **D18** (1978) 2199.

4. S. Glashow, *Nucl. Phys.* **22** (1961) 579; S. Weinberg, *Phys. Rev. Lett.* **19** (1967) 1264; A. Salam, in *Proc. of the Nobel Symposium on Elementary Particle Theory*, ed. N. Svartholm (Stockholm, 1968).

5. M. Kobayashi and T. Maskawa, *Prog. Theor. Phys.* **49** (1973) 652.

6. S. Adler, *Phys. Rev.* **177** (1969) 2426; J. Bell and R. Jackiw, *Nuovo Cimento* **51** (1969) 47; W.A. Bardeen, *Phys. Rev.* **184** (1969) 1848.

7. R. Jackiw and C. Rebbi, *Phys. Rev. Lett.* **37** (1976) 172; C. Callan, R. Dashen and D. Gross, *Phys. Lett.* **B63** (1976) 334.

8. A. Belavin, A. Polyakov, A. Schwarz and Yu. Tyupkin, *Phys. Lett.* **B59** (1975) 85.

9. N. Manton, *Phys. Rev.* **D28** (1983) 2019; F. Klinkhamer and N. Manton, *Phys. Rev.* **D30** (1984) 2212.

10. R. Dashen, B. Hasslacher and A. Neveu, *Phys. Rev.* **D10** (1974) 4138; J. Boguta, *Phys. Rev. Lett.* **50** (1983) 148; J. Burzlaff, *Nucl. Phys.* **B233** (1984) 262.

11. V. Kuzmin, V. Rubakov and M. Shaposhnikov, *Phys. Lett.* **B155** (1985) 36.

12. H. Aoyama and H. Goldberg, *Phys. Lett.* **B188** (1987) 506.

13. P. Arnold and L. McLerran, *Phys. Rev.* **D36** (1987) 581; **D37** (1988) 1020; A. Ringwald, *Phys. Lett.* **B201** (1988) 510.

14. M. Shaposhnikov, *Nucl. Phys.* **B287** (1987) 757; **B299** (1988) 797; A. Bochkarev, S. Khlebnikov and M. Shaposhnikov, *Nucl. Phys.* **B329** (1990) 490; L. McLerran, *Phys. Rev. Lett.* **62** (1989) 1075; N. Turok and P. Zadrozny, *Phys. Rev. Lett.* **65** (1990) 2331; A. Cohen, D. Kaplan and A. Nelson, *Nucl. Phys.* **B349** (1991) 727; M. Dine, P. Huet, R. Singleton and L. Susskind, *Phys. Lett.* **B257** (1991) 351; L. McLerran, M. Shaposhnikov, N. Turok and M. Voloshin, TPI-MINN-90/64-T.

15. O. Espinosa, *Nucl. Phys.* **B343** (1990) 310.

16. L. McLerran, A. Vainshtein and M. Voloshin, *Phys. Rev.* **D42** (1990) 171.

17. V.I. Zakharov, *Classical corrections to instanton induced interactions*, TPI-MINN-90/7-T (1990).

18. L. Yaffe, in *Proc. of the Santa Fe Workshop on Baryon number violation at the SSC?*, eds. M. Mattis and E. Mottola (World Scientific, Singapore, 1990).

19. S. Khlebnikov, V. Rubakov and P. Tinyakov, *Nucl. Phys.* **B350** (1991) 441.

20. A.H. Mueller, *Nucl. Phys.* **B348** (1991) 310; **B353** (1991) 44.

21. P. Arnold and M. Mattis, *Phys. Rev.* **D42** (1990) 1738.

22. H. Lehmann, K. Symanzik and W. Zimmermann, *Nuovo Cimento* **1** (1955) 205.

23. R. Peccei and H. Quinn, *Nuovo Cimento* **A 41** (1977) 309.

24. I. Affleck, *Nucl. Phys.* **B191** (1981) 445.

25. V.V. Khoze and A. Ringwald, *Nucl. Phys.* **B355** (1991) 351.

26. V.V. Khoze and A. Ringwald, *Phys. Lett.* **B259** (1991) 106.

27. I.I. Balitsky and A.V. Yung, *Phys. Lett.* **B168** (1986) 113.

28. A.V. Yung, *Nucl. Phys.* **B297** (1988) 47.

29. V.V. Khoze and A. Ringwald, *Valley Trajectories in Gauge Theories*, CERN-TH.6082/91.

31. L. Yaffe, *Nucl. Phys.* **B151** (1979) 247.

30. V.N. Gribov, in *Proc. of the 12th Winter School of the Leningrad Nuclear Physics Institute* (1977).

32. D.I. Diakonov and V.Yu. Petrov, in *Proc. of the 26th Winter School of the Leningrad Nuclear Physics Institute* (1991).

33. A.H. Mueller, *On Higher Order Semiclassical Corrections to the High Energy Cross Sections in the One Instanton Sector*, CU-TP-512.

34. P. Arnold and M. Mattis, *Mod. Phys. Lett.* **A6** (1991) 2059.

35. P. Arnold and M. Mattis, *Peace in the valley: Concordant Approach to Distorted Instantons*, LA-UR-91-1858 (1991).

36. J. Verbaarschot, *Nucl. Phys.* **B362** (1991) 33.

37. I.I. Balitsky and V.M. Braun, in preparation.

38. V.V. Khoze and A. Ringwald, in preparation.

39. M. Voloshin, *Nucl. Phys.* **B359** (1991) 301.

40. Xu Li, L. Mclerran, M. Voloshin and Rang-tai Wang, *Corrections to High Energy Particles Interacting Through an Instanton as Quantum Fluctuations in the Position of the Instanton*, TPI-MINN-91/16-T.

41. S. Khlebnikov and P. Tinyakov, *Constraint Dependence of the Instanton Calculations and Exponentiation of Hard-Soft Corrections at High Energies* CERN-TH.6146/91.

42. D.I. Diakonov and M.V. Polyakov, *Baryon Number Non-Conservation at High Energies and Instanton Interactions*, Leningrad NPI preprint (1991).

43. V.I. Zakharov, *Nucl. Phys.* **B353** (1991) 683; *High-energy vs large-order perturbative behaviour of weak interactions*, MPI-PAE/PTh 11/91.

44. M. Maggiore and M. Shifman, *Non-perturbative processes at high energies in weakly coupled theories: multi-instantons set an early limit*, TPI-MINN-91/24-T (1991).

45. V.V. Khoze, J. Kripfganz and A. Ringwald, in preparation.

46. A. Ringwald, *Baryon-Numver Vilation at the Electroweak Scale*, CERN-TH.6135/91.

PATTERN RECOGNITION IN HIGH ENERGY PHYSICS WITH NEURAL NETWORKS

Carsten Peterson[1]

Department of Theoretical Physics, University of Lund
Sölvegatan 14A, S-22362 Lund, Sweden

Abstract

Artificial neural networks (ANN) are introduced in the context of analyzing particle physics data. The power of these techniques are demonstrated in applications ranging from off-line jet identification to tracking. Among other things very successful results are presented for b-quark identification using hadronic information only. Also a novel approach for tracking using deformable templates is shown.

1 Introduction

Particle physics contains many challenging feature recognition problems ranging from off-line data analysis to low-level experimental triggers. Flavour tagging and Higgs detection are obvious examples. With powerful algorithms at our hands, signal-to-background ratios could be substantially reduced. In particular for the next generation of accelerators (LHS, SSC) the availability of such algorithms that can be executed in real-time will be crucial. The event rate at these machines is of the order of one event per 10-100 ns. Another class of feature recognition problems is track finding. Again, with the high luminosity expected at LHS and SSC real-time track reconstruction would be advantageous, if not crucial.

It turns out that proof-of-concept for using artificial neural networks (ANN) in particle physics can be established already in off-line analysis of data from existing accelerators. In the domain of quark-gluon separation and heavy quark tagging results have been achieved with ANN that are superior to conventional approaches [1, 2, 3, 4, 5]. Also promising approaches to use ANN for solving the optimization problem of track finding have been suggested [6, 7], which have recently been successfully confronted with realistic data [8]. Alternative ANN inspired algorithms based deformable templates have also been successfully explored for tracking [9, 10].

There are two basic kinds of neural network architectures, *feed-forward* and *feed-back*. In feed-forward networks (see fig 1a) information is processed from bottom to top in a layered strucure. These networks are most commonly used for feature recognition problems in an adaptive manner. Feed-back networks are characterized by an arbitrary connection structure (see fig. 1b) and that the information is processed in all directions. These networks have been successfully to optimization problems. Here the networks are "programmed" once and for all and are hence not adaptive.

---

[1] thepcap@seldc52 (bitnet); carsten@thep.lu.se (internet).

QCD at 200 TeV, Edited by L. Cifarelli
and Y. Dokshitzer, Plenum Press, New York, 1992

 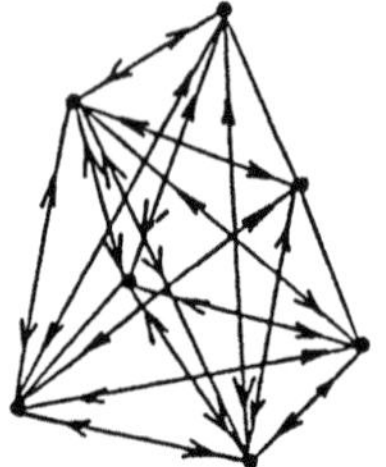

Figure 1. Feed-forward and feed-back architectures

The talk is organized as follows: In Section 2 quark-gluon separation in both e^+e^- annihilation and hadron-induced reactions is discussed using multi-layer perceptron (MLP) networks. Also in this Section b-quarks tagging using both MLP and so-called topological maps can be found. Using ANN to reconstruct invariant masses is another successful application, which is described in Section 3. Feedback networks and deformable networks are used in Section 4 to solve the track-finding problem. Section 5 contains a brief summary and outlook.

1.1   Feed-forward Networks

When analyzing experimental data it is a standard procedure to make various cuts in observed kinematical variables x_k in order to single out desired features. A specific choice of cuts corresponds to a particular set of feature functions $o_i = F_i(x_k)$ in terms of the kinematical variables x_k. This procedure is often not very systematic and quite tedious. Ideally one would like to have an automated optimal choice of the functions F_i. This is exactly what feature recognition artificial neural networks (ANN) aim at. For feed-forward ANN the following form of F_i is often chosen

$$F_i(x_k) = g[\sum_j \omega_{ij} g(\sum_k \omega_{jk} x_k)] \tag{1}$$

where ω_{ij} and ω_{jk} are the parameters to be fitted to the distributions and

$$g(x) = 0.5[1 + \tanh(x)] \tag{2}$$

Eq. (1) corresponds to the feed-forward architecture of fig. 2. The bottom layer (input) corresponds to the measured kinematical variables x_k and the top layer to the features o_i. The mission of the so-called hidden layer is to build up an internal representation of the observed data. Eq. (1) and fig. 2 are easily generalized to more than one hidden layer. Each unit or neuron has the threshold behaviour given by $g(x)$. Fitting to a given data set (or learning) takes place with gradient descent on e.g. a summed square error

$$E = \frac{1}{2} \sum_i (o_i - t_i)^2 \tag{3}$$

with respect to the weights ω_{ij} and ω_{jk}, where t_i are the desired feature values. In this process, which is called *back-propagation (BP)* [11], the *training patterns* are presented over and over again with successive adjustments of the weights. Once this iterative learning has reached an acceptable level in terms of a low error E the weights are frozen and the ANN is ready to be used on patterns it has never seen before. The capability of the network to correctly characterize these *test patterns* is called *generalization* performance.

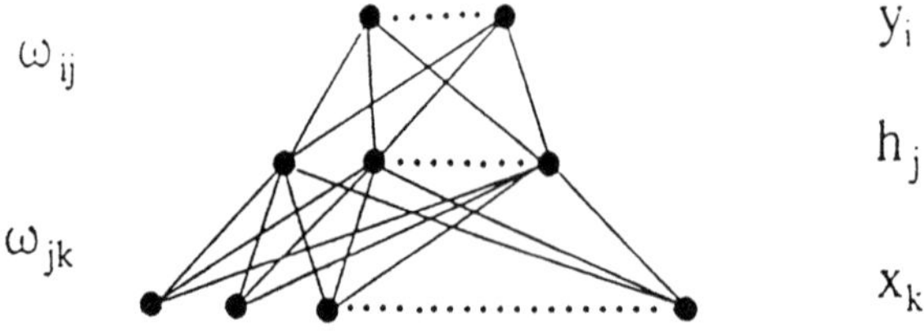

Figure 2. A one hidden layer feed-forward neural network architecture.

150

This back-propagation procedure assumes the knowledge about what features (o_i) are relevant from the outset and separates the data accordingly. There is also an alternative approach, *self-organization* where the network organizes the data into features without any external teacher (no output units) [12]. The underlying architecture consists of an input layer (x_k) and a layer of feature nodes denoted h_j (see fig. 3), where

$$h_j = g(\sum_k \omega_{jk} x_k) = g(\vec{\omega}_j \cdot \vec{x}) \tag{4}$$

For all patterns presented the weights are updated such that the angles between $\vec{\omega}_j$ and $\vec{x}$ are minimized. Also, topological ordering between the feature nodes h_j are introduced with a "mexican-hat" potential, such that neighbouring nodes in a plane react to similar features. Such a system has no "teacher" like the feed-forward BP network above where answers are compared with correct values t_i. The results in this approach are extremely easy to analyze; the weight vectors $\vec{\omega}_j$ for the different feature nodes point in the direction of typical data in $\vec{x}$-space.

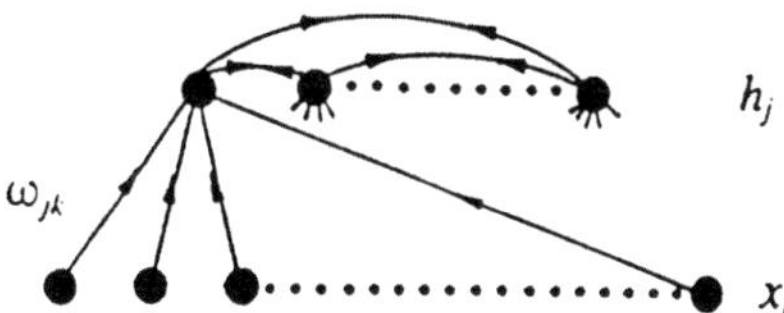

Figure 3. A one-layer self-organizing network. Lateral interactions between the feature nodes correspond to the "mexican-hat" potential.

The feed-forward ANN approach have turned out to be very profitable for a wide range of pattern recognition problems ranging from identification of handwritten numerals [13], transformation of written text to speech [14] to quark identification in particle physics [1][2][3][4][5]. The latter domain of applications is the forcus of this paper. Most applications described here are done with the BP algorithm since it seems to give best performance in terms of classification percentages.

1.2 Feed-back Networks

Finding good solutions to difficult combinatorial optimization problems is another area where ANN has shown great power (see e.g. ref. [15]). The basic approach is as follows: Consider an energy function

$$E = -\frac{1}{2} \sum_{ij} \omega_{ij} S_i S_j \tag{5}$$

where the binary neurons S_i represent the decisions and the connections ω_{ij} encode the various constraints. Using mean field techniques [16] to minimize eq. (5) corresponds to iterative updating of eq. (2). In Section 4 this technique will be used for the track-finding problem.

2 Identifying the Origin of a Hadronic Jet

2.1 Background

In high energy lepton-lepton, lepton-hadron, hadron-hadron and nucleus-nucleus collisions quarks and gluons are produced. These basic quanta of Quantum Chomodynamics (QCD) can never be observed due to the confinement mechanism. They fragment into jets of hadrons, which are observed in the laboratory. This fragmentation process can presently not be calculated from first principles, but theoretically plausible Monte Carlo Markov models [17][18][19] have been developed that reproduce the data extremely well. The main problem here is given a pattern of hadrons in terms of their kinematical variables unfold their origin in terms of quark species or gluon identity. In e^+e^- - reactions the kinematical information is typically given in terms of energy and momenta, whereas in hadron-induced reactions calorimeter detectors pick up transverse energy deposits (E_T) in different cells. The cell positions are given by azimuth angle ϕ and pseudorapidity η.

2.2 Quark-gluon Separation

Being able to distinguish whether a jet of hadrons originates from a quark or a gluon is important from many perspectives. It can shed light on the hadronization mechanism. For example experimental studies of the so-called string effect needs identification of the gluon jet. Also a fairly precise identification of the gluon jet is required for establishing the existence of the 3-gluon coupling in e^+e^- -annihilation reactions. In refs. [1][2][3] the back-propagation (BP) algorithm for MLP networks was used to do the separation both in e^+e^- reactions and in hadron-induced large p_T processes.

e^+e^- **reactions** [1][2]. In order not to reveal to much about the MC model dependent low momentum part of the jet, four-momenta $(\vec{p}_k, E_k)$ of the four leading particles in the jet were used as inputs to the network. One output node representing the quark/gluon option was used and 10 hidden units. Training and test sets were generated at 2 different energies (92 GeV and 29 GeV) with 3 different MC generators; JETSET [17] , ARIADNE [19] and HERWIG [18]. The inputs were either single jets defined by the LUCLUS clustering algorithm in JETSET [17] or entire 3-jet events. After training was completed the network was tested with a middle point succes criteria where > 0.5 for the output node is interpretted as a gluon jet and < 0.5 as a quark jet.

On the average the network was able to correctly classify 85% of the test set jets. The MC model independence of the results were demonstrated by training on MC data generated by one model and tested on MC data from another. Almost no deterioration in performance was obeserved. Also runs where detector acceptance effects were included showed only $O(2\%)$ degradation.

The QCD matrix element suppresses gluon jet production as $1/E_{gluon}$. A fair part of the ANN discrimination orininates from this property. In order to factor out this matrix element dependence from the intrinsic differences between quark and gluon jets one should train different networks with quark and gluon jets in different energy intervals and combine the answers with the appropriate matrix element predicitions. Indeed with such a procedure the classification power increases to 92% [5].

Large p_T processes [4]. In this case the network gets no lead from QCD matrix element information, since the kinematics of the incoming quarks is unknown. Hence we expect a lower classification performance. Another difference is that in hadron-hadron collisions the momenta and energies of the produced hadrons are available in a "raw" form in terms of towers in a calorimeter representing the transverse energies E_T, as mentioned above. In ref. [3] a set of $p\bar{p}$ events at 630 GeV were generated with the PYTHIA MC [20]. The transverse energy of the fragmentation products was mapped into a calorimeter with a granularity of $\Delta\eta = 0.20$ in pseudorapidity ($\approx$ longitudinal velocity) and $\Delta\phi = 0.26$ in azimuth. This calorimeter had a complete coverage in ϕ and extended down to $|\eta| \leq 2$. The set-up corresponds to the UA2 calorimeter at CERN. Jet transverse energies were collected in cones of radius 0.8 in η, ϕ space. The calorimeter information was presented to the network as follows: Take the E_T of the leading cell in the 7×7 matrix and assign it to the first node x_1. Assign the η and ϕ coordinates relative to the center of the jet to x_2 and x_3 respectively. Then take the second leading cell and assign its E_T, η and ϕ to x_4, x_5 and x_6 and so on for the first 15 cells. This corresponds to 45 input nodes. The reason for choosing this representation of the input data rather than the 7×7 cells directly is that in this way invariances of the data is more efficiently incorporated.

After training the the network correctly classifies $70 - 72\%$ of the jets using the 0.5-critera as above. Rather than having this success criteria one should vary the *cut* of the output node and choose a value corresponding to an optimal efficiency and signal-to-background ratio. In fig. 4 the signal-to-background ratio as a function of the signal efficiency is shown. In ref. [3] a similar encoding was successfully used for separating jets steeming from the intermediate vector boseon W from those originating from QCD collision processes. Such a network is able to reduce the QCD background to $W/Z^0 \rightarrow jets$ by factors 20-30.

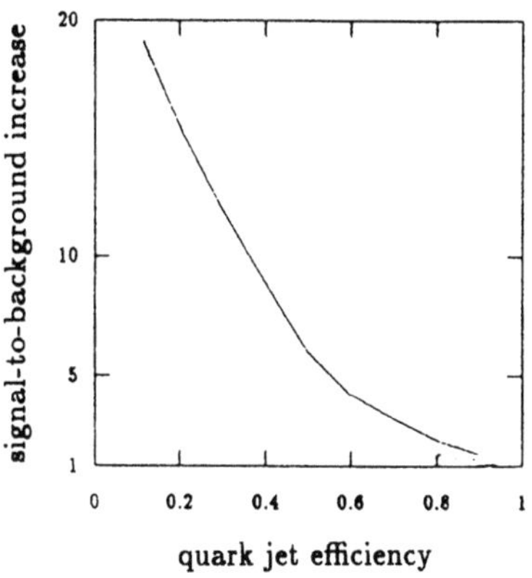

Figure 4. The increase in signal-to-backround ratio versus the efficiency for a neural network quark trigger.

2.3 Heavy Quark Jet Tagging

So-called heavy quarks (c and b) are produced at high energies. In contrast to u- and d- quarks they are unstable and deacay weakly emitting leptons. The conventional way of identifying heavy quarks in e^+e^--reactions is either through leptonic tagging or secondary vertex with efficiency/purity levels of approximately (5%/95%) and (25%/80%) respectively. We have designed a network that identifies heavy quarks [2], which is entirely based upon hadronic information. As input variables we use the total jet energy and momentum along with the energy and direction for the 6 leading particles in the jet, giving a total of 20 input units. Again one layer of 10 hidden units is used and a single output neuron (b=1, non-b=0). As in the large p_T application above the option of varying the cut on the output node can be used to select a desirable efficiency versus purity. The network is then able to produce efficiency/purity numbers comparable with what is expected from vertex detectors.

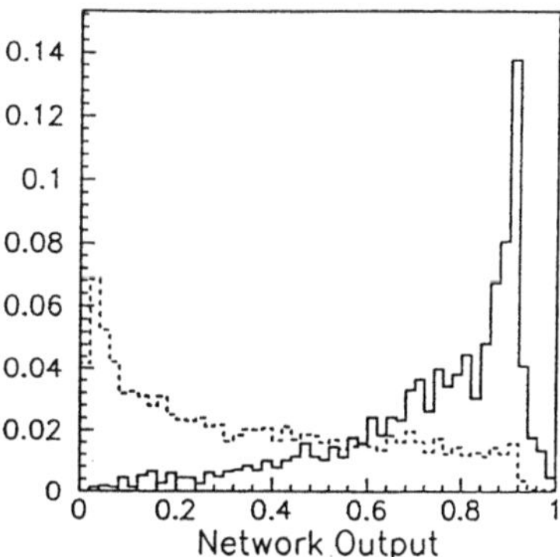

Figure 5. Output distributions for b (full line) and non-b (dashed line) hadronic jets (from ref. [21]).

This remarkable result implies that a general purpose hadronic detector could be very efficient also for heavy flavour tagging. Even better efficiency/purity ratios were obtained in subsequent work [21] by preprocessing the kinematic variables in terms of different shape variables. In fig. 5 the distribution of events for the output node is shown from [21]. It is clear from this figure how chosing the output threshold governs the efficiency/purity ratio.

It is very illuminating to study heavy quark tagging in a self-organizing network. In ref. [4] such a network consisting of a plane of 7 × 7 feature nodes was used to disentangle b-, c- and light (uds) quarks. The input layer has 12 nodes, corresponding to p_z and p_T for the 6 leading particles in the jet. The resulting distributions of the feature nodes are shown in fig. 6 and in fig. 7 the mapping is shown in terms of dominating quarks. Not surprisingly the two extremes in terms of quark

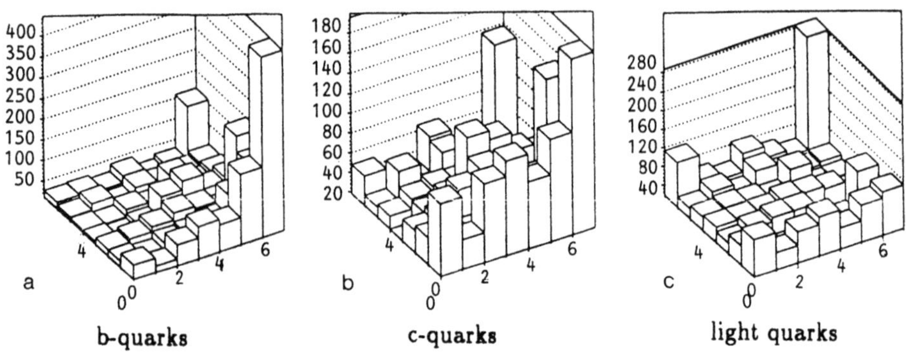

Figure 6. Distribution of b-quarks (**a**), c-quarks (**b**) and uds-quarks (**c**) over the self-organized 7×7 feature nodes.

masses occupy two distinct areas separated by the c-quarks. It is very interesting to inspect the corresponding weight vectors $\vec{\omega}_j$. For example the nodes in the lower right corner have $\vec{\omega}_j$'s with an even distribution of momenta (see fig. 8), which is exactly what one expects for b-quarks since they decay more isotropically.

3 Mass Reconstruction

When hunting for new particles or resonances one often encounters the problem of computing invariant masses of expected decay products. For example, in the case of the intermediate vector boson W (produced in $\bar{p}p$ collisions) in its hadronic decay channel, $W \rightarrow q\bar{q} \rightarrow$ hadrons, M_W is reconstructed relativistically from the momenta and assumed masses of the produced hadrons. The problem here is that the q and $\bar{q}$ jets are not the only hadrons in the collision - there are also remnants from projectile hadrons. An additional complication is that the q and/or $\bar{q}$ jets can give rise to additional jets with bremsstrahlung. So identfying the appropriate jets is crucial for a good

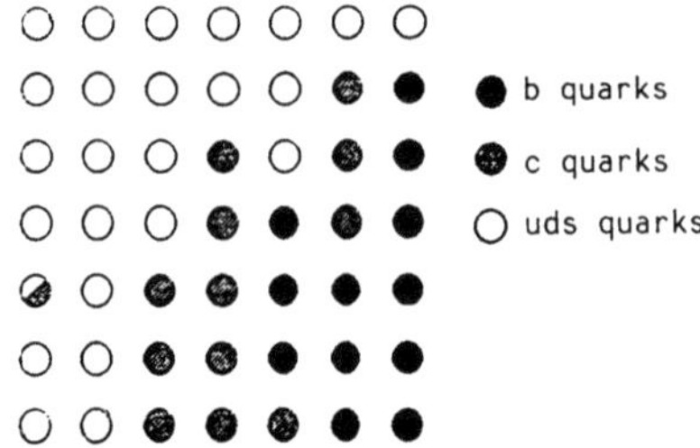

Figure 7. The resulting map for uds- c- and b-quarks. The shading indicates the dominant flavor for the units. The units are numbered as in fig. 6.

reconstruction of the mass. The "standard" procedure [22] for doing this is by sweeping through the calorimeter with a "window" of a certain (ϕ, η) size. The two windows with largest total E_T are selected as containing the two jets and the hadrons in these jets are used to compute M_W.

The neural network approach to this problem is as follows [23]. Since this is not a classification task, a linear rather than non-linear output node is used for the answer (M_W). As inputs the 30 largest towers from the calorimeter are used (cf. quark/gluon separation above) together with the total E_T. A network with two hidden layers with 36 and 15 hidden units respectively is trained with the BP algorithm. As a training set MC generated data with a flat distribution of M_W in the range [50,150] GeV is used. When tested on "real data" in terms of MC generated realistic W-masses the ANN approach produces a distibution which is more narrow and symmetric than the one using conventional methods (see fig. 9). The main reason why the ANN method does better than the conventional method is that it captures the gluon bremsstrahlung tails well.

Figure 8. The weight vectors corresponding to the 4 leading hadrons for b-sensitive (a) and uds-sensitive (b) units in the self-organizing network. The p_T component has been multiplied by a factor of 5 relative to the p_z component.

4 Track Finding

The neural network paradigm has also shown great promise in solving difficult optimization problems [24], in particular with K-valued so-called Potts neurons [16, 15]. In this case feed-back networks are used. In refs. [6][7] this approach was pursued for the track finding problem with encouraging results with respect to solution quality.

The track finding problem is to construct smooth curves through a set of signal points subject to application specific requirements. Typical applications are determination of moving target trajectories and tracks in high energy physics experiments. Common for these applications is the existence of a set of N signal points. The task is to connect the N signal points with continuous smooth tracks.

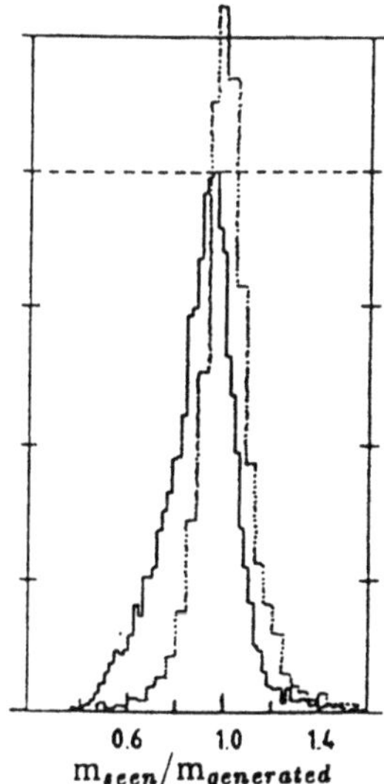

Figure 9. The reconstructed mass (M_W) divided by the true mass (M_W^0) using the ANN method (dotted line) and the conventional "window" method (full line).

4.1 A "Conventional" Neural Approach

In refs. [6][7] binary neurons S_{ij} were introduced to represent whether the decision is to connect points i with j or not. Consider an energy function of the form

$$E = E_1 + E_2 \tag{6}$$

where the "cost" part E_1 is chosen such that short adjacent segments with small relative angles are favoured. In ref. [7] the following choice was made

$$E^{(cost)} = -\frac{1}{2} \sum_{ijkl} \delta_{jk} \frac{\cos^m \theta_{ijl}}{r_{ij} + r_{jl}} S_{ij} S_{kl} \tag{7}$$

where m is an odd integer and the line segments r_{ij} and angles θ_{ijl} are defined in fig. 10. The constraint term in eq. (6) has two parts $E_2^{(a)}$ and $E_2^{(b)}$, where $E_2^{(a)}$ takes care of the requirement that there should be no bifurcating tracks

$$E_2^{(a)} = -\frac{\alpha}{2}[\sum_{ik} S_{ij}S_{kj} + \sum_{jl} S_{ij}S_{il}] \tag{8}$$

The other part, $E_2^{(b)}$, ensures that the number of neurons that are "on" is roughly equal to the number of signals N. It takes the form

$$E_2^{(b)} = \frac{\beta}{2}[\sum_{ij} S_{ij} - N]^2 \tag{9}$$

In eqs. (8,9) α and β are a Lagrange multipliers.

Performing gradient descent on eq. (6) corresponds to changing S_{ij} according to a step-function updating equation

$$S_{ij} = sign(-\frac{\Delta E}{\Delta V_{ij}}) \tag{10}$$

With such a procedure one will very likely end up in local minima, which might represent bad solutions to the problem. A standard way out of this is by introducing noise to the system in a cost-effective deterministic way; the *mean field approximation (MFT)*. It amounts to replacing the step-function of eq. (10) with a sigmoid function

$$V_{ij} = \frac{1}{2}[1 + \tanh(-\frac{\partial E}{\partial V_{ij}} \frac{1}{T})] \tag{11}$$

where the temperature T represents the noise is and V_{ij} are the so called mean field variables, $V_{ij} =< S_{ij} >_T$. Eqs. (11) are solved iteratively until a stable state is obtained. This MFT approximation works particularly well on problems that can be mapped onto neural systems. Also eqs. (11) are isomorfic to VLSI circuitry [24] , which facilitates real-time implementations. In refs. [6][7] these equations produced good solutions to modest sized problems with appropriate choice of parameters (α, β and T). In figs. 11 and 12 we show typical a evolution of the solutions and the energy behaviour with repect to number of iterations (T).

There are in principle N^3 operations to be carried out at each iteration. However, due to the local nature of most track finding problems, this number can be substantially reduced. Neurons need only to be defined within an interaction radius R_c. With on the average m "active" partners within R_c on then has $O(Nm^2)$ computations.

In ref. [8] the number of active neurons was further reduced under realistic circumstances using data from the cylindrical ALEPH TPC detector at CERN. Only neurons connecting points satisfying the following cuts were included:

- $\Delta\phi < 0.15$ rad

- $\Delta\theta < 0.10$ rad

- difference in pad-row^2 <4

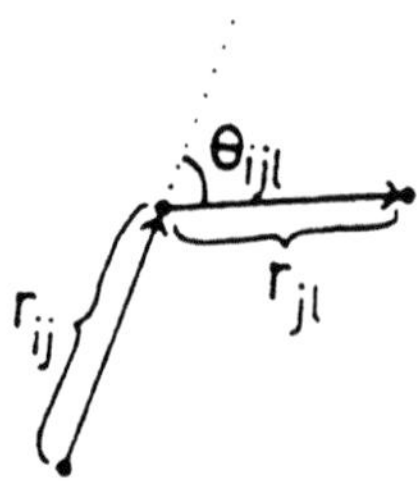

Figure 10. Definition of segment lengths r_{ij} and angles θ_{ijl} between segments

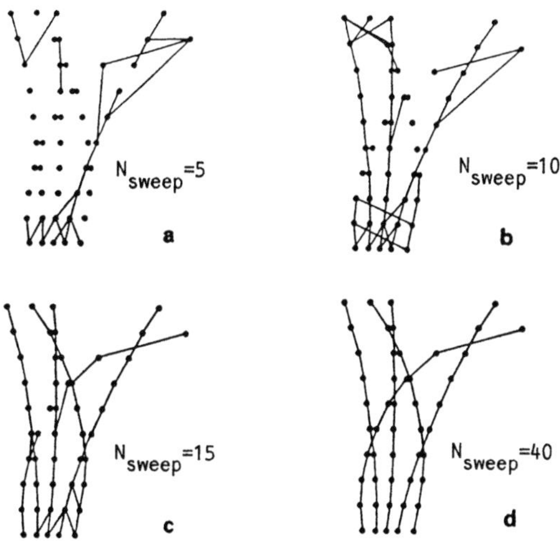

Figure 11. Segments with $V_{ij} > 0.1$ at different evolution stages of the MFT equations.

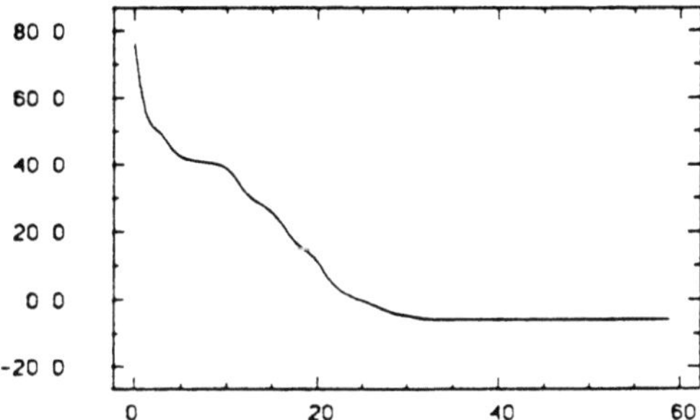

Figure 12. E(T) for the problem in fig. 11.

Also only those neurons correponding to segments pointing inwards towards the origin exist. In fig. 13a segments corresponding to a $Z^0 \rightarrow$ *hadrons* event is shown prior to convergence. In fig. 13b the surviving $S_{ij}=1$ neurons are shown for the same event. In general the quality of the solutions are very good. The efficiency of the ANN alogorithm is around 99% which is comparable with the conventional tracking method used in ALEPH (99.7%) [8]. In ref. [8] the authors also perform simulation studies with track densities corresponding to LHC and SSC detectors ($\sim$200 tracks/event). Fig. 14 shows the time consumption consumption of the ANN algorithm and the conventinal method [8]. Two things emerge from fig. 14. One is that the ANN method scales better with the number of tracks than the conventional method. One also notes that about 60% of the ANN time consumption is due to initialization of the network - computing weights from the observed coordinates.

A few variations of the neuronic approach are possible:

- If one knew the number of tracks in advance from e.g. histogramming (see next Section) one could have chosen a different neural encoding where a neuron S_{ia} is "on" if point i is assigned to track a.

- One of the soft constraints in eq. (8) could have been realized in a "hard" way with a so-called Potts encoding [16], in which case one has

$$\sum_j S_{ij} = 1 \tag{12}$$

---

[2] In a cylindrical detector concentrical *pad-rows* around the origin detect the signals.

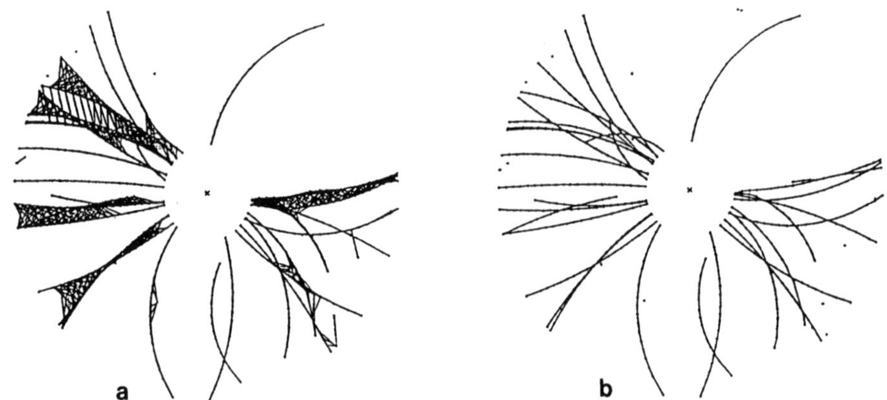

Figure 13. (a) Active neuron segments for a real $Z^0 \rightarrow hadrons$ event (x-y projection) from ref. [8]. (b) Final surviving $S_{ij}=1$ neurons for the same event.

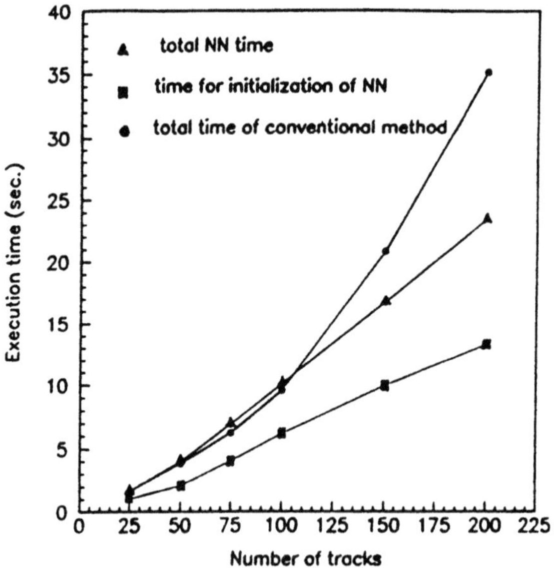

Figure 14. Execution time as a function of the number of tracks (from ref. [8]).

Both of these items will be involved in development of the deformable templates method in the next section.

4.2 Deformable Templates Approach

Even though the ANN approach above seems to work very well it may not be the optimal way to proceed for the particle physics track finding problem:

1. It only solves the combinatorial optimization part of the problem; assigns signals to tracks. In reality one also needs to know the momenta corresponding the tracks. In the neural approach one then has to augment the algorithm with some fitting procedure. It would be nice to have a algorithm that does both things simultaneously.

2. The neural approach is presumable more powerful than what is needed for this problem. The parametric form of the tracks are known in advance - helices. The network has no prior information about this. The fact that the problem is fairly easy to solve for the ANN algorithm is reflected in a very smooth phase transition (see fig. 12). However, for other applications with no prior knowledge of the parametric form of the tracks, the very versatile ANN approach is the way to go.

3. The number of degrees of freedom needed to solve the a N signal problem is large even with the connectivity restrictions inposed in refs. [7, 8]. For a problem with N signals and M tracks one should only need N×M degrees of freedom.

4. As demonstrated in ref. [9] the neural approach is somewhat sensitive to noise. Again with prior knowledge of the parametric form one should be more robust with respect to noise.

All these issues can be accounted for in a novel approach [10, 26] based on so-called deformable templates or elastic nets [25]. A similar approach was independently pursued in ref. [9]. The strategy is to try to match the observed events to simple parameterized models where the form of the models contains the *a priori* knowledge about the possible tracks - circles passing through the origin (the collision point). In addition, this formulation allows for some data points, hopefully those corresponding to sensor noise, to be unmatched. The mechanism involved is closely related to redescending M-estimators used in Robust Statistics [27].

Let us for simplicity limit the discussion to two dimensions, where θ_a and κ_a are the emission angle and the curvature for the a:th track. A Hough transform [28] is used to give initial conditions for the templates and to specify the number of templates required. Hough transforms are essentially variants of "histogramming" or "binning" techniques which have previously been applied to particle tracking.

Assume that we are armed with a set of M deformable templates (θ_a, κ_a) from the Hough transform. A fitness measure is defined between the measurement points and the deformable templates as

$$E[V_{ia}, \theta_a, \kappa_a] = \sum_{i,a} V_{ia} M(\theta_a, \kappa_a, \vec{x}_i) + \lambda \sum_i \{\sum_a V_{ia} - 1\}^2, \tag{13}$$

where: (i) $\vec{x}_i$ labels the positions of the measurement points $(i = 1, ..., N)$, (ii) θ_a and κ_a are the angle of orientation at the origin and the curvature of the circles with $a = 1, ..., M$ indexing the circles, (iii) $M(\theta_a, \kappa_a, \vec{x}_i)$, which we will abbreviate to M_{ia}, is a measure of distance between the ith point and the ath circle, and (iv) V_{ia} is a binary decision unit (neuron) such that

$$V_{ia} = 1 \tag{14}$$

if the ath circle goes through the ith point and is zero otherwise. We want to minimize $E[V_{ia}, \theta_a, \kappa_a]$ with respect to $V_{ia}, \theta_a, \kappa_a$ subject to the global constraint that each point is either matched to a unique circle or not matched. More precisely, given i there exist a unique a such that $V_{ia} = 1$. The second term in eq. (13) imposes a penalty λ if a specific point is unmatched to any circle. Note that with eq. (13) the problem is parametrized in two ways - with the "neurons" V_{ia} and with the template coordinates θ_a and κ_a.

The Boltzmann distribution for $E[V_{ia}, \theta_a, \kappa_a]$ in eq. (13) reads

$$P[V_{ia}, \theta_a, \kappa_a] = \frac{e^{-\beta E[V_{ia}, \theta_a, \kappa_a]}}{Z} \tag{15}$$

with $\beta = 1/T$ as the inverse temperature.

We can now define so-called marginal distributions by either integrate out the neuronic or the template cordinate degrres of freedom. If we choose the latter alternative we end up with a pure neuronic description of the problem similar to the one in the previous Section. Let us use the former alternative, defining the marginal distribution P_M as

$$P_M[\theta_a, \kappa_a] = \sum_V P[V_{ia}, \theta_a, \kappa_a] \tag{16}$$

ensuring that we sum only over configurations of the V_{ia}'s which satisfy the global constraints defined in eq. (14). One gets [10]

$$P[\theta_a, \kappa_a] = \frac{1}{Z} \prod_i \{ e^{-\beta\lambda} + \sum_a e^{-\beta M_{ia}} \} \tag{17}$$

which we can write as

$$P[\theta_a, \kappa_a] = \frac{1}{Z} e^{-\beta E_{eff}[\theta_a, \kappa_a]} \tag{18}$$

where the *effective energy* is given by

$$E_{eff}[\theta_a, \kappa_a] = \frac{-1}{\beta} \sum_i \log\{ e^{-\beta\lambda} + \sum_a e^{-\beta M_{ia}} \} \tag{19}$$

Gradient descent on eq. (19) gives

$$\frac{d\theta_a}{dt} = -\frac{\partial E_{eff}}{\partial \theta_a} = -\sum_i \frac{e^{-\beta M_{ia}}}{e^{-\beta\lambda} + \sum_b e^{-\beta M_{ib}}} \frac{\partial M_{ia}}{\partial \theta_a} \tag{20}$$

$$\frac{d\kappa_a}{dt} = -\frac{\partial E_{eff}}{\partial \kappa_a} = -\sum_i \frac{e^{-\beta M_{ia}}}{e^{-\beta\lambda} + \sum_b e^{-\beta M_{ib}}} \frac{\partial M_{ia}}{\partial \kappa_a} \tag{21}$$

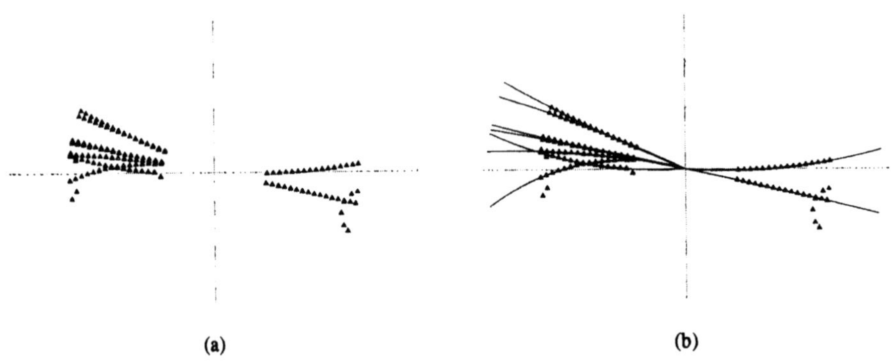

(a)   (b)

Figure 15. (a) Simulated data from DELPI TPC. (b) Result from hybrid Hough/deformable template system.

These equations have an interesting structure since decision elements

$$V_{ia} = \frac{e^{-\beta M_{ia}}}{e^{-\beta\lambda} + \sum_b e^{-\beta M_{ib}}} \tag{22}$$

are mixed with parameter fitting parts proportional to $\partial M_{ia}/\partial\theta_a$ and $\partial M_{ia}/\partial\kappa_a$.

In ref. [10] it is shown how the Hough transform is a $\beta \mapsto \infty$ and $\lambda \mapsto 0$ limit of eq. (13). In ref. [10] it is discussed how this approach is related to that of ref. [9], where also Hough transforms are used specify the number of tracks and initial parameters values. The template "arms" corresponding to the different tracks are fitted to the data in a serial manner one by one. In other words one elastic net is used per track. An explicit repulsive force is introduced between the tracks to ensure only one "net" per track. The approach here is somewhat different. First of all the elastic net consists of many "arms" finding the solution simultanously. Repulsive forces between the tracks is implicitly present through the winner-takes-all mechanism of eq. (22).

The approach of ref. [10] has been explored with simulated DELPHI TPC events with encouraging results (see fig. 15). We are presently investigating this algorithm in depth for a variety of applications [26].

5 Summary and Outlook

This talk have displayed of the power of the neural network paradigm for high energy physics applications. Needless to say the ANN approach looks very promising. The technology is here to stay.

The neural network method is very efficient for extracting features in hadronic data. World record performance can be obtained for quark/gluon separation. With respect to heavy quark tagging the results are in parity with those expected from a vertex detector. A similar network is also able to reduce the QCD background to $W/Z^0 \to jets$ by a factor 20-30.

Most results reported here (refs. [1, 2, 3, 4]) were obtained by using JETNET F77 software package [29].

Ongoing projects [30] include reproducing the theoretical "QCD calorimeter" given the experimentaly measured hadronic and electromagnetic showers. In other words finding the inverse to the QCD cascading, fragmentation and detector simulation processes. Another challenge [30] is to improve the signal/noise ratio for separating $H^0 \to W^+W^-$ at LHC/SSC energies where one of the W's decays into hadronic jets from a background consisting of $W+jets$, $tt \to W+jets$ and $bb \to l+jets$.

It should be emphasized that the feed-forward ANN approach is nothing but fitting data with layers of sigmoids. For high-dimensional problems this is more effecient than the more common approach of using Gaussian expansions. Furthermore sigmoidal amplifiers are easily available VLSI devices, which facilitates hardware implementations.

The neural network approach is in general very noise- and damage-resistant. Hence it is suitable for high energy experiments where various parts of an detector may malfunction. Also with its inherent concurrency and simple structure fast execution custom made hardware could be an asset for real-time triggering. Such systems are already underway [31].

In addition to feature recognition ANN will very likely also play an important role in track finding since feed-back networks easily lend themselves to real-time hardware implementations.

References

[1] L. Lönnblad, C. Peterson and T. Rögnvaldsson, "Finding Gluon Jets with a Neural trigger", *Physical Review Letters* **65**, 1321 (1990).

[2] L. Lönnblad, C. Peterson and T. Rögnvaldsson, "Using Neural Networks to Identify Jets", *Nuclear Physics* **B 349**, 675 (1991).

[3] P. Bhat, L. Lönnblad, K. Meier and K. Sugano, "Using Neural Networks to Identify Jets in Hadron-Hadron Collisions", LU TP 90-13 (to appear in *Proc. of the 1990 DPF Summer Study on High Energy Physics Research Directions for the Decade, Colorado, 1990*).

[4] L. Lönnblad, C. Peterson, H. Pi and T. Rögnvaldsson, "Self-organizing Networks for Extracting Jet Features", *Computer Physics Communications* **67**, 193 (1991).

[5] I. Scabai, F. Czakó and Z. Fodor, "Quark and Gluon Jet Separation using Neural Networks", ITP Budapest Report 477 (1990).

[6] B. Denby, "Neural Networks and Cellular Automata in Experimental High Energy Physics", *Computer Physics Communications* **49**. 429. 1988.

[7] C. Peterson, "Track Finding with Neural Networks", *Nuclear Instruments and Methods* **A279**, 537 (1989).

[8] G. Stimpfl-Abele and L. Garrido, "Fast Track Finding with Neural Nets", UAB-LFAE 90-06 (1990) (submitted to Computer Physics Communications).

[9] M. Gyulassy and H. Harlander, "Elastic Tracking and Neural Network Algorithms for Complex Pattern Recognition", *Computer Physics Communications*, **66** 31 (1991).

[10] A. Yuille, K. Honda and C. Peterson, "Particle Tracking by Deformable Templates", *Proceedings of 1991 IEEE INNS International Joint Conference on Neural Networks*, Vol. 1, pp 7-12, Seattle, WA (July 1991).

[11] D. E. Rumelhart, G. E. Hinton and R. J. Williams, "Learning Internal Representations by Error Propagation", in D. E. Rumelhart and J. L. McClelland (eds.) *Parallel Distributed Processing: Explorations in the Microstructure of Cognition (Vol. 1)*, MIT Press (1986).

[12] T. Kohonen, "Self Organized Formation of Topologically Correct Feature Maps", *Biolological Cybernetics* **43**, 59 (1982);
T. Kohonen, *Self-organization and Associative Memory*, third edition, Springer-Verlag, Berlin, Heidelberg (1990).

[13] Y. LeCun et. al., "Backpropagation Applied to Handwritten Zip Code Recognition", *Neural Computation* **1**, 541 (1989).

[14] T.J. Sejnowski and C.R. Rosenberg, " Parallel Networks that learn to Pronounce English Text", *Complex Systems* **1**, 145 (1987).

[15] C. Peterson, "Parallel Distributed Approaches to Combinatorial Optimization", *Neural Computation* **2**, 261 (1990).

[16] C. Peterson and B. Söderberg, "A New Method for Mapping Optimization Problems onto Neural Networks", *International Journal of Neural Systems* **1**, 3 (1989).

[17] T. Sjöstrand, JETSET 7.2 program and manual. See B. Bambah et al., QCD Generators for LEP, CERN-TH.5466/89.

[18] G. Marchesini and B. R. Webber, *Nucl. Phys.* **B310** (1988) 461;
I. G. Knowles, *Nucl. Phys.* **B310**, 571 (1988).

[19] L. Lönnblad, "ARIADNE-3, A Monte Carlo for QCD Cascades in the Colour Dipole Formulation", Lund preprint LU TP 89-10.

[20] H-U. Bengtsson, T. Sjöstrand, *Computer Physics Communications* **46**, 43 (1987).

[21] L. Bellantoni, J. Conway, J. Jacobsen, Y.B. Pan and S.L. Wu, "Using Neural networks with Jet Shapes to Identify b Jets in e^+e^- Interactions", CERN-PPE/91-90 (submitted to Nuclear Instruments and Methods A).

[22] J. Alitti et al., "Measurements of the Transverse Momentum Distributions of W and Z bosons at the CERN $\bar{p}p$ Collider", *Zeitschrift für Physik* **C 47**, 523 (1990).

[23] L. Lönnblad, C. Peterson and T. Rögnvaldsson, "Mass Reconstruction with a Neural Network", *LU TP 91-25* (to appear in Physics Letters B).

[24] J.J. Hopfield and D.W. Tank, "Neural Computation of Decisions in Optimization Problems", *Biological Cybernetics* **52**, 141 (1985).

[25] R. Durbin, and D. Willshaw, "An Analog Approach to the Travelling Salesman Problem Using an Elastic Net Method", *Nature.* **326**, 689 (1987).

[26] M. Ohlsson, C. Peterson and A. Yuille, "Track Finding with Deformable Templates - The Elastic Arms Approach", *LU TP 91-27* (to appear in Computer Physics Communications).

[27] P.J. Huber, *Robust Statistics*, John Wiley and Sons (New York 1981).

[28] R.O. Duda and P.E. Hart, *Pattern Classification and Scene Analysis*, John Wiley and Sons (New York, 1973).

[29] L. Lönnblad, C. Peterson and T. Rögnvaldsson, "Pattern Recognition in High Energy Physics with Artificial Neural Networks - JETNET 2.0", *LU TP 91-18* (to appear in Computer Physics Communications) [Program and manual available via Email request].

[30] L. Lönnblad, C. Peterson and T. Rögnvaldsson, work in progress.

[31] B. Denby et. al., "Neural Networks for Triggering", FERMILAB-CONF-90/20.

FINAL STATES IN SMALL-x PROCESSES AT VERY HIGH ENERGIES

B.R. Webber

Cavendish Laboratory, University of Cambridge
Madingley Road, Cambridge CB3 0HE, U.K.

1. INTRODUCTION

Much progress has been made over the past few years in the understanding of small-x processes. One of the key developments [1-3] has been the proof of k_t-*factorization* for leading logarithms of x. This implies that the cross section for a generic small-x process like heavy quark hadroproduction at a c.m. energy $\sqrt{S}$ much greater than the quark mass M is expressible in the form

$$\sigma(AB \to Q\bar{Q}X) \sim \int \frac{dx}{x}\frac{dy}{y} d^2k_t d^2q_t \mathcal{F}_A(x, k_t)\mathcal{F}_B(y, q_t)\hat{\sigma}(\hat{s}, M; k_t, q_t) . \tag{1}$$

Here $\mathcal{F}_{A,B}(x, k_t)$ are generalized structure functions for the incoming hadrons A and B, giving the probability (per unit of $\ln x$) of finding a gluon at longitudinal momentum fraction x and transverse momentum k_t. We can take the momentum k of a small-x gluon in hadron A to be of the form $k = xp_A + k_t$, so that $k^2 \simeq -k_t^2$ and k_t also describes the off-shellness of the gluon. Similarly, for a constituent of hadron B with small momentum fraction y the momentum is approximately $q = yp_B + q_t$ and $q^2 \simeq -q_t^2$. When integrated over transverse momenta up to some limit μ, the generalized structure functions become the usual structure functions giving the momentum fraction distribution at scale μ:

$$\int_0^\mu d^2k_t \mathcal{F}(x, k_t) = F(x, \mu^2) . \tag{2}$$

The object $\hat{\sigma}$ in Eq. (1) is a generalized subprocess cross section for the two off-shell gluons to produce a heavy quark-antiquark pair with c.m. energy-squared $\hat{s}$. Such a quantity can be defined in an unambiguous, gauge-invariant way in the small-x region [1-3]. At $k_t = q_t = 0$ it becomes the usual on-shell $gg \to Q\bar{Q}$ cross section.

The generalized structure functions $\mathcal{F}_{A,B}$ and subprocess cross section $\hat{\sigma}$ provide a complete description of the fully inclusive heavy quark production cross section at small x, but for many purposes we would like to know the more exclusive properties of the final state. In particular, we note from Eq. (2) that a subprocess with characteristic scale μ may be expected to involve gluons with transverse momenta $k_t \lesssim \mu$. These transverse momenta arise from the emission of gluons in the course of the evolution of the structure functions from the typical hadronic scale up to scale μ. After hadronization, this "initial-state radiation" leads to an associated hadron multiplicity and transverse energy flow that should be taken into account in studies of the final state. It is especially important to develop Monte Carlo event generators[1] that include this component of hadron production correctly, in addition to that coming directly from the heavy quark production subprocess.

A general theoretical framework for the simulation of initial-state radiation in small-x processes is provided by the coherent parton branching algorithm formulated in Refs. [5-7]. This approach generates a structure function whose asymptotic behaviour, at very small x and very high energy, satisfies the Lipatov evolution equation [8]. The corresponding anomalous dimension is given by the Lipatov expression, which sums singular contributions from all orders in perturbation theory, and therefore we shall call this the *all-loop* formulation of the branching process. In contrast, the *one-loop* formulation will signify the conventional treatment corresponding to the leading-order Altarelli-Parisi equation [9] with coherence [10], in which the structure function evolution is given by the one-loop anomalous dimension, even for $x \to 0$.

The simulation of exclusive final states in small-x processes is much more complicated than that at larger x values, so for definiteness and simplicity we shall concentrate on the electroproduction of heavy quarks via photon-gluon fusion, as depicted in Fig. 1. This process involves only one incoming gluon instead of the two in Eq. (1), and the cross section takes the simpler form

$$\sigma(ep \to Q\bar{Q}X) \sim \frac{\alpha}{\pi} \int \frac{dx}{x} \frac{dy}{y} \frac{dQ^2}{Q^2} d^2 k_t \mathcal{F}_p(x, k_t) \hat{\sigma}(\hat{s}, M; k_t, q_t) , \tag{3}$$

where $x \equiv x_n$, $k \equiv k_n$, y is the electron momentum fraction transferred to the photon and $Q^2 = -q^2 \simeq q_t^2$ is the photon virtuality. The Bjorken variable is $x_B = Q^2/yS$ and the c.m. energy-squared of the fusion subprocess is given by

$$\hat{s} = (x/x_B - 1)Q^2 - k_t^2 - 2k_t \cdot q_t . \tag{4}$$

Notice that the gluon momentum fraction x becomes equal to x_B in the limit that Q^2 is much larger than $\hat{s}$ and k_t^2. However, the factor of $1/Q^2$ from the photon propagator favours much smaller values of Q^2, close to the photoproduction limit, and thus x is usually unrelated to x_B.

In Sect. 2 the coherent parton branching algorithm is reviewed in both the conventional one-loop (Altarelli-Parisi) and the improved all-loop (Lipatov) formulations. Details of the way in which both formulations have been implemented in a Monte Carlo program [11] are given in Sect. 3. Next, in Sect. 4, the off-shell subprocess is considered in more detail, especially the question of how the relevant upper limit μ on the gluon virtuality, which determines the amount of initial-state radiation, emerges naturally from the dynamics of the subprocess. Sect. 5 presents the preliminary results of a Monte Carlo simulation of heavy quark electroproduction at small x. Here we compare and contrast the conventional prescription of one-loop branching plus on-shell subprocess with the improved all-loop plus off-shell formulation. Finally in Sect. 6 some conclusions are drawn and directions for future work are suggested.

---

[1] For a review and earlier references, see Ref. [4].

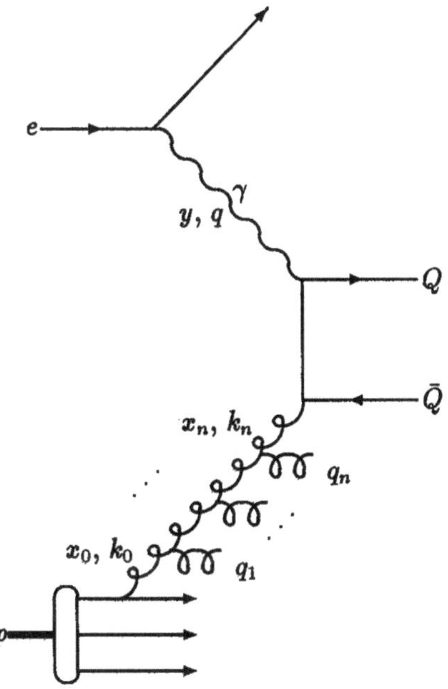

Fig.1. Electroproduction of a heavy quark-antiquark pair.

2. SMALL-x PARTON BRANCHING

In an initial-state gluon cascade like that in Fig. 1, a gluon of low virtuality k_0^2 evolves towards higher spacelike virtuality and lower energy via multigluon emission $q_1,\ldots,q_n$. This is in contrast to the situation in final-state cascades, where a parton of high timelike virtuality evolves to lower energy and virtuality. Thus the ordering of energy and virtuality is in opposite directions in initial-state, spacelike, branching but in the same direction in the final-state, timelike, case. In both cases, however, the coherence properties of the emission process imply that the ordering of virtuality is more precisely an *angular ordering* of the gluon emissions.

By angular ordering we mean that for finite x, to leading collinear and infrared order, the initial-state parton cascade in Fig. 1 takes place in the phase space of θ_i-ordering, where θ_i is the angle between the incoming proton momentum p and that of the emitted gluon q_i. Outside this region there is destructive interference and the multiparton distributions vanish to leading order [10].

The energy and angular distributions along the cascade are determined by the Altarelli-Parisi splitting function and by the appropriate Sudakov form factor, which sums the leading-order virtual corrections. For a purely gluonic cascade as shown, the splitting function is given to one-loop order by

$$P_g(z,\alpha_S) = \bar{\alpha}_S \left[\frac{1}{1-z} + \frac{1}{z} - 2 + z(1-z) \right] , \qquad \bar{\alpha}_S = \frac{C_A \alpha_S}{\pi} , \qquad (5)$$

where $(1-z)$ is the momentum fraction of the emitted gluon. Only the first two terms are relevant for $x \to 0$ and for soft gluon coherence. At the inclusive level, the infrared

singularity due to the term $1/(1 - z)$ is cancelled by the Sudakov form factor, to be specified in more detail shortly.

2.1. One-loop Formulation

As will be discussed below, angular ordering can be extended to the small x region only by taking into account additional *non-Sudakov* virtual corrections. However, even if these additional virtual terms are neglected, the correct $x \to 0$ behaviour of highly inclusive quantities such as the structure function can be reproduced, at least to one-loop accuracy, provided one at the same time reduces the emission phase space from θ_i- to q_{ti}-ordering, where q_{ti} is the transverse momentum of the emitted gluon [10,12]. In this formulation, the correct behaviour of the structure function to one loop is obtained by q_{ti}-ordering at $x \to 0$ and θ_i-ordering at finite x, including $x \to 1$. Therefore in the one-loop formulation the following ordered phase space is used for the initial-state radiation in all regions of x:

$$q'_{ti} > q'_{t\,i-1} \, , \tag{6}$$

where the variable q'_{ti} is related to the transverse momentum and angle of the emitted parton by

$$q'_{ti} \equiv q_{ti}/(1 - z_i) \simeq E_{i-1}\theta_i \, , \tag{7}$$

$(1 - z_i)$ being the momentum fraction of q_i, and E_i the energy component of the spacelike momentum k_i (see Fig. 1). Ordering in q'_{ti} is thus essentially equivalent to θ_i-ordering for finite x and to q_{ti}-ordering for $x \to 0$ $(z_i \to 0)$.

The spacelike gluon anomalous dimension $\gamma_N(\alpha_S)$, defined by

$$\int dx \, x^{N-2} F(x, \mu^2) \sim (\mu^2/\mu_0^2)^{\gamma_N(\alpha_S)} \, , \tag{8}$$

where μ_0 is a collinear momentum cutoff, is known at the 2-loop level [13], and for $N \to 1$ is given by

$$\gamma_N(\alpha_S) = \frac{\bar{\alpha}_S}{N - 1}(1 + c\,\alpha_S) \, , \tag{9}$$

with c a known number. Notice the absence of singular terms of the type $\alpha_S^2/(N - 1)^3$ and $\alpha_S^2/(N - 1)^2$. For $N \to 1$, q'_{ti}-ordering gives an anomalous dimension which agrees with (9) apart from the $c\,\alpha_S$ correction [10].

The structure function corresponding to the anomalous dimension (9) has the small-x behaviour (for fixed α_S)

$$F(x, \mu^2) \sim \exp \sqrt{a \ln(1/x)} \, , \qquad a = 4\bar{\alpha}_S \ln(\mu^2/\mu_0^2) \, . \tag{10}$$

2.2. All-loop Formulation

The main results of Refs. [5-7] concern the coherence structure of real emission and the presence of non-Sudakov virtual corrections. We may summarize the important features as follows:

(i) *Coherence.* For $x \to 0$, interference among soft gluons gives multiparton distributions for the initial-state branching in Fig. 1 which are confined to the θ_i-ordered region:

$$\theta_i > \theta_{i-1} \quad \Rightarrow \quad q'_{ti} > z_{i-1}q'_{t\,i-1} \, . \tag{11}$$

This is the same region given by soft gluon interference [10] at finite x, including $x \to 1$. Notice that, due to the rescaling by z_{i-1}, the phase space given by θ_i-ordering is much larger than that given by q'_{ti}-ordering used in the one-loop formulation:

$$\{\theta_i > \theta_{i-1}\} \sim \{q_{ti} > z_{i-1}q_{ti-1}\}$$
$$\supset \quad \{q_{ti} > q_{ti-1}\}, \quad x \to 0. \tag{12}$$

This implies that for $x \to 0$ angular ordering gives essentially no constraint on the transverse momenta of emitted partons.

(ii) *Non-Sudakov form factor.* In the structure function to any order in α_S, the integrations over the θ_i-ordered region produce powers of $\ln x$ which are not present in either the exact 2-loop result [13] or the leading-order calculation to all loops [5-7]. The origin of these $\ln x$ contributions is that in the θ_i-ordered region the lower bound on q_{ti} vanishes either for $q_{ti-1} \to 0$ or for $z_{i-1} \to 0$. Thus the q_{ti} integrations generate $\ln z_{i-1}$ contributions that give rise to powers of $\ln x$. These $\ln x$ singular terms are cancelled by additional virtual corrections of non-Sudakov type. Like the usual Sudakov virtual contributions they factorize and exponentiate, to give the following non-Sudakov form factor:

$$\Delta_{ns}(z_i, q_{ti}^2, k_{ti}^2) = \exp\left[-\bar{\alpha}_S(k_{ti}^2) \ln\left(\frac{1}{z_i}\right) \ln\left(\frac{k_{ti}^2}{z_i q_{ti}^2}\right)\right] \tag{13}$$

for the i-th branching in Fig. 1. Notice that this expression has a Gaussian behaviour for large $\ln 1/z_i$. Thus $\Delta_{ns}^i/z_i \to 0$ for $z_i \to 0$.

For finite x (including $x \to 1$), the non-Sudakov form factor gives a non-leading correction ($\Delta_{ns} \simeq 1$) and can be neglected. The branching process then becomes *local* in the variables z_i and q_{ti}, and one obtains the same structure as in the Altarelli-Parisi equation with coherence, i.e. with θ_i-ordering.

In the $x \to 0$ region, the branching process is *non-local* and the Altarelli-Parisi equation is not valid. This is because the non-Sudakov form factor (13) is not only a function of the branching variables z_i and q_{ti}, but also of the transverse momentum k_{ti}, which depends on the previous development of the cascade:

$$k_{ti} = |q_{t1} + \cdots + q_{ti}|. \tag{14}$$

In this new formulation the gluon structure function at small x satisfies the Lipatov equation [8] asymptotically.

Since for $x \to 0$ there is no ordering in q_{ti} (or k_{ti}), the transverse momenta generated during a small-x parton cascade perform a random walk and can in principle vanish. However, the region of vanishing transverse momenta is screened by the non-Sudakov form factor, corresponding to the order-by-order cancellation between real emission and virtual contributions in this region.

These cancellations are crucial for the generation of the correct powers of $\ln x$ in the expansion of the structure function and the correct $N \to 1$ behaviour of the spacelike gluon anomalous dimension. From the new branching formulation one finds [5-7] that $\gamma_N(\alpha_S)$ is given by

$$N = 1 + \bar{\alpha}_S \chi(\gamma_N), \tag{15}$$

where $\chi(\gamma)$ is the Lipatov characteristic function [8]

$$\chi(\gamma) = 2\psi(1) - \psi(\gamma) - \psi(1 - \gamma), \tag{16}$$

ψ being the logarithmic derivative of the gamma function. Thus γ_N is given by the expansion

$$\gamma_N(\alpha_S) = \frac{\bar{\alpha}_S}{N - 1} + 2\zeta_3 \left(\frac{\bar{\alpha}_S}{N - 1}\right)^4 + 2\zeta_5 \left(\frac{\bar{\alpha}_S}{N - 1}\right)^6 + \cdots, \tag{17}$$

where ζ_i is the Riemann zeta function. Notice that the α_S^2, α_S^3, and α_S^5 terms are absent. This is consistent with the 2-loop result (9) for $N \to 1$, apart from the subleading term $c\,\alpha_S$. The one-loop approximation $\gamma_N = \bar{\alpha}_S/(N-1)$ corresponds to taking $\chi(\gamma) = 1/\gamma$, which is valid for small γ.

Although each term is singular only at $N = 1$, the expansion (17) develops a square root singularity at $N = 1 + (4\ln 2)\bar{\alpha}_S$. The presence of this singularity at $N > 1$ implies that the behaviour of the structure function for $x \to 0$ is more singular than that given by any finite number of loops. Using the full anomalous dimension given by (15), one finds for fixed α_S the simple power behaviour

$$F(x,\mu^2) \sim x^{-p}, \qquad p = (4\ln 2)\bar{\alpha}_S . \tag{18}$$

If the value of α_S is taken to run in the expected way, the small-x behaviour is no longer given by a simple inverse power of x [14-16], but in either case one finds a violation of the Froissart bound. Ref. [14] discusses how this violation can be overcome by a consistent unitarization procedure, but this goes beyond the scope of the analysis described here.

3. MONTE CARLO METHOD

In the Monte Carlo simulation of the heavy quark production process in Fig. 1, emission of gluonic radiation takes place in two stages. First there is the stage of initial-state branching shown explicitly, in which gluons are emitted from the struck gluon into the angular ordered region (11) and distributed according to the proper Sudakov and non-Sudakov form factors. In the second stage (not shown), these primary emitted gluons and the produced heavy quarks undergo timelike cascading and radiate secondary gluons according to the coherent branching formalism [10,17-20] with angular ordering and the usual Sudakov form factor.

The following subsections first describe the all-loop and one-loop Monte Carlo algorithms and compare the two formulations, then conclude by recalling the algorithm for the final-state emission.

3.1. All-loop Algorithm

The initial-state emission takes place in the angular ordered phase space (11). The branching distribution is given in Ref. [6] as

$$d\mathcal{P}_i = \tilde{P}_g^i(z_i, q_{ti}^2, k_{ti}^2)\, \Delta_s(q_{ti}', z_{i-1}q_{t\,i-1}')\Theta(q_{ti}' - z_{i-1}q_{t\,i-1}')\Theta(1 - z_i - Q_0/q_{ti}')\, dz_i\, dq_{ti}'^2/q_{ti}'^2 \tag{19}$$

where

$$\tilde{P}_g^i(z_i, q_{ti}^2, k_{ti}^2) = \frac{\bar{\alpha}_S(q_{ti}^2)}{1 - z_i} + \frac{\bar{\alpha}_S(k_{ti}^2)}{z_i}\Delta_{ns}(z_i, q_{ti}^2, k_{ti}^2) . \tag{20}$$

The two Θ-functions enforce the angular ordering constraint (11) and an infrared cutoff on the emitted transverse momenta, $q_{ti} > Q_0$, which corresponds to $1 - z_i > Q_0/q_{ti}'$.

The Sudakov form factor Δ_s is given by

$$\Delta_s(q_{ti}', z_{i-1}q_{t\,i-1}') = \exp\left[-\int_{(z_{i-1}q_{t\,i-1}')^2}^{q_{ti}'^2} \frac{dq_t'^2}{q_t'^2} \int_0^{1-Q_0/q_t'} \frac{dz}{1-z}\bar{\alpha}_S(q_t^2)\right] . \tag{21}$$

Note that the q_t' integration region corresponds to the angular ordering constraint (11). The argument of α_S is the square of the transverse momentum $q_t = (1 - z)q_t'$ [21], and the infrared cutoff $(1 - z) > Q_0/q_t'$ is the same as in the phase space for real emission.

These features ensure that the soft singular term $1/(1 - z_i)$ of $\tilde{P}_g^i$ is properly cancelled at the inclusive level by corresponding contributions from the Sudakov form factor, and all the Q_0 dependence disappears from inclusive quantities. The form factor (21) can be interpreted as the probability for not emitting any gluons (with transverse momenta above the cutoff Q_0) within the angular region $\theta_{i-1} < \theta < \theta_i$.

The splitting function $\tilde{P}_g^i$ differs from the usual form (5) in the following three respects:

(i) As explained in Sect. 2.2, the $1/z_i$ term is modified by the virtual corrections of non-Sudakov type, which screen the singularity ($\Delta_{ns}^i/z_i \to 0$ for $z_i \to 0$). This screening is a small-x dynamical effect which is absent in the one-loop formulation.

(ii) The second modification is to the arguments of the running couplings α_S in $\tilde{P}_g^i$. In general [21], the appropriate hard scale for the argument of α_S is $(1 - z_i)|k_i^2|$, which tends to k_{ti}^2 as $z_i \to 0$ and to q_{ti}^2 as $z_i \to 1$. Therefore k_{ti}^2 is taken as the argument of α_S in the non-Sudakov form factor (13) and in the $1/z_i$ term, while the argument is q_{ti}^2 for the soft gluon emission term $1/(1 - z_i)$ and in the Sudakov form factor.

(iii) Finally, the finite terms $-2 + z_i(1 - z_i)$ in (5), which are unimportant for both $x \to 0$ and $x \to 1$, are neglected. This simplification allows both the form factors and the distributions of q_{ti}' and z_i to be calculated analytically in terms of simple functions.

The following procedure is used in the Monte Carlo program for generating an initial-state parton shower like that in Fig. 1 at a given hard scale μ^2. We start at a low scale μ_0^2, generating an x-value x_0 according to a given input structure function $F(x_0, \mu_0^2)$. The corresponding initial transverse momentum k_{t0} of the struck gluon is given a Gaussian distribution with width μ_0. Next the gluon emissions $q_1, \ldots, q_n$, if any, are generated sequentially. Thus the scheme adopted is one of forward evolution from μ_0 to the higher scale μ, rather than backward evolution [18,22] from μ to μ_0, which would be more convenient for phenomenology but has yet to be formulated for small x.

The emission of the i-th gluon takes place according to (19) and proceeds as follows. First q_{ti}' is selected according to the Sudakov form factor

$$\Delta_s(q_{ti}', z_{i-1}q_{t\,i-1}') \, dq_{ti}'^2/q_{ti}'^2 \; , \tag{22}$$

in the region $q_{ti}' > z_{i-1}q_{t\,i-1}'$ (angular ordering) and $q_{ti}' > Q_0$ (infrared cutoff). The first emission takes place in $q_{t1}' > \mu_0$. Notice that at this stage the emitted transverse momentum q_{ti} is not determined since it depends on both q_{ti}' and z_i.

Next z_i is selected according to the distribution $\tilde{P}_g^i$ in (19). For the first term the argument of α_S is $q_{ti}^2 = (1 - z_i)^2 q_{ti}'^2$, which means that the distribution of z_i is affected by the running of α_S. For the second term, both α_S and Δ_{ns} depend on the non-local variable $k_{ti} = k_{t\,i-1} + q_{ti}$. However, the use of this variable only becomes crucial at small z_i, where we may make the approximation $q_{ti} \simeq q_{ti}'$. Therefore in the second term k_{ti} may be taken to be fixed as far as the selection of z_i is concerned. After the selection of z_i, and the choice of a random azimuthal angle, the emitted momentum q_i and the true value of k_{ti} can be computed.

For evolution up to a hard scale μ^2, branching continues until a q_{ti}' value that would violate the bound

$$q_{ti}' < \mu \tag{23}$$

is selected, say for $i = n + 1$. In that case we have an event with n initial-state gluons corresponding to Fig. 1. We shall discuss in Sect. 4 how the appropriate value of μ in Eq. (23) is related to the kinematics of the heavy quark production subprocess.

3.2. One-loop Algorithm

In the one-loop formulation the phase space is given by the q'_{ti}-ordering (6) and the branching distribution is

$$dP_i = P_g^i(z_i, q_{ti}^2, k_{ti}^2)\,\Delta_s(q'_{ti}, q'_{t\,i-1})\Theta(q'_{ti} - q'_{t\,i-1})\Theta(1 - z_i - Q_0/q'_{ti})\,dz_i\,dq_{ti}'^2/q_{ti}'^2 \qquad (24)$$

where P_g^i is obtained from $\tilde{P}_g^i$ by setting $\Delta_{ns} = 1$:

$$P_g^i(z_i, q_{ti}^2, k_{ti}^2) = \frac{\bar{\alpha}_S(q_{ti}^2)}{1 - z_i} + \frac{\bar{\alpha}_S(k_{ti}^2)}{z_i}\,. \qquad (25)$$

Notice that in this scheme there is again a relation between the phase space of the branching and the range of integration of the Sudakov form factor, which implies that inclusive quantities are infrared finite. The event is generated in the same way as before, with the same constraint (23) on the evolution variable.

The important differences between the all-loop and one-loop branching schemes are thus:

- The different phase space for the variable q'_{ti} specified by the arguments of the Θ-functions in (19) and (24);

- The corresponding difference in the range of integration of the Sudakov form factor;

- The presence in (19) of the non-Sudakov form factor.

3.3. Final-state Emission

In both the one-loop and all-loop initial-state branching processes each primary emitted gluon q_i in Fig. 1 undergoes a timelike cascade according to the algorithm of Ref. [17]. The cascade is confined within an angular cone around the spatial direction of q_i of aperture θ_i, which is obtained via Eq. (7) from the value of q'_{ti} generated in the initial-state branching. The cascade continues until the evolution variables of all final-state gluons fall below the cutoff scale Q_0. This treatment takes soft gluon coherence in timelike parton evolution fully into account to leading order.

In general there will also be gluon radiation from the heavy quarks produced in the hard subprocess, which could be handled in a similar way to that from the primary emitted gluons [19]. This is expected to be important when the heavy quarks move relativistically, with large relative transverse momentum. However, the photon-gluon fusion cross section favours the kinematic region in which the heavy quarks are produced near threshold and have little phase space for gluon emission. We shall therefore neglect this component of the final state in the present treatment.

4. OFF-SHELL SUBPROCESS

For the photon-gluon fusion subprocess, the explicit form of the small-x off-shell differential cross section is [1,2]

$$\frac{d\hat{\sigma}}{d\Omega} = \frac{1}{4}e_Q^2\alpha\alpha_S xyS\sqrt{1 - \frac{4M^2}{\hat{s}}}\left[\frac{1}{(\hat{t} - M^2)(\hat{u} - M^2)} - \frac{1}{k_t^2 q_t^2}\left(1 + \frac{x_Q y_{\bar{Q}} S}{\hat{t} - M^2} + \frac{y_Q x_{\bar{Q}} S}{\hat{u} - M^2}\right)^2\right]$$

$$(26)$$

where $d\Omega$ represents the element of solid angle in the subprocess c.m. frame, $\hat{t} = (p_Q - p_\gamma)^2$ and $\hat{u} = (p_{\bar{Q}} - p_\gamma)^2$ are the usual Mandelstam variables, and we have written the heavy quark and antiquark momenta as

$$p_Q = x_Q p_p + y_Q p_e + p_{tQ} \,, \qquad p_{\bar{Q}} = x_{\bar{Q}} p_p + y_{\bar{Q}} p_e + p_{t\bar{Q}} \,. \tag{27}$$

If, for example, the gluon momentum k goes on-shell, $k_t^2 \to 0$, then $\hat{t} \to M^2 - x y_{\bar{Q}} S$ and $\hat{u} \to M^2 - x y_Q S$, so the $1/k_t^2$ singularity is cancelled and we recover the usual on-shell expression. On the other hand, when k_t^2 becomes much larger than M^2 and Q^2 then the cross section is suppressed.

This dynamical cutoff on the gluon virtuality is illustrated in Fig. 2, which shows the (unnormalized) value of the partially integrated cross section

$$\sigma(k_t^2) \equiv \int \frac{dx}{x} d\phi_{qk} \hat{\sigma}(\hat{s}, M; k_t, q_t) \,, \tag{28}$$

where ϕ_{qk} represents the azimuthal angle between the photon and gluon transverse momenta. Two different sets of values of M, Q^2 and $x_B = Q^2/yS$ are shown at the energy of HERA ($\sqrt{S} = 314$ GeV), corresponding to: (a) bottom quark production, and (b) deep inelastic charm production. We see that when $M^2 \gg Q^2$, as in case (a), then the natural cutoff is around $k_t^2 \sim 4M^2$, while for deep inelastic production with $Q^2 \gg M^2$, as in (b), the cutoff becomes Q^2. In general, the important region is $k_t^2 \lesssim Q^2 + 4M^2$. This is much smaller than the kinematically allowed region $k_t^2 \lesssim W^2 \sim Q^2/x_B$ ($= 10^3$ GeV^2 in Fig. 2).

Thus at small x the dynamics of the hard subprocess provide a natural upper limit on the evolution of the initial-state gluon cascade,

$$\mu^2 \sim Q^2 + 4M^2 \tag{29}$$

indicated by the arrows in Fig. 2, beyond which the contribution to the cross section is suppressed.

In the conventional method of calculation, based on the factorization of inclusive hard cross sections at finite x, the on-shell ($k_t^2 = 0$) value of the subprocess cross section would be multiplied by the gluon structure function $F(x, \mu^2)$ evaluated at some appropriate scale μ^2, such as that given by Eq. (29). This is equivalent to approximating Fig. 2 by a step-function which drops sharply from the on-shell value to zero at $k_t^2 = \mu^2$. It can be seen that this procedure has two main effects on the value obtained for the heavy quark production cross section:

(i) The on-shell subprocess cross section is larger in the region $k_t^2 < \mu^2$, so the cross section there is over-estimated;

(ii) The 'tail' of the cross section at $k_t^2 > \mu^2$ is ignored, which leads to an under-estimate.

Asymptotically, the second effect dominates, so the true asymptotic cross section is expected to be larger than the conventional Born approximation. However, at sub-asymptotic energies the first effect still plays a role and the expectation is not so clear.

5. RESULTS ON HEAVY QUARK ELECTROPRODUCTION

Figs. 3–13 show the preliminary results of our new Monte Carlo simulation based on the ideas outlined above. Since the program is at present limited to lepton-hadron collisions, we have chosen to concentrate on bottom quark production at HERA as a good testing-ground for this new approach to small-x physics. The two parts of each figure correspond to two different choices of input gluon structure function at a starting scale of $\mu_0^2 = 5$ GeV^2:

(a) $F(x, \mu_0^2) = 3(1 - x)^5$ ('flat' input distribution);

(b) $F(x, \mu_0^2) = 0.677(1 - x)^5/\sqrt{x}$ ('steep' input distribution).

Both functions are normalized to a gluon momentum fraction of 0.5. Since quarks are neglected in our treatment, the gluon momentum fraction is fixed throughout the evolution of the structure function.

For each choice of input structure function, the simulation was carried out in both the 'one-loop' and the 'all-loop' approaches described above. The results are shown in Figs. 3–13 by the crosses and circles, respectively. For the 'one-loop' simulations, in addition to using the one-loop (Altarelli-Parisi) evolution algorithm of Sect 3.2, we replaced the off-shell subprocess cross section by its on-shell value multiplied by a step-function, as explained in Sect. 4.

For the timelike evolution of the primary emitted gluons into final-state jets, a cutoff $Q_0 = 1$ GeV was used. Study of jet fragmentation in e^+e^- annihilation suggest that for this value of the cutoff approximately 2.5 charged particles are produced for each gluon in the partonic final state. Therefore, in the spirit of local parton-hadron duality [25,26], we simply multiply the number of gluons by this factor in order to present the hadron multiplicity distributions shown in Figs. 6 and 7.

The preliminary results shown indicate that, in the case of bottom quark production at HERA, the results from the improved all-loop evolution plus off-shell subprocess are not very different from the those of the conventional one-loop treatment. The most significant differences occur in the associated multiplicity of produced hadrons, Figs. 6 and 7. These difference arise from the additional phase space available for primary gluon emission in the all-loop evolution: the region of disordered transverse momenta is forbidden in one-loop evolution, while in the new treatment it is allowed, although suppressed at very small momentum fractions by the non-Sudakov form factor.

The curves in Figs. 4, 9, 12 and 13 show the corresponding predictions of Ellis and Kunszt [23], based on the on-shell Born diagram (γg) or on the Born contribution plus gluon bremsstrahlung ($\gamma g + \gamma gg$) as indicated. The predictions are not strictly comparable since we use simplified input gluon distributions and neglect quarks, whereas they used the structure functions of Eichten et al. [24] (EHLQ set 1), which take quarks into account in the (one-loop) evolution equations. However the EHLQ set 1 structure functions are similar to our 'flat' parametrization (a), and quarks are not very important, so the left-hand Figures are broadly similar to the results of Ellis and Kunszt where they can be compared.

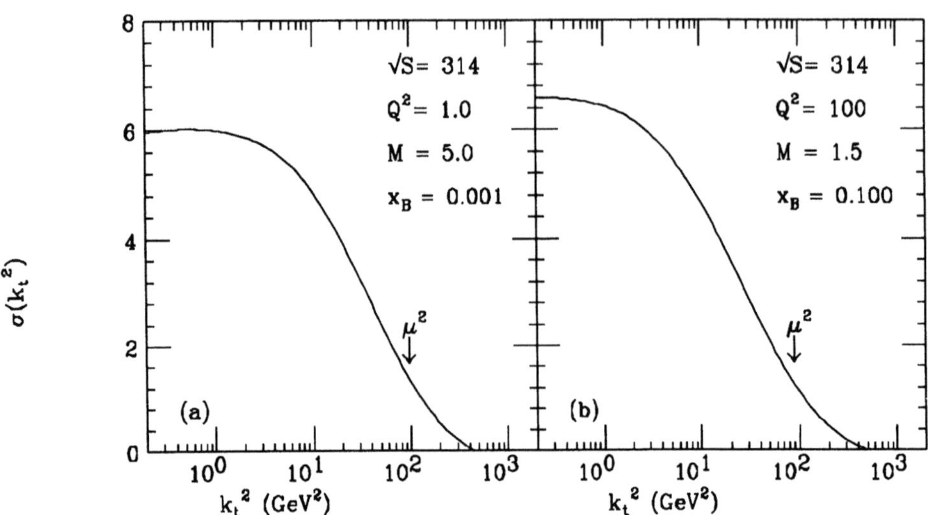

Fig. 2. Dependence of $\gamma g \to Q\bar{Q}$ subprocess cross section on gluon off-shellness k_t^2.

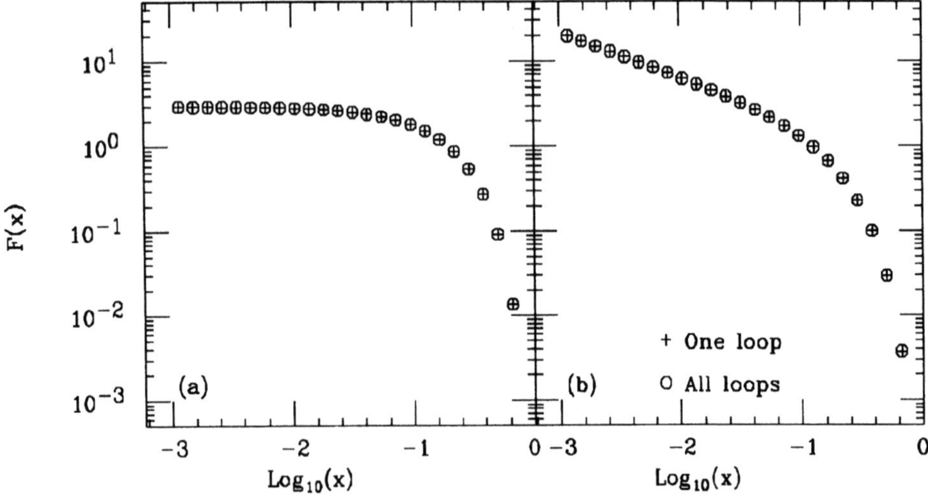

Fig. 3. Input gluon structure functions. Figs. 4–13 give results for
(a) 'flat' input, (b) 'steep' input as shown.

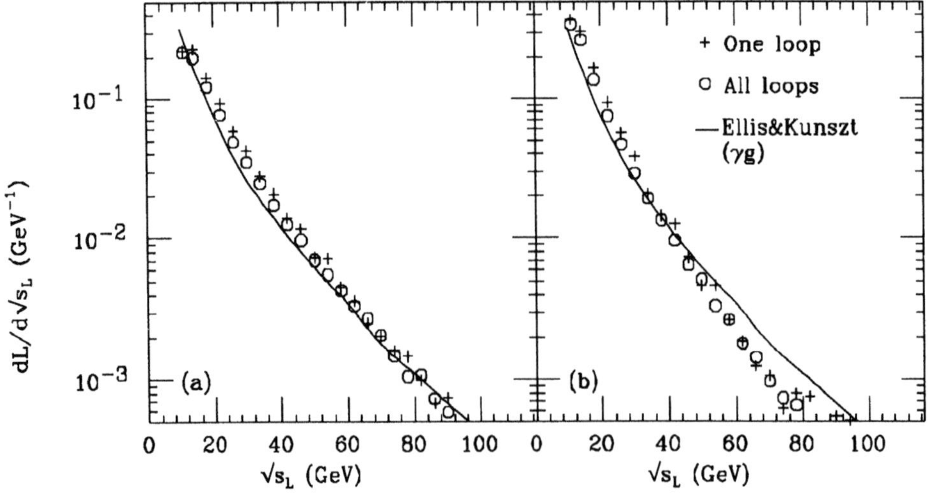

Fig. 4. Luminosity function for γg interactions, as a function of $\sqrt{s}_L$ where $s_L = xyS$.

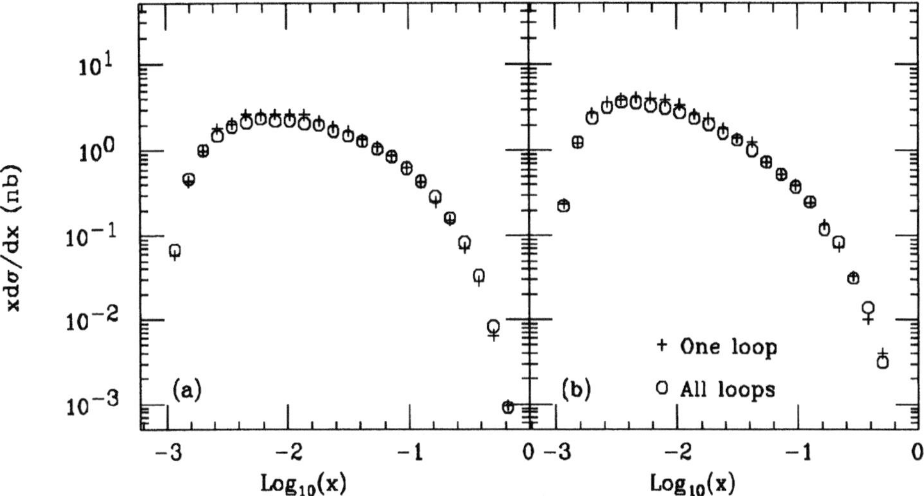

Fig. 5. Momentum fraction distribution of the interacting gluon.

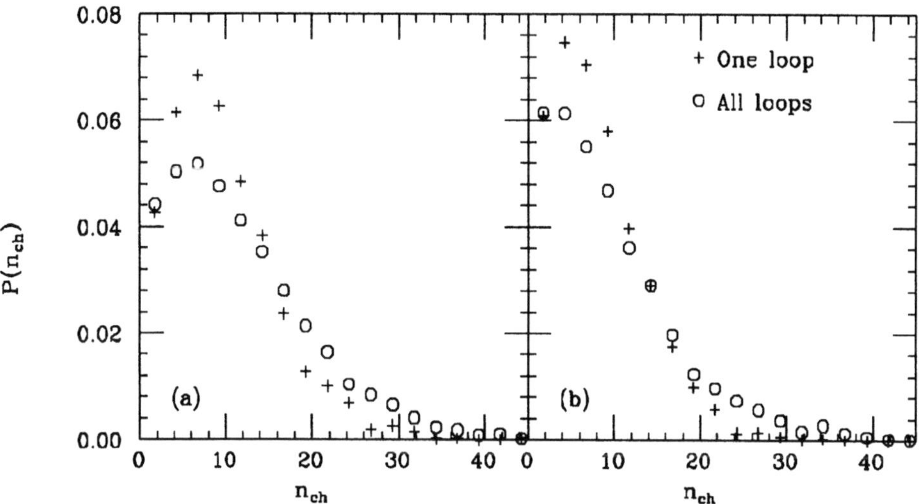

Fig. 6. Multiplicity distribution of charged hadrons from initial-state radiation.

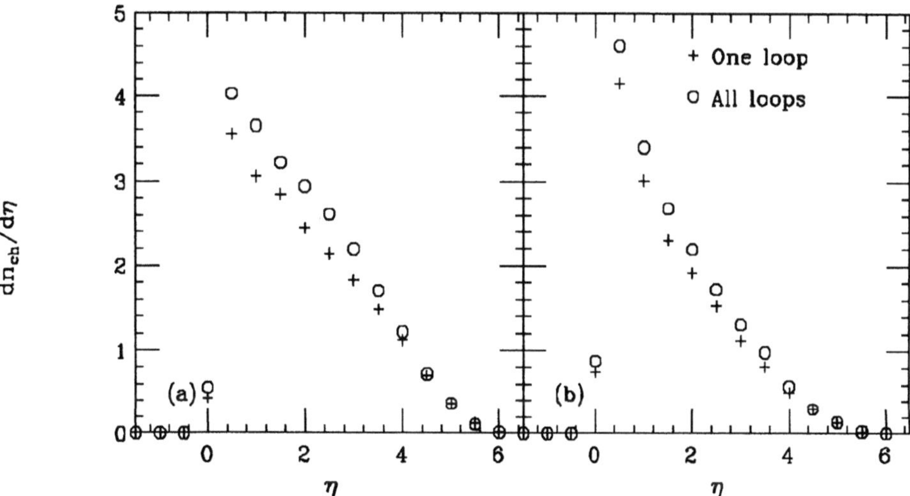

Fig. 7. Pseudorapidity distribution of charged hadrons from initial-state radiation.

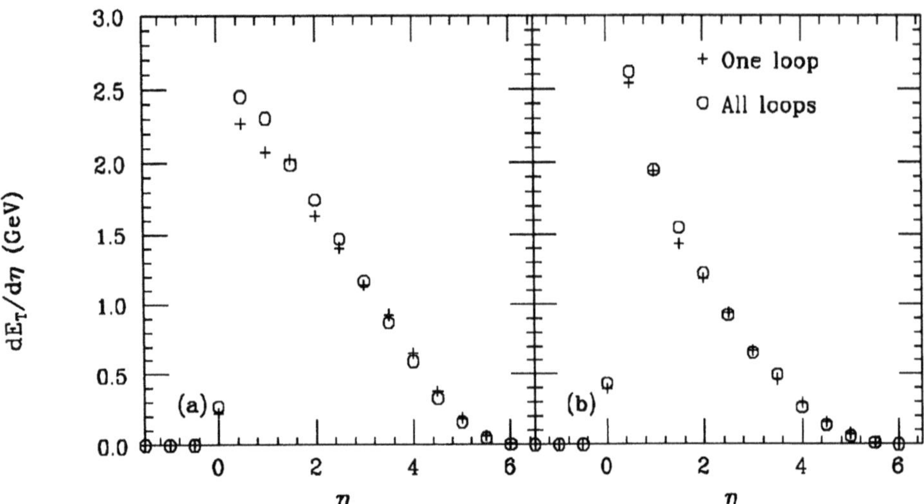

Fig. 8. Pseudorapidity distribution of transverse energy from initial-state radiation.

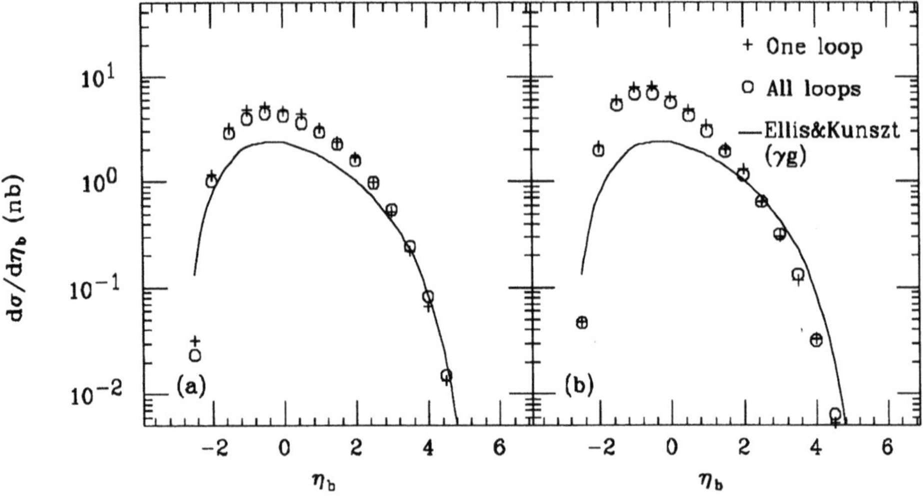

Fig. 9. Pseudorapidity distribution of produced b and $\bar{b}$ quarks.

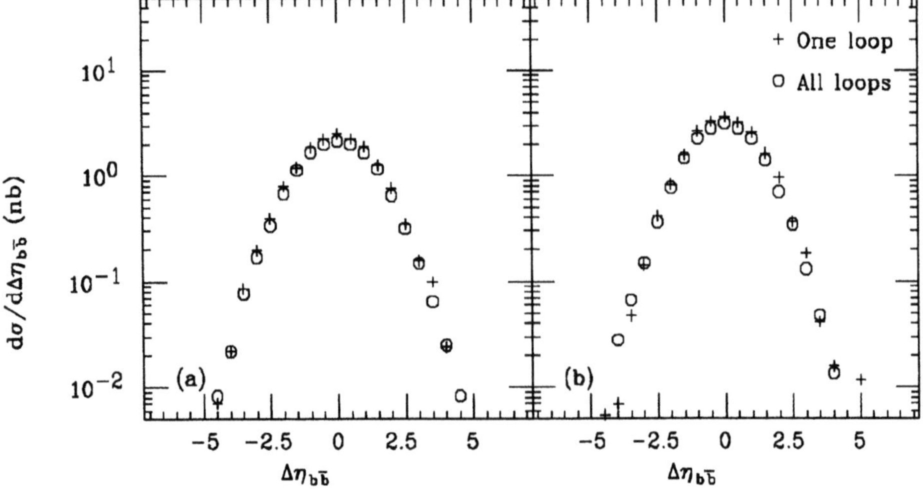

Fig. 10. Pseudorapidity difference between produced b and $\bar{b}$ quarks.

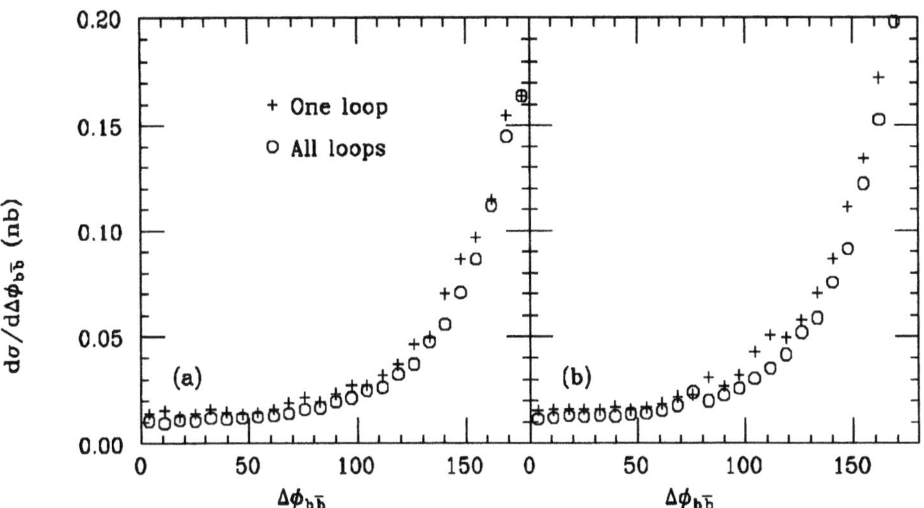

Fig. 11. Difference in azimuth between produced b and $\bar{b}$ quarks.

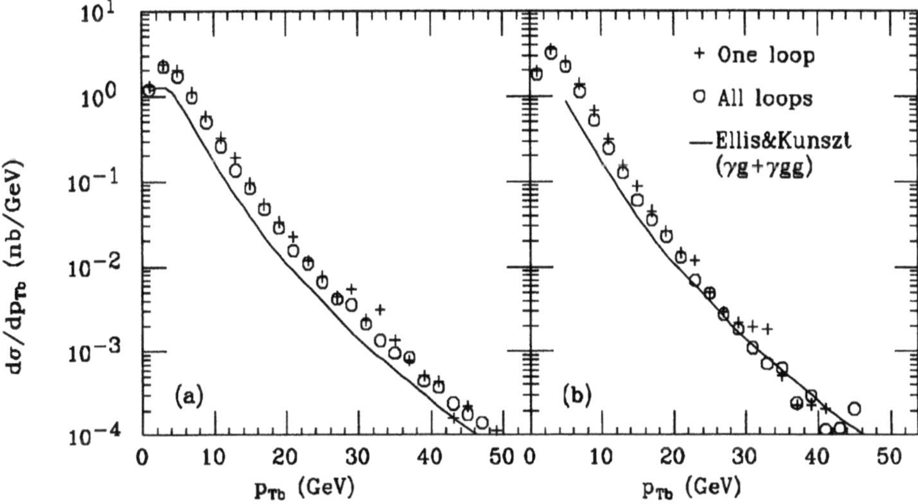

Fig. 12. Transverse momentum distribution of produced b and $\bar{b}$ quarks.

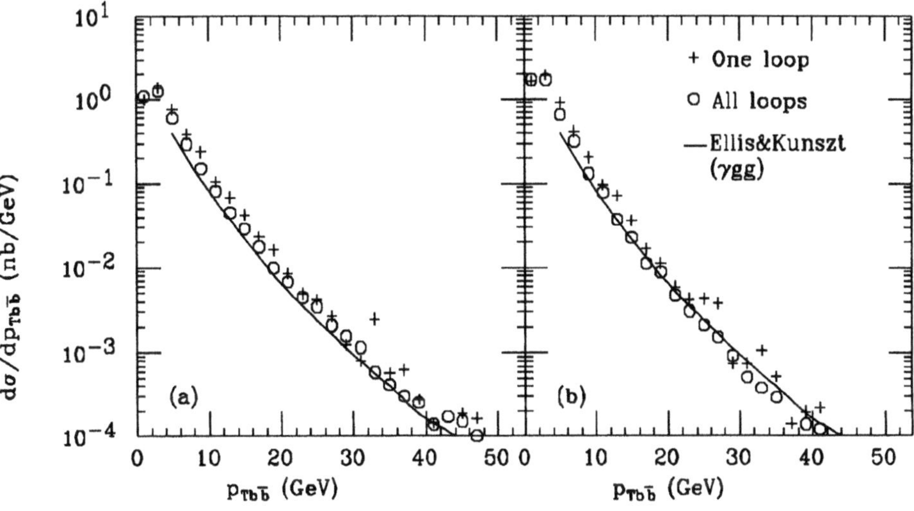

Fig. 13. Transverse momentum distribution of produced $b\bar{b}$ pairs.

6. CONCLUSIONS AND FUTURE DIRECTIONS

The preliminary results presented here on heavy quark electroproduction confirm those obtained earlier in studies of initial-state QCD radiation at higher Q^2 [11]. The one-loop and all-loop evolution algorithms give similar results for highly inclusive quantities, such as the structure function and the momentum distributions of produced heavy quarks, over a wide range of x and Q^2. This can be understood as follows. In the expansion (17) of the anomalous dimension, there are no corrections to the one-loop expression until the order α_S^4. Correspondingly, the all-loop characteristic function $\chi(\gamma)$ in Eq. (16), which governs the evolution of the structure function, remains close to the one-loop form $\chi(\gamma) = 1/\gamma$, except at large values of γ ($\gamma \sim \frac{1}{2}$). This implies that, except for the beginning of the evolution, the effective anomalous dimensions in the two formulations are nearly equal, leading to similar structure functions. For a more detailed analysis, see Ref. [16].

On the other hand, on comparing Figs. 4(a) and 4(b), etc., one sees significant differences depending on whether one uses a flat or steep form for the input gluon structure function. This underlines the importance of determining the input gluon distribution experimentally down to the lowest possible values of x. In this respect new data, especially from HERA, will play a crucial rôle.

The similarity of the inclusive distributions generated by one-loop and all-loop evolution is due to a cancellation of leading higher-order corrections in inclusive observables. In less inclusive quantities such as the associated multiplicity and distributions of emitted hadrons, we find significant differences between the results of the two evolution algorithms, which are explicable in terms of the two new dynamical features of the all-loop algorithm, namely the enlarged phase space due to angular ordering and the screening of the $1/z$ singularity in the gluon splitting function by the non-Sudakov form factor.

We describe the results given here as preliminary because there are several improvements that have yet to be made to our small-x Monte Carlo simulation program to improve its scope and reliability. Apart from technical developments to improve efficiency at x values below 10^{-3}, where the difference between Altarelli-Parisi and Lipatov evolution becomes more significant [2], we need to include light quarks in the initial- and final-state evolution, radiation from the produced heavy quarks, more exact kinematics, and a more detailed hadronization model. We then plan to study hadroproduction processes, in which smaller values of x should be accessible with higher heavy quark cross sections.

ACKNOWLEDGEMENTS

I am most grateful to Drs. L. Cifarelli and Yu. Dokshitzer, and to the Director and staff of the Ettore Majorana Centre, for organizing such a stimulating and informative workshop.

All of the work reported here was done in collaboration with G. Marchesini. We benefited greatly from discussions with S. Catani, M. Ciafaloni, F. Fiorani, E.M. Levin, A.H. Mueller and M.G. Ryskin.

This research was supported in part by the U.K. Science and Engineering Research Council and the Italian Ministero della Pubblica Istruzione.

REFERENCES

1. S. Catani, M. Ciafaloni and F. Hautmann, Phys. Lett. 242B (1990) 97; in *Proc. DESY Topical Meeting on the Small-x Behaviour of Deep Inelastic Structure Functions in QCD*, ed. A. Ali and J. Bartels [Nucl. Phys. B (Proc. Suppl.) 18C (1990) 220]; Cambridge preprint Cavendish–HEP–90/27, to be published in Nucl. Phys. B.

2. J.C. Collins and R.K. Ellis, Nucl. Phys. B360 (1991) 3.

3. E.M. Levin, M.G. Ryskin, Yu.M. Shabelski and A.G. Shuraev, DESY preprints 91–054, 91–065.

4. B.R. Webber, Ann. Rev. Nucl. Part. Sci. 36 (1986) 253.

5. M. Ciafaloni, Nucl. Phys. B296 (1987) 249.

6. S. Catani, F. Fiorani and G. Marchesini, Phys. Lett. 234B (1990) 339.

7. S. Catani, F. Fiorani, G. Marchesini and G. Oriani, Nucl. Phys. B361 (1991) 645.

8. L.N. Lipatov, Yad. Fiz. 23 (1976) 642 [Sov. J. Nucl. Phys. 23 (1976) 338];
E.A. Kuraev, L.N. Lipatov and V.S. Fadin, Zh. Eksp. Teor. Fiz. 72 (1977) 373 [Sov. Phys. JETP 45 (1977) 199];
Y.Y. Balitskii and L.N. Lipatov, Yad. Fiz. 28 (1978) 1597 [Sov. J. Nucl. Phys. 28 (1978) 822];
J. Bartels, Nucl. Phys. B151 (1979) 293;
T. Jaroszewicz, Acta Phys. Pol. B11 (1980) 965; Phys. Lett. 116B (1982) 291.

9. G. Altarelli and G. Parisi, Nucl. Phys. B126 (1977) 298;
Yu.L. Dokshitzer, Sov. Phys. JETP 73 (1977) 1216.

10. A. Bassetto, M. Ciafaloni and G. Marchesini, Phys. Rep. 100 (1983) 201;
Yu.L. Dokshitzer, V.A. Khoze, S.I. Troyan and A.H. Mueller, Rev. Mod. Phys. 60 (1988) 373.

11. G. Marchesini and B.R. Webber, Nucl. Phys. B349 (1991) 617.

12. N. Mitra, Nucl. Phys. B218 (1983) 145.

13. G. Curci, W. Furmanski and R. Petronzio, Nucl. Phys. B175 (1980) 27; W. Furmanski and R. Petronzio, Zeit. Phys. C11 (1982) 293;
J. Kalinowski, K. Konishi, P.N. Scharbach and T.R. Taylor, Nucl. Phys. B181 (1981) 253;
E.G. Floratos, C. Kounnas and R. Lacaze, Phys. Lett. 98B (1981) 89;
I. Antoniadis and E.G. Floratos, Nucl. Phys. B191 (1981) 217.

14. L.V. Gribov, E.M. Levin and M.G. Ryskin, Phys. Rep. 100 (1983) 1;
E.M. Levin and M.G. Ryskin, in *Proc. DESY Topical Meeting on the Small-x Behaviour of Deep Inelastic Structure Functions in QCD*, ed. A. Ali and J. Bartels [Nucl. Phys. B (Proc. Suppl.) 18C (1990) 92].

15. J. Kwiecinski, Zeit. Phys. C29 (1985) 561.

16. E.M. Levin, G. Marchesini, M.G. Ryskin and B.R. Webber, Nucl. Phys. B357 (1991) 167.

17. G. Marchesini and B.R. Webber, Nucl. Phys. B238 (1984) 1;
    B.R. Webber, Nucl. Phys. B238 (1984) 492.

18. G. Marchesini and B.R. Webber, Nucl. Phys. B310 (1988) 461.

19. G. Marchesini and B.R. Webber, Nucl. Phys. B330 (1990) 261;

20. S. Catani, G. Marchesini and B.R. Webber, Nucl. Phys. B349 (1991) 635.

21. D. Amati, A. Bassetto, M. Ciafaloni and G. Veneziano, Nucl. Phys. B173 (1980) 429.

22. T. Sjöstrand, Phys. Lett. 157B (1985) 321.

23. R.K. Ellis and Z. Kunszt, Nucl. Phys. B303 (1988) 653.

24. E. Eichten, I. Hinchliffe, K. Lane and C. Quigg, Rev. Mod. Phys. 56 (1984) 579;
    ibid. 58 (1986) 1065.

25. D. Amati and G. Veneziano, Phys. Lett. 83B (1979) 87.

26. Yu.L. Dokshitzer and S.I. Troyan, Leningrad Nuclear Physics Institute preprint N922 (1984);
    Ya.I. Azimov, Yu.L. Dokshitzer, V.A. Khoze and S.I. Troyan, Phys. Lett. 165B (1985) 147; Zeit. Phys. C27 (1985) 65.

STRUCTURE FUNCTION FOR LARGE AND SMALL X[1]

Giuseppe Marchesini

Dipartimento di Fisica, Università di Parma
INFN, Gruppo Collegato di Parma, Italy

Abstract

We describe generalized soft factorization theorems and the resulting recurrence relation which we use to derive for the structure function a) the evolution equation for large x, to double logarithms; b) the Lipatov equation for small x, to all collinear logarithms; c) a general initial state branching process and d) an integral equation valid for all values of x.

1. Introduction

A characteristic feature of the new (and future) hadron colliders is that data for hard cross sections will be available also for very small values of $x \simeq Q^2/S$, with Q the hard scale and $\sqrt{S}$ the center of mass energy. Examples of these cross sections are heavy quark production discussed at this meeting [1,2,3].

Since many years one knows how to compute in perturbative QCD hard cross sections as long as x is finite. The leading (and the next to leading) collinear logarithms are resummed by the Altarelli–Parisi equation of Ref. [4] for the structure function. Near the kinematical boundary $x \to 1$ the structure function expansion contains also power of $\ln(1-x)$ of infrared type. These terms are correctly resummed [5], at least to double logarithmic accuracy, by this evolution equation provided one uses as evolution parameter the emission angle with respect to the incoming hadron. Similarly in the region of small x the structure function involves also power of $\ln x$. The difference with respect to the $x \to 1$ case is that for $x \to 0$ one finds huge cancellations among collinear singularities so that double logarithmic accuracy is inadequate. Actually the accurate calculations of the leading $x \to 0$ contributions requires one not to make any collinear approximation.

Hard processes in the small x region should be smoothly connected to soft processes in which one does not have any hard scale ($Q \sim m$). The variable $x \to 0$ corresponds then

[1]Research supported in part by the Ministero della Università e della Ricerca Scientifica.

Talk given at the 17th Workshop: "QCD at 200 TeV", Erice, June 1990

to $m^2/S \to 0$ and actually the high energy total cross section is given by an expansion in $\ln(S/m^2)$. Since long time it is known that the resummation of the leading terms is given by the Lipatov equation [6]. Although this does not have the form of an evolution equation of the renormalization group type, as one should expect for hard processes, it has been suggested [5,7] that the $x \to 0$ structure function could be described by the Lipatov equation.

In a recent analysis [8,9,10] of hard perturbative QCD for $x \to 1$ and $x \to 0$ it has been shown that this assumption is correct. One finds that the Altarelli–Parisi equation breaks down for $x \to 0$ and should be substituted by the Lipatov integral equation. To study this transition we used a unified treatment of hard collisions near the two boundaries $x \to 1$ and 0. This calculation is based on the use of *generalized soft gluon factorisation theorems* which allow us to compute both real emission and virtual correction contributions near these phase space boundaries. These theorems do not involve any collinear approximation.

In a first paper [9] we performed an exclusive analysis and computed, to double logarithmic accuracy, the multi-gluon distributions in deep inelastic process in both the singular regions $x \to 0$ and $x \to 1$. We showed (see also Ref. [8]) that multi-gluon emission is described by a branching process which can be used to construct a Monte Carlo program for numerical studies [11]. This program has been used to compute heavy flavour cross sections and their associated QCD radiation [2]. At the inclusive level, i.e. neglecting the associated final state radiation, one finds that such initial state branching for finite x corresponds to the Altarelli–Parisi equation, while for $x \to 0$ corresponds to the Lipatov equation for the structure function.

In a second paper [10] we used the fact that the soft factorisation theorems do not involve any collinear approximation and performed the direct inclusive calculation of the structure function for $x \to 0$. We showed that the Lipatov equation is obtained without the introduction of any collinear approximation.

In this talk I would like to generalize this last inclusive study and deduce an iterative equation for the structure function in both regions $x \to 0$ and $x \to 1$. In the last region this iterative equation is obtained to double logarithmic accuracy and corresponds to the inclusive initial state branching obtained in Refs. [8,9]. For $x \to 0$ one recovers the result of Ref. [10] with no collinear approximations. This inclusive study is much simpler than the exclusive one of Ref. [9,3] and allows us a better control of the approximations. Therefore, in the course of the present description, I hope to elucidate the important role of coherence in this type of physical problems. On the technical point of view the systematic treatment of soft gluon emission from bounces of fast charges allows one to simplify the calculation of sums of Feynman diagrams. In particular, as shown in Ref. [10], calculating the structure function for $x \to 0$ one finds that in those phase space regions where amplitudes cannot be computed without collinear approximations the distribution vanishes in the soft limits due to destructive interference. This allows one to reach an all-loop accuracy by using a technique which is typical of double logarithmic accuracy.

In Sect. 2 we describe the general soft factorization theorems, which are valid to all collinear logarithms, and deduce the inclusive recurrence relation. In Sect. 3 we show that this relation gives the structure function in both regions of small x (to all loops) and large x (to double logarithms). We deduce also a unified integral equation for all values of x.

2. Soft gluon factorization and recurrence relation

We consider deep inelastic scattering which to parton level represented in Fig. 1

184

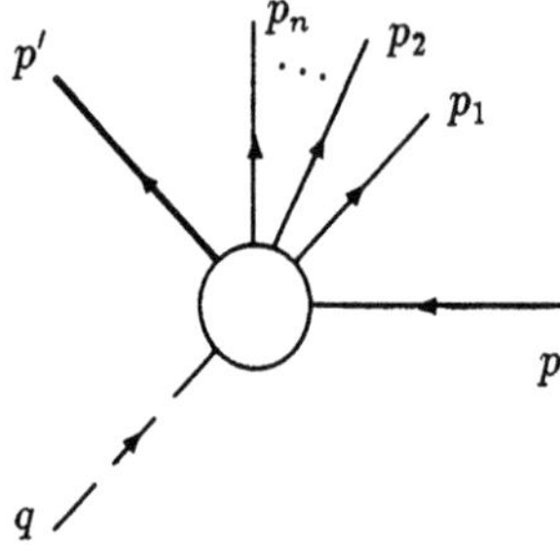

Figure 1

where q is the hard colour singlet probe and p' represents the recoiling system of partons, such as, for instance, a pair of heavy quark-antiquark. For simplicity we assume that the incoming p and the outgoing partons $p_1 \ldots p_r$ are gluons. We take $p = (E, 0, 0, E)$ and introduce a light-like vector $\bar{p}$ in the opposite z-direction. The general emission momentum can be written as

$$p_i = y_i p + \bar{y}_i \bar{p} + p_{ti}, \quad \xi_i \equiv \frac{\bar{y}_i}{y_i} = \frac{p_{ti}^2}{s\, y_i^2}, \quad s = 2p\bar{p}, \tag{1}$$

where $\ln \xi_i$ is the (pseudo-)rapidity.

The cross section of this process factorizes into the elementary heavy flavour distribution and the structure function. As discussed at this meeting [1], for $x \to 0$ this factorization theorem can be generalized to include higher order corrections if we introduce the structure function at fixed total transverse momentum Q_t, which is given by

$$\tilde{F}(x, \mathbf{Q}_t) = \frac{1}{x} \sum_r \int \prod_{i=1}^{r} (dp_i) \left| M^{(r)}(p_1 \ldots p_r) \right|^2 \theta_{12\ldots r}^Y \delta(1 - x - Y_r) \delta^2(\mathbf{Q}_t - \mathbf{Q}_{rt}),$$

$$(dp) = \frac{d^3 p}{(2\pi)^3 2\omega}; \quad Y_r = \sum_1^r y_i; \quad \mathbf{Q}_{rt} = \sum_{i=1}^{r} \mathbf{p}_{it}; \quad \Theta_{12\ldots n}^Y = \theta(y_1 - y_2) \cdots \theta(y_{n-1} - y_n);$$

$$\tag{2}$$

with $\left| M^{(r)}(p_1 \ldots p_r) \right|^2$ the spin and colour average of the multi-gluon squared subamplitudes. Notice that we have substituted the usual symmetrization factor $1/r!$ by y-ordering. The structure function is given by

$$F(x, Q) = \int d^2 Q_t \theta(Q - Q_t) \tilde{F}(x, \mathbf{Q}_t). \tag{3}$$

For $x \to 0$ this Q_t factorization is explicit in the gauge $\eta = \bar{p} \equiv q + xp \sim p'$. Here the amplitude of Fig. 1 can be factorized [9] as follows

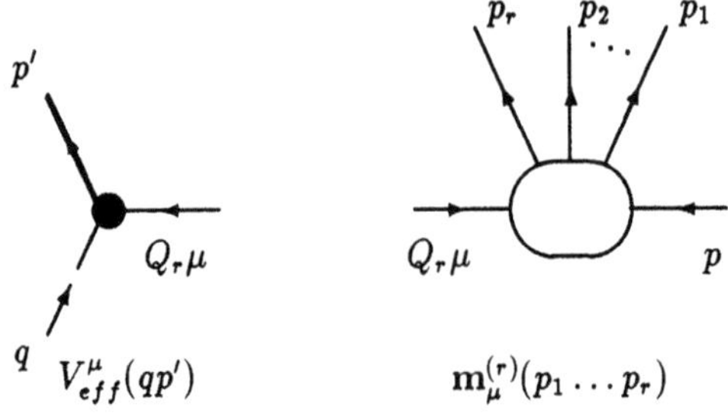

Figure 2

where the subamplitude $\mathbf{m}_\mu^{(r)}$ is a matrix in the colour indices of the external $p, p_1, \cdots p_r$ and internal Q_r gluons. In this gauge the elementary heavy quark amplitude $V_{eff}^\mu(qp')$, is proportional to $\bar{p}^\mu$ where μ is the Lorentz index of the internal gluon Q_r. The subamplitudes in (2) are then defined by

$$\mathbf{M}^{(r)}(p_1 \ldots p_r) = \frac{\bar{p}^\mu}{2p\bar{p}} \mathbf{m}_\mu^{(r)}(p_1 \ldots p_r) \, . \tag{4}$$

To the perturbative order α_S^n, the distribution $\left| M^{(r)} \right|^2$ contains $(n - r)$ virtual loops. In order to explicitly show the cancellation of real and virtual infrared singularities it is convenient to treat the virtual gluons on the mass-shell, as in old fashion perturbation theory so that to the perturbative order α_S^n the distribution $\left| M^{(r)} \right|^2$ can be expressed as an integral over the $n - r$ on-shell momenta $v_1 \ldots v_{n-r}$. We denote by $\mathbf{f}(p_1 \ldots p_r; v_1 \ldots v_{n-r})$ the integrand which depends on the real and virtual momenta, on the helicity and colour of $\{p_i\}$ and on the colour of Q_r (see Ref. [10] for details).

The perturbative expansion of $\tilde{F}(x, \mathbf{Q}_t)$ can be written as

$$\tilde{F}(x, \mathbf{Q}_t) = \frac{1}{x} \sum_{n=1}^\infty \alpha_S^n \int \prod_{i=1}^n (dk_i) \sum_R Tr\{\mathbf{f}_n^{(R)}(k_1 \ldots k_n)\} \, \theta_{12 \ldots n}^Y \, \delta_3(p - q - Q^{(R)}) \, , \tag{5}$$

where we introduced the short notation

$$\delta_3(p - q - Q^{(R)}) \equiv \delta\left(1 - x - Y^{(R)}\right) \delta^2\left(\mathbf{Q}_t - \mathbf{Q}_t^{(R)}\right) \, , \tag{6}$$

and where $k_1, \ldots, k_n$ are the n real $p_1 \ldots p_r$ or the $n - r$ on-shell virtual $v_1 \ldots v_{n-r}$ momenta. We denote by R the given assignment of real momenta $\{p_i\}$ into the set $\{k_i\}$. Therefore $\mathbf{f}_n^{(R)}$ is the multi-gluon integrand which corresponds to this assignment and we have $Y^{(R)} = Y_r$ and $\mathbf{Q}_t^{(R)} = \mathbf{Q}_{rt}$ The structure function is obtained by summing over all different assignments.

2.1) *Phase space for $x \to 0$ and $x \to 1$.*

To compute $\tilde{F}(x, \mathbf{Q}_t)$ we follow the method of Ref. [9]. We first show that the dominant $x \to 1$ and $x \to 0$ contributions are given by the following soft phase space for both real and virtual gluons

$$\begin{aligned} 1 \simeq y_1 \gg \cdots \gg y_n \gg x, \quad & x \to 0 \\ x \gg y_1 \gg \cdots \gg y_n, \quad & x \to 1 \, . \end{aligned} \tag{7}$$

The fact that for $x \to 1$ the strong ordered phase space gives the dominant contribution is well known [5]. The fact that this gives also the dominant contribution for $x \to 0$ is due to the infrared regularities of the inclusive distributions and then of the integrand in (5). As a result the phase space region in which the real or virtual gluon k_n is softer than x ($y_n < x$) gives a contribution to the structure function of order x which can be neglected for $x \to 0$. Therefore in this case x plays the role of an infrared cutoff.

In the regions (7) we have that in each three gluon vertex, one of the gluons is much softer than the other two. We can then use the soft gluon techniques [5,12] in which one approximates the three gluon vertex by taking only the contribution in which the Lorentz index is conserved between the two hard gluons (helicity conservation of soft gluon emission).

2.2) *Soft gluon factorization*

In order to compute $f_n^{(R)}$ we follow an iterative procedure: the Feynman graphs for $f_n^{(R)}$ are obtained from the graphs of $f_{n-1}^{(R')}$ by inserting a softer real or virtual gluon k_n in all possible ways. We have the following cases.

a) Real emission contribution for $x \to 1$. In this case all external and internal lines are harder than k_n and, as well known, the emission of k_n factorizes and is given by the eikonal current, without any collinear approximation. We have the recurrent relation

$$\alpha_S \sum_R Tr\left\{ \mathbf{f}_n^{(R)}(k_1 \dots k_n)\right\} \delta_3\left(p - q - Q^{(R)}\right) \simeq$$
$$\simeq -g_S^2 \sum_{R'} Tr\left\{ \mathbf{f}_{n-1}^{(R')}(k_1 \dots k_{n-1}) \left[\mathbf{J}_{eik}^{(R')}(k_n)\right]^2\right\} \delta_3\left(p - q - Q^{(R')} - k_n\right) , \tag{8}$$

where $R \equiv (p_1 \dots p_r k_n)$, $R' \equiv (p_1 \dots p_r)$ and the eikonal current is given by

$$J_{eik}^{(R)} = -\frac{p}{p k_n}\mathbf{T}_p + \sum_1^r \frac{p_i}{p_i k_n}\mathbf{T}_{p_i} + \frac{p'}{p' k_n}\mathbf{T}_{p'} , \tag{9}$$

with $\mathbf{T}_p$ and $\mathbf{T}_{p_i}$ the colour matrix charges of incoming and outgoing gluons. Because of charge conservation we have $\mathbf{T}_{p'} = \mathbf{T}_p - \sum_1^r \mathbf{T}_{p_i}$.

b) Real emission contribution for $x \to 0$. In this case the emitted gluon k_n is softer than all lines in the graphs but the last internal gluon Q_r. In Ref. [9] we showed that also in this case the emission of k_n factorizes. One has two different classes of emission: k_n emitted from all external and internal lines but Q_r; and k_n emitted from Q_r. The first class gives rise to the same eikonal current contribution in eq. (9). The second class gives a non eikonal contribution. Even in this case the emission factorizes in terms of the following non-eikonal current

$$J_{ne}^{\mu_n}(k_n, Q_r) = -\frac{2(Q_r - xp)_\mu}{Q_r^2}\left(g^{\mu\mu_n} - k_n^\mu \frac{p'^{\mu_n}}{p' k_n}\right)\mathbf{T}_{p'} . \tag{10}$$

The corresponding recurrence relation is given by

$$\alpha_S \sum_R \delta_3\left(p - q - Q^{(R)}\right) Tr\left\{\mathbf{f}_n^{(R)}(k_1 \dots k_n)\right\} \simeq g_S^2 \frac{-Q_r^2}{(Q_r + k_n)^2}\delta_3\left(p - q - Q^{(R')} - k_n\right)$$
$$\sum_{R'} Tr\left\{\mathbf{f}_{n-1}^{(R')}(k_1 \dots k_{n-1})\left[\mathbf{J}_{eik}^{(R')}(k_n) + \mathbf{J}_{ne}(k_n, Q_r)\right]^2\right\} . \tag{11}$$

The first factor is the ratio of propagators with and without the emission of k_n. This factor was not present in (8) since it was already taken into account within the eikonal current itself (see Ref. [10]).

It is an important consequence of coherence that (11) is equivalent to (8). Without any collinear approximation, one shows [10] that in the region (7)

$$\frac{Q_r^2}{(Q_r + k_n)^2}\left[\mathbf{J}_{eik}^{(R')}(k_n) + \mathbf{J}_{ne}^{(R')}(k_n, Q_r)\right]^2 \simeq \left[\mathbf{J}_{eik}^{(R')}(k_n)\right]^2 . \tag{12}$$

The presence of the non-eikonal current and the ratio of the internal propagators compensate to give simply the square of the eikonal current.

c) Virtual emission contribution for $x \to 1$ and $x \to 0$. Taking on-shell the softest virtual gluon k_n, we can use the factorization theorems of previous cases and obtain

$$\alpha_S \sum_R Tr\left\{\mathbf{f}_n^{(R)}(k_1 \dots k_n)\right\} \delta_3\left(p - q - Q^{(R)}\right) \simeq g_S^2 \sum_{R'} \delta_3\left(p - q - Q^{(R')}\right)$$
$$Tr\left\{\mathbf{f}_{n-1}^{(R')}(k_1 \dots k_{n-1})\left[\left(\mathbf{J}_{eik}^{(R')}(k_n)\right)^2 + 2\theta(y_n - x)\mathbf{J}_{eik}^{(R')}(k_n)\mathbf{J}_{ne}(k_n, Q_r + k_n)\right]\right\} , \tag{13}$$

with $R = R' \equiv (p_1 \ldots p_r)$. Notice that in the case of $x \to 1$ the non-eikonal emission is absent and one recovers the usual result [5]. For $x \to 0$ one has an additional non-eikonal contribution from the softest internal line Q_r.

The virtual non-eikonal contribution has a simple structure in the colour space. Without any collinear approximation in the region (7) and for finite Q_t one shows [10]

$$g_s^2 \int (dk_n) 2\mathbf{J}_{eik}^{(R')}(k_n)\mathbf{J}_{ne}(k_n, Q_r + k_n) \simeq \bar{\alpha}_S \int \frac{dy_n}{y_n} \frac{d^2 k_{nt}}{\pi k_{nt}^2} \Theta\left(k_{nt} - Q_{rt}\right) \qquad \bar{\alpha}_S \equiv \frac{\alpha_S C_A}{\pi} , \tag{14}$$

with transverse momenta taken with respect to the incoming parton. This result is obtained observing that for finite Q_t there are no collinear singularities. Therefore the system of external gluons is seen as a single parton of total charge $\mathbf{T}_{p'}$ and the colour algebra gives simply $(\mathbf{T}_{p'})^2 = C_A$. For vanishing Q_t the jet is resolved and the collinear singularity $\xi_n \to 0$ is screened by the jet angular spreading.

2.3) *Recurrence relation*

Even with the simplifications in (12) and (14) the recurrence relations (8), (11) and (13) have a complex colour algebra due to the presence of the various different terms in $\mathbf{J}_{eik}^{(R)}(k_n)$. However we can take advantage of large cancellations between real and virtual contributions. This is evident observing that the eikonal current $\mathbf{J}_{eik}^{(R)}$ in the real and virtual terms are the same, but have opposite signs and different phase space δ-functions. Taking together the eikonal contributions for the real emission and virtual correction one can approximate [10]

$$-g_S^2 \int (dk_n) \left[\mathbf{J}_{eik}^{(R)}(k_n)\right]^2 \left\{\delta_3(p - q - Q^{(R)} - k_n) - \delta_3(p - q - Q^{(R)})\right\} \simeq$$
$$\bar{\alpha}_S \int \frac{dy_n}{y_n} \frac{d\xi_n}{\xi_n} \left\{\delta_3(p - q - Q^{(R)} - k_n) - \delta_3(p - q - Q^{(R)})\right\} , \quad k_{tn} = y_n\sqrt{s\xi_n} . \tag{15}$$

The introduction of the angular variable here makes explicit the effects of coherence. For the case $x \to 1$, as well known this approximation is correct at least to double logarithmic level. For the case $x \to 0$ we have shown that, at the inclusive level, this is accurate to all-loops. Again this is due to coherence in the phase space (7) and to real-virtual cancellation. The jet of external gluons in R are seen by the softest gluon k_n as a single parton of total charge $\mathbf{T}_{p'}$. The colour algebra is then simply done ($\mathbf{T}_{p'}^2 = C_A$) and we obtain the result (15).

Finally using (14) and (15) the emission factors become diagonal in colour and we can trivially perform the colour algebra by introducing the functions $f_n^R \equiv Tr\{\mathbf{f}_n^R\}$. Therefore, in the strongly y-ordered region (7) one obtains the following general recurrence relation

$$\alpha_S \sum_R f_n^R(k_1 \ldots k_n) (dk_n) \delta_3\left(p - q - Q^{(R)}\right) \simeq \bar{\alpha}_S \sum_R f_{n-1}^{(R)}(k_1 \ldots k_{n-1}) \frac{dy_n}{y_n} \frac{d\xi_n}{\xi_n}$$
$$\left\{\delta_3\left(p - q - Q^{(R)} - k_n\right) - \delta_3\left(p - q - Q^{(R)}\right) + \delta_3\left(p - q - Q^{(R)}\right) \theta(y_n - x)\theta\left(k_{nt} - Q_t^{(R)}\right)\right\} . \tag{16}$$

The three δ-functions in the square bracket have the following origin: the first corresponds to the real emission (8) or (11); the second and third correspond to the eikonal and non-eikonal virtual correction (13) respectively. The third term can be neglected for $x \to 1$ and one recovers the Altarelli–Parisi equation. For $x \to 0$ this term is essential and one obtains the Lipatov equation. We show this explicitly in the next section. Let us recall again that this inclusive recurrence relation for soft emission is accurate for $x \to 0$ to all collinear orders, while for $x \to 1$ to double logarithms.

3. Iterative equation for the structure function

In this Section we deduce from (16) the structure function in both regions of large and small x. Moreover we describe the corresponding inclusive initial state branching [8,9] for all values of x which is used in a recent Monte Carlo simulation [2,11], and a general iterative equation for the structure function.

3.1 Evolution equation for large x

As a first application of the recurrence relation (16) we now deduce the well known result [5] that the structure function for large x satisfies the Altarelli–Parisi equation with an angular evolution variable. In eq. (16) we neglect the last δ-function with the factor $\theta(y_n - x)$ which corresponds to the non-eikonal virtual corrections. Observe that, without the non-eikonal contribution, the kernel of (16) gives a result symmetric with respect to $\{y_i\}$ and $\{\xi_i\}$. Therefore we exchange y- with ξ-ordering and introduce the variables z_i

$$y_i = (1 - z_i)x_{i-1}, \quad x_i = z_i x_{i-1}, \quad x = \prod_{i \in R} z_i, \tag{17}$$

where now the range of z_i is between 0 and 1. The y_n integration in (16) is then

$$\frac{dy_n}{y_n}\left[\frac{1}{z_n}\delta_3(p - Q^{(R)} - k_n) - \delta_3(p - Q^{(R)})\right] = dz_n\left\{\frac{1}{(1-z)_+} + \frac{1}{z_n}\right\}\delta_3(p - Q^{(R)} - k_n), \tag{18}$$

where $k_n = 0$ for $z_n = 0$. The $1/z_n$ factor in the real emission δ-function of the right hand side is due to the $1/x$ factor in the definition of the structure function in (2). We recognize in (18) the gluon splitting function apart from the finite terms $-2 + z_n(1 - z_n)$ which are not obtained in this approach. Notice that neglecting the non-eikonal contributions the recurrence relation can be integrated over Q_t and one obtains directly the kernel of the Altarelli–Parisi equation for the structure function $F(x, Q)$ with ξ as evolution variable.

3.2 Lipatov equation for small x.

For $x \to 0$ the non-eikonal virtual corrections become relevant. Due to the factor $\theta(k_n - Q_t^{(R)})$ in the last δ-function in (16), we cannot integrate the recurrence relation over the total transverse momentum Q_t, thus one needs to work with the Q_t unintegrated structure function. As discussed in [10], for $x \to 0$ the phase space can be approximated as follows

$$\Theta_{12...n}^Y \delta(1 - x - Y^R) \;\to\; \Theta_{12...n}^Y \delta(1 - y_1)\theta(y_n - x). \tag{19}$$

This is due to the fact that for the leading contribution x plays the role of an infrared cutoff (see Ref. [10] for details). By using this approximation for $\delta(1 - x - Y^R)$ we find that (16) can be written in the form

$$\begin{aligned}
\alpha_S \int_x^{y_{n-1}} (dk_n) \sum_R f_n^R(k_1 \ldots k_n)\delta^2\left(\mathbf{Q}_t - \mathbf{Q}_t^{(R)}\right) &\simeq \bar{\alpha}_S \sum_R f_{n-1}^{(R)}(k_1 \ldots k_{n-1}) \\
&\times \int_x^{y_{n-1}} \frac{dy_n}{y_n} \frac{d^2 k_{nt}}{\pi k_{nt}^2}\left[\delta^2\left(\mathbf{Q}_t - \mathbf{Q}_t^{(R)} - \mathbf{k}_{nt}\right) - \delta^2\left(\mathbf{Q}_t - \mathbf{Q}_t^{(R)}\right)\Theta\left(Q_t^{(R)} - k_{nt}\right)\right].
\end{aligned} \tag{20}$$

From this recurrence relation and the definition of $\bar{F}(x, Q_t^2)$ in (5) we obtain the integral equation for the N-moment of the Q_t-structure function

$$\bar{F}_N(Q_t^2) = \frac{\bar{\alpha}_S}{N - 1}\left\{\int_0^\infty \frac{d^2 Q_t'}{\pi\left(\mathbf{Q}_t' - \mathbf{Q}_t\right)^2}\left[\bar{F}_N(Q_t'^2) - \bar{F}_N(Q_t^2)\right] + \int_{Q_t^2}^\infty \frac{dk_t^2}{k_t^2}\bar{F}_N(Q_t^2)\right\}, \tag{21}$$

where we have introduced $\mathbf{Q}'_t = \mathbf{Q}_t - \mathbf{k}_t$ with $\mathbf{k}_t$ the soft gluon transverse momentum. This is equivalent to the Lipatov equation.

3.3 Initial state branching for large and small x.

In order to obtain the inclusive initial state branching we need to evaluate the contribution to the structure function with a fixed number of emitted gluons and with any number of virtual loops. Of course each contribution is infrared divergent and we need a cutoff Q_0 in the emitted gluon transverse momenta. This distribution can be directly obtained by solving the recurrence relation (16). One finds that the Q_t structure function is given by

$$\tilde{F}(x, \mathbf{Q}_t) =$$
$$\sum_r \frac{(\bar{\alpha}_S)^r}{x} \int_0^1 \prod_1^r \frac{dy_i}{y_i} \int_{\xi_0}^{\bar{\xi}} \prod_1^r \frac{d\xi_i}{\xi_i} \, \delta(1 - x - \sum y_i) \, \delta^2(\mathbf{Q}_t - \mathbf{Q}_{tr}) \, \Theta^Y_{12\ldots r} \, V_r(p_1 p_2 \ldots p_r) , \tag{22}$$

where ξ_0 and $\bar{\xi}$ are the minimum and maximum allowed angles (or rapidities), and the transverse momenta p_{ti} are related to ξ_i by the usual relation in (1) $p_{ti} = y_i \sqrt{s\xi_i}$. The function $V_r(p_1 p_2 \ldots p_r)$ sums all the virtual corrections and is given by

$$V_r(p_1 p_2 \ldots p_r) = \exp\left\{ -\bar{\alpha}_S \int_0^1 \frac{dy}{y} \int_{\xi_0}^{\bar{\xi}} \frac{d\xi}{\xi} \right\} \prod_{i=1}^r \exp\left\{ \bar{\alpha}_S \int_{x_i}^{x_{i-1}} \frac{dy}{y} \int_{\xi_0}^{\bar{\xi}} \frac{d\xi}{\xi} \, \theta\left(q - Q_{ti} \right) \right\}$$

$$\mathbf{Q}_{ti} = \sum_{j=1}^i \mathbf{p}_{tj} , \qquad x_i = 1 - y_1 - \ldots - y_i , \qquad q = y\sqrt{s\xi} . \tag{23}$$

The first exponential is the eikonal form factor which sums all virtual corrections corresponding to the second δ-function in (16). The $y_i \to 0$ soft singularity in the real emission (22) is cancelled by the $y \to 0$ singularity in this form factor. The contribution of the third δ-function in (16), corresponding to the non-eikonal virtual corrections, are summed by the other r non-eikonal form factors. Notice that only in the case of a soft exchange ($x_i \ll x_{i-1}$) the corresponding non-eikonal form factor gives a singular contribution.

In order to exploit the coherence properties of these distributions and to find the relation with the equations of previous subsections, it is convenient to exchange y- with ξ-ordering. We proceed as in Ref. [9]. Taking into account that the non-eikonal form factors are singular only for soft exchanges and that fast emissions can be ordered both in angle and energy, we can make in all regions (7) the replacement (see Ref. [9] for details)

$$\Theta^Y_{12\ldots r} V_r(p_1 p_2 \ldots p_r) \quad \to \quad \Theta^\xi_{r \ldots 21} V_r(p_1 p_2 \ldots p_r) , \tag{24}$$

so that the gluons $\{p_1 \ldots p_r\}$ are now emitted in the angular ordered phase space

$$\bar{\xi} > \xi_r > \ldots > \xi_1 > \xi_0 . \tag{25}$$

Since the y-variables are not any more ordered we use (17) so that the y-integration factors in (22) can be written

$$\frac{(\bar{\alpha}_S)^r}{x} \prod_1^r \frac{dy_i}{y_i} \, \delta(1 - x - \sum y_i) = \prod_1^r dz_i \, \bar{\alpha}_S \left\{ \frac{1}{1 - z_i} + \frac{1}{z_i} \right\} \delta(x - z_1 \ldots z_r) , \tag{26}$$

and we recover the gluon splitting function apart from the finite contribution $-2 + z_i(1 - z_i)$. Following Ref. [9] we introduce the rescaled transverse momenta

$$q_i \equiv x_{i-1}\sqrt{s\xi_i} = p_{ti}/(1 - z_i) , \quad \bar{Q} \equiv x\sqrt{s\bar{\xi}} , \tag{27}$$

so that the angular ordered region (25) becomes

$$\left\{\bar{\xi} > \xi_r \,,\ \xi_i > \xi_{i-1} \,,\ \xi_1 > \xi_0\right\} \quad \rightarrow \quad \left\{\bar{Q} > z_r q_r \,,\ q_i > z_{i-1} q_{i-1} \,,\ q_1 > z_0 q_0 \equiv Q_s\right\} \qquad (28)$$

with Q_s the collinear cutoff.

Next, from the first exponential in (23), we extract the Sudakov form factors which regularize all the soft singularities for $z_i \rightarrow 1$ and normalize the distribution (26). Following [9] we can write (23) in the form involving the variable in (27)

$$V_r(p_1 p_2 \ldots p_r) = \Delta_s(\bar{Q}, z_r q_r) \prod_1^r \Delta_s(q_i, z_{i-1} q_{i-1}) \Delta_{ns}(z_i, q_i, Q_{ti}) \,, \qquad (29)$$

where the Sudakov form factor Δ_s is given by

$$\Delta_s(q_i, z_{i-1} q_{i-1}) = \exp\left[-\int_{(z_{i-1} q_{i-1})^2}^{(q_i)^2} \frac{dq^2}{q^2} \int_0^{1-Q_0/q} dz \frac{\bar{\alpha}_S}{1-z}\right] \,. \qquad (30)$$

Note that the q integration region corresponds to the angular ordering constraint (28). We have introduced the infrared cutoff $(1 - z) > Q_0/q$ which corresponds to a minimum transverse momentum given by Q_0. In order to obtain an infrared finite result we assume the same cutoff in the real emission distribution in (26).

The non-Sudakov form factor is given by

$$\Delta_{ns}(z_i, q_i, Q_{ti}) = \exp\left\{-\bar{\alpha}_S \int_{z_i}^1 \frac{dz}{z} \int_{(z q_i)^2}^{Q_{ti}^2} \frac{dq^2}{q^2}\right\} \,. \qquad (31)$$

The upper limit Q_{it} in the q integration comes from the non-eikonal form factor in (23), while the lower limit $z q_i$ corresponds to the angular ordering $\xi > \xi_i$ ($q = z\, x_{i-1} \sqrt{s\,\xi} > z q_i$) and comes from a part of the eikonal form factor not included into Δ_s. For very small and large values of Q_t this expression needs corrections. Recall that for $Q_{ti} \rightarrow 0$ the collinear singularities in the non-eikonal virtual contribution (14) is screened by the jet angular spreading. This implies that $\Delta_{ns} \rightarrow 1$ for $Q_{ti} \rightarrow 0$. Moreover the kinematical boundary $Q_{ti} < z x_{i-1} \sqrt{s\xi_i}$, corresponding to $\xi < \bar{\xi}$, implies that for very large Q_{ti} one has $\Delta_{ns} \rightarrow 1$

By using the z-distribution in (26) and the form factors in (29) one obtains [9] for the inclusive initial state emission the branching distribution (see also [8])

$$\begin{aligned}
dP_i &= \tilde{P}_g^i \, dz_i \, \Delta_s(q_i, z_{i-1} q_{i-1}) \frac{dq_i^2}{q_i^2} \, \Theta\left(q_i - z_{i-1} q_{i-1}\right) \Theta\left(1 - z_i - Q_0/q_i\right) \,, \\
\tilde{P}_g^i &= \tilde{P}_g z_i, q_i, Q_{ti}) = \left[\frac{\bar{\alpha}_S}{1 - z_i} + \frac{\bar{\alpha}_S}{z_i} \Delta_{ns}(z_i, q_i, Q_{ti})\right] \,.
\end{aligned} \qquad (32)$$

We have included the non-Sudakov form factors Δ_s only in the $z_i \rightarrow 0$ singular contribution of the gluon splitting function. This is because for finite z_i the function Δ_s is regular and gives a non leading correction. The appropriate hard scale for the argument of α_S is $(1 - z_i)|Q_i^2|$ (see Ref. [13]), which tends to Q_{ti}^2 as $z_i \rightarrow 0$ and to q_{ti}^2 as $z_i \rightarrow 1$.

It is easy to include in $\tilde{P}_g^i$ the finite terms $-2 + z_i(1 - z_i)$ which are relevant in the region of x not large or small. The branching distribution (32) is the basis for the new Monte Carlo program which simultaneously takes into account coherence for large x (to double logarithms) and small x (to all loops). For the description of this Monte Carlo see the talk by B. Webber [2].

We conclude by observing that from the factorized branching structure in (32) we can deduce an iterative equation for the Q_t structure function. To take into account the ξ-ordering we need to introduce explicitly into the Q_t-structure function the dependence on the variable $\bar{Q}$ of the maximum angle defined in (27). From (32) we obtain the following integral equation valid for both large and small x

$$\tilde{F}(x, \mathbf{Q}_t, \bar{Q}) = \delta(1 - x)\delta(\mathbf{Q}_t)\Delta_s(\bar{Q}, Q_s) +$$
$$\int \Delta_s(\bar{Q}, zq)\theta(\bar{Q} - zq)\frac{dz}{z}\frac{d^2q}{q^2}\theta(1 - z - \frac{Q_0}{q})\tilde{P}_g(z, q, Q_t)\tilde{F}(\frac{x}{z}, \mathbf{Q}_t - (1 - z)\mathbf{q}, q), \tag{33}$$

where the distribution $\tilde{P}_g$ is given in (32). For finite x one can neglect the non-Sudakov form factor Δ_{ns} so that the distribution $\tilde{P}_g$ becomes the usual gluon splitting function. In this case (33) is the integral equation version of the Altarelli–Parisi equation, as one can simply check by differentiating with respect to $\bar{Q}$ and using (30).

For small x we can neglect in P_g the $1/(1 - z)$ contribution and (33) becomes the Lipatov equation. This result is directly obtained by showing that for small x the Q_t structure function becomes independent of $\bar{Q}$. For $Q_t < \bar{Q}$ this is a consequence of the fact that $\Delta_{ns} \to 1$ for vanishing Q_t. The analysis of the other region is more complex. One has to take into account the kinematical boundary $\xi < \bar{\xi}$ in the non-Sudakov form factor which implies that $\Delta_{ns} \to 1$ for very large Q_t. As a result the Q_t dependence for $\bar{Q} < Q_t$ turns out to be non leading for small x [9]. The explicit form of this dependence will be described elsewhere.

Acknowledgements

Most of the results here described have been obtained together with S. Catani, F. Fiorani and G. Oriani. I am most grateful for valuable discussions with M. Ciafaloni, Yu.L. Dokshitzer, E.M. Levin, A.H. Mueller, M.G. Ryskin and B.R. Webber.

References

1. M. Ciafaloni, talk at this meeting.

2. B.R. Webber, talk at this meeting.

3. S. Catani, M. Ciafaloni and F. Hautmann, Phys. Lett. 242B (1990) 97; Nucl. Phys. B366 (1991) 135;
   J.C. Collins and R.K. Ellis, Nucl. Phys. B360 (1991) 3;
   E.M. Levin, M.G. Ryskin, Yu.M. Shabelski and A.G. Shuraev, DESY preprints 91-054, 91-065.

4. V.N. Gribov and L.N. Lipatov, Yad. Fiz. 15 (1972) 781,1218 [Sov. J. Nucl. Phys. 15 (1972) 78];
   G. Altarelli and G. Parisi, Nucl. Phys. B126 (1977) 298;
   Yu.L. Dokshitzer, Sov. Phys. JETP 73 (1977) 1216.

5. A. Bassetto, M. Ciafaloni and G. Marchesini, Phys. Rep. 100 (1983) 201;
   Yu.L. Dokshitzer, V.A. Khoze, S.I. Troyan and A.H. Mueller, Rev. Mod. Phys. 60 (1988) 373.

6. L.N. Lipatov, Yad. Fiz. 23 (1976) 642 [Sov. J. Phys. 23 (1976) 338]; E.A. Kuraev, L.N. Lipatov and V.S. Fadin, Zh. Eksp. Teor. Fiz. 72 (1977) 373 [Sov. Phys. JETP 45 (1977) 199]; Ya. Balitskii and L.N. Lipatov, Yad. Fiz. 28 (1978) 1597 [Sov. J. Nucl. Phys. 28 (1978) 822].

7. L.V. Gribov, E.M. Levin and M.G. Ryskin, Phys. Rep. 100 (1983) 1.

8. M. Ciafaloni, Nucl. Phys. B296 (1987) 249.

9. S. Catani, F. Fiorani and G. Marchesini, Phys. Lett. 234B (1990) 339; S. Catani, F. Fiorani and G. Marchesini, Nucl. Phys. B336 (1990) 18.

10. S. Catani, F. Fiorani, G. Marchesini and G. Oriani, Nucl. Phys. B361 (1991) 645.

11. G. Marchesini and B.R. Webber, Phys. Lett. 349B (1991) 617.

12. S. Catani and M. Ciafaloni, Nucl. Phys. B249 (1985) 301; S. Catani, M. Ciafaloni and G. Marchesini, Nucl. Phys. B264 (1986) 558.

13. D. Amati, A. Bassetto, M. Ciafaloni and G. Veneziano, Nucl. Phys. B173 (1980) 429.

QCD RESULTS FROM OPAL AT LEP

Siegfried Bethke

Physikalisches Institut
Universität Heidelberg
D-6900 Heidelberg, Germany

ABSTRACT

Recent results on jet physics and on tests of perturbative QCD in hadronic final states of e^+e^- annihilations around the Z^0 pole are presented. Various jet algorithms are discussed in detail. Studies of 3- and 4-jet final states provide significant tests of the abelian nature of QCD, in particular for the gluon self coupling and for the running coupling constant α_s (asypmptotic freedom). The coupling α_s is determined, from studies of event topologies, jet production rates and from Z^0 line shapes and decay asymmetries, to be $\alpha_s(M_{Z^0}) = 0.119 \pm 0.008$. Studies of soft hadron and gluon coherence phenomena are shortly reviewed.

1. INTRODUCTION

Hadronic final states of high energy e^+e^- annihilations have proven to be a significant testing ground for Quantum Chromodynamics (QCD) [1], the nonabelian gauge theory of the strong interactions between quarks and gluons [2,3,4]. The exact knowledge of the quantum numbers and the energy of the initial state, the absence of underlying events due to target remnants, the clean signature of hadronic final states and the clear jet structures of such events, which are interpreted as footprints of the underlying quarks and gluons, are outstanding advantages of e^+e^- annihilation reactions, both in view of QCD studies and for precision tests of the standard model of electroweak interactions [5].

LEP, the large e^+e^- collider with the four experiments ALEPH, DELPHI, L3 and OPAL [6,7,8,9] at the European Centre for High Energy Physics (CERN) in Geneva, operates at centre of mass energies around the resonance of the Z^0 gauge boson and thus provides a significant increase of both the available centre of mass energy and event statistics. This is illustrated in Fig. 1, where the cross sections of the reactions $e^+e^- \rightarrow$ hadrons, $e^+e^- \rightarrow \mu^+\mu^-$ and $e^+e^- \rightarrow \gamma\gamma$, as measured by OPAL [10] at LEP and by previous experiments at lower c.m. energies (E_{cm}), are plotted as a function of E_{cm}. While at lower energies the cross sections are dominated by the $1/E_{cm}^2$ behaviour of the reaction $e^+e^- \rightarrow \gamma \rightarrow f\bar{f}$ ($f \equiv$ fermion), at $E_{cm} > 60$ GeV the process $e^+e^- \rightarrow Z^0 \rightarrow f\bar{f}$ takes over, leading to resonant cross sections at LEP which are more than 2 orders of magnitude larger than those experienced at the PETRA, PEP and TRISTAN e^+e^-

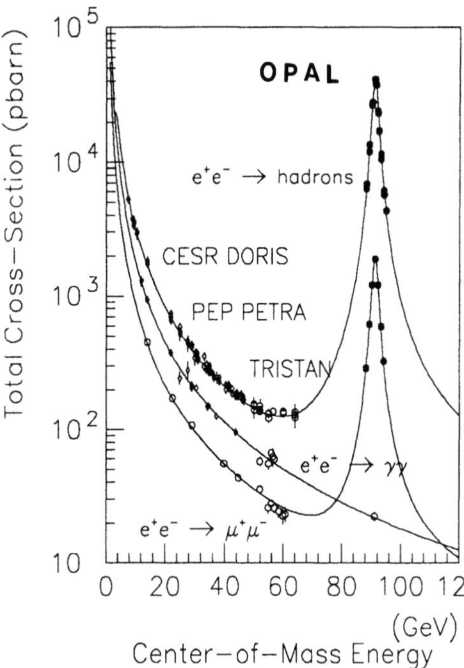

Fig. 1. Total cross sections of various processes in e^+e^- annihilation.

colliders. The large number of hadronic Z^0 decays detected so far, approaching 500.000 for OPAL alone, and the increase in centre of mass energy which leads to a decrease of (noncalculable) hadronization effects and thus to better jet identifications, make LEP the ideal laboratory for precision tests of QCD.

Studies of hadronic event shapes are summarized in section 2. Algorithms for quantifying jet production in e^+e^- annihilation and their corresponding applications in experimental analyses are presented in section 3. Among the most important results are tests of the gluon spin and of the gluon self coupling in angular correlations within 3-jet and and 4-jet final states, and the experimental evidence for asymptotic freedom through the energy dependence of 3-jet event production rates. Measurements of $\alpha_s(M_{Z^0})$ are summarized in section 4. Studies of soft hadron and gluon coherence phenomena are presented in section 5, and some aspects of nonperturbative hadron production as Intermittency and Bose-Einstein correlations are discussed in section 6.

2. Hadronic Event Shapes

Hadronic event shapes are tools to study both the amount of gluon radiation and details of the hadronization process. Since the laboratory frame in which the events are measured is (almost always) identical to the centre of mass system of the annihilation reaction, events from $q\bar{q}$ final states without any hard gluon radiation predominantly result in two collimated, back-to-back jets of hadrons. The emission of one hard, energetic gluon leads to planar 3-jet events, while the emission of two or more energetic gluons can cause nonplanar multi-jet like event structures. Additional radiation of soft, low-energetic gluons together with the hadronization process blurs this naive classification of event classes, however. Observables to classify hadronic events according to their overall jet structure are therefore defined, which are insensitive to the radiation of soft and collinear

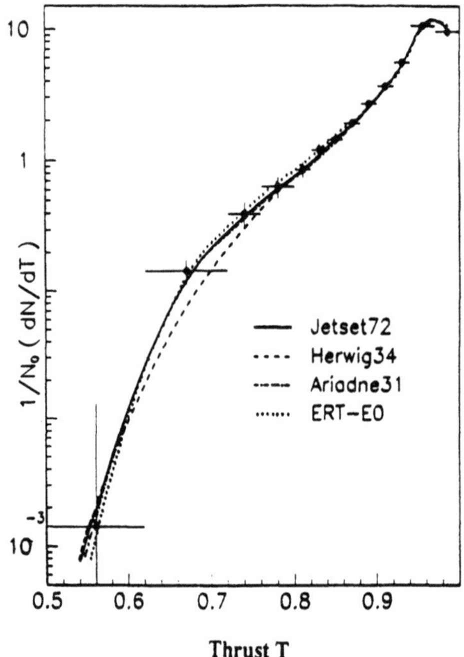
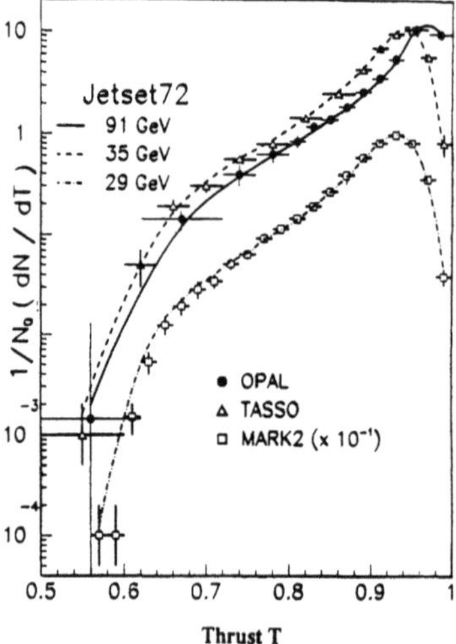

Fig. 2. Thrust distribution measured at $E_{cm} = 91$ GeV, compared to various QCD plus hadronization models tuned to describe hadronic Z^0 decays.

Fig. 3. The Thrust distributions measured at various c.m. energies, compared to model predictions with parameters tuned to the 91 GeV data.

gluons. Such observables can be calculated and predicted both by perturbative QCD [11] and by QCD plus hadronization models.

A typical event shape observable is Thrust, T, which is defined as the normalized sum of the momentum components of all particles of a given event with respect to a certain event axis; this axis is chosen such that T is maximized. Ideal 2-jet events result in $T = 1$, while for planar 3-jet events $2/3 < T < 1$ and for a completely spherical event $T = 1/2$. In Fig. 2 the Thrust distribution of hadronic Z^0 decays, measured around c.m. energies of 91 GeV [12], is compared to the predictions of several QCD plus hadronization models. The data are corrected for detector acceptance and resolution, and the effects of initial state photon radiation, which are small at the Z^0 pole, are unfolded.

Models studied in this comparison are the Jetset QCD shower plus string hadronization model [13], the Herwig QCD shower plus cluster hadronization model [14], the Jetset $O(\alpha_s^2)$ QCD model plus string hadronization [13] and the Ariadne QCD colour dipole plus string hadronization model [15]. QCD shower models are based on the development of quark and gluon cascades down to invariant parton masses of about 1 GeV, calculated in leading log approximations (LLA) of perturbative QCD. At Z^0 energies, these models result in partonic final states of typically 10 quarks and gluons. The main event charcteristics are largely determined by the parton cascade and the available phase space for hadronization is relatively small. $O(\alpha_s^2)$ QCD models are based on complete next-to-leading order calculations, which predict the production of up to 4-parton final states of massless quarks and gluons; hadronization sets in at much larger energy scales (typically 10 GeV at Z^0 energies).

For the models shown in Fig.2, the QCD scale parameter Λ and several parameters specifying the hadronization process were optimized to describe the measured event shape distributions of T_{major} and of H_2/H_0 [12]. From studies similar as presented in Fig. 2 it is evident that most of the global event shape distributions can be well described by the QCD plus hadronization models studied so far; QCD shower models are somewhat superior compared to $O(\alpha_s^2)$ QCD models, presumably because the latter cannot generate events with more than four parton jets. The overall best description of hadronic events is provided by the JETSET parton shower model [13].

QCD predicts scaling violations for observables which do not depend on absolute energies or momenta, as for example the T-distribution. These scaling violations are caused by the energy dependence of α_s, determining that the relative amount of gluon radiation, which in leading order is proportional to α_s, decreases with increasing energy. It is thus expected that fewer 3- and multijet like events are observed at higher c.m. energies, and event shape distributions should become more 2-jet like. The OPAL collaboration compared their measured event shape distributions with similar measurements done at lower c.m. energies. As seen in Fig. 3 for the T-distribution, small but significant differences between these distributions are indeed observed. In order to disentangle the signal for QCD violations from similar effects which might be caused by a possible, energy dependend impact of the hadronization process, OPAL also compared the expectation of QCD shower models with the data at 29 and 35 GeV. The model predictions, with no parameter changed but the c.m. energy, describe the various data distributions well. The energy dependence of these distributions can thus be explained by QCD scaling violations plus an energy independend parametrization of hadronization.

3. Physics with Jets

Studies of jet production provide the most intuitive tests of the underlying parton structure of hadronic events at high energies [4,16]. While hadron jets and the corresponding jet multiplicity of individual events can often be inferred from graphical displays of measured hadronic events, quantitative studies require the exact definition of *resolvable* jets within a given event. In sections 3.1 and 3.2, jet algorithms will be defined and some of their features will be discussed. Physics results from studies which involve jet finding algorithms will be presented in sections 3.3 to 3.5 and in 4.1.

3.1 Jet Algorithms and Recombination Schemes

The most commonly used algorithm to define and reconstruct jets of hadrons was introduced by the JADE collaboration [17,18]: the scaled pair mass of two resolvable jets i and j, $y_{ij} = M_{ij}^2/E_{vis}^2$, is required to exceed a threshold value y_{cut}; E_{vis} is the total visible energy of the event. In a recursive process, the pair of particles n and m (or clusters of particles) with the smallest value of y_{nm} is replaced by (or "recombined" into) a single jet (or cluster) k with four-momentum $p_k = p_n + p_m$, as long as $y_{nm} < y_{cut}$. The procedure is repeated until all pair masses y_{ij} are larger than the jet resolution parameter y_{cut}, and the remaining clusters of particles are called jets.

Several definitions of resolution criteria y_{ij} exist; in addition, since $O(\alpha_s^2)$ QCD calculations are available for massless partons only and adding two (massless) 4-vectors leads - in general - to a vector which has nonzero mass, the recombination of jets can be performed in many ways which finally ensure massless jets [11,19,16,20]. The most commonly used schemes are summarized in Table 1. The E0 and the original JADE scheme are equivalent to second order perturbation theory. While the E, the E0, p and p0 schemes were used and compared to each other in one of our previous studies ([19]; see

Table 1. Definition of resolution measures y_{ij} and of combination schemes for various jet algorithms; s is the total centre of mass energy or - in experimental implementations - the total visible energy squared, $\vec{p}_i$ denotes a 3-vector and $p_i \equiv (E_i, \vec{p}_i)$ is the corresponding 4-vector.

Algorithm	Resolution	Combination	Remarks
E	$\dfrac{(p_i+p_j)^2}{s}$	$p_k = p_i + p_j$	Lorentz invariant
JADE	$\dfrac{2E_iE_j(1-\cos\theta_{ij})}{s}$	$p_k = p_i + p_j$	conserves $\sum E, \sum \vec{p}$
E0	$\dfrac{(p_i+p_j)^2}{s}$	$E_k = E_i + E_j;$ $\vec{p}_k = \dfrac{E_k}{|\vec{p}_i+\vec{p}_j|}(\vec{p}_i + \vec{p}_j)$	conserves $\sum E$, but violates $\sum \vec{p}$
p	$\dfrac{(p_i+p_j)^2}{s}$	$\vec{p}_k = \vec{p}_i + \vec{p}_j;$ $E_k = |\vec{p}_k|$	conserves $\sum \vec{p}$, but violates $\sum E$
p0	$\dfrac{(p_i+p_j)^2}{s}$	$\vec{p}_k = \vec{p}_i + \vec{p}_j;$ $E_k = |\vec{p}_k|$	as p-scheme, but E_{vis} updated after each recomb.
D	$\dfrac{2\cdot\min(E_i^2,E_j^2)\cdot(1-\cos\theta_{ij})}{s}$	$p_k = p_i + p_j$	conserves $\sum E, \sum \vec{p};$ avoids exp. problems
G	$\dfrac{8E_iE_j(1-\cos\theta_{ij})}{9(E_i+E_j)^2}$	$p_k = p_i + p_j$	conserves $\sum E, \sum \vec{p};$ avoids exp. problems

also section 4.1 and [21]), the recently proposed "Durham" (D) [22] and "Geneva" (G) schemes [20] were not utilized in an experimental analysis so far. These schemes were invented in order to allow for resummations of leading and next-to-leading logarithms to all orders of QCD perturbation theory; see [21,23,20] for further information.

In $O(\alpha_s^2)$ QCD calculations, jet production rates defined for the algorithms listed in table 1 are described as quadratic functions of $\alpha_s(\mu)$:

$$
\begin{aligned}
R_2 &\equiv \frac{\sigma_{2-jet}}{\sigma_{tot}} = 1 + C_{2,1}(y_{cut}) \cdot \alpha_s(\mu) + C_{2,2}(y_{cut}, f) \cdot \alpha_s^2(\mu) \\
R_3 &\equiv \frac{\sigma_{3-jet}}{\sigma_{tot}} = C_{3,1}(y_{cut}) \cdot \alpha_s(\mu) + C_{3,2}(y_{cut}, f) \cdot \alpha_s^2(\mu) \\
R_4 &\equiv \frac{\sigma_{4-jet}}{\sigma_{tot}} = C_{4,2}(y_{cut}) \cdot \alpha_s^2(\mu),
\end{aligned}
\tag{1}
$$

where σ_{tot} is the total hadronic cross section, $\sigma_{n}-jet$ are the cross sections for n-parton event production, μ is the renormalization scale at which α_s is evaluated and $f = \mu^2/E_{cm}^2$ is the renormalization scale factor. The kth order QCD coefficients for n-jet production, $C_{n,k}$, depend on the jet resolution parameter y_{cut}; in addition, the next-to-leading order coefficients $C_{2,2}$ and $C_{3,2}$ are recombination scheme dependent and exhibit an explicit dependence on the renormalization scale factor f; see [11,20] for numerical results of these coefficients. The coupling constant $\alpha_s(\mu)$ can be written as a function of $ln(\mu^2/\Lambda_{\overline{MS}}^2)$ [24], where $\Lambda_{\overline{MS}}$ is the QCD scale parameter which must be determined by experiment:

$$
\alpha_s(\mu) = \frac{12\pi}{(33 - 2 \cdot N_f) \cdot ln(\frac{\mu}{\Lambda_{\overline{MS}}})^2} \cdot \left(1 - 6 \cdot \frac{153 - 19 \cdot N_f}{(33 - 2 \cdot N_f)^2} \cdot \frac{ln(ln(\frac{\mu}{\Lambda_{\overline{MS}}})^2)}{ln(\frac{\mu}{\Lambda_{\overline{MS}}})^2}\right),
\tag{2}
$$

with the number of active quark flavours N_f equal to 5.

3.2 Hadronization Corrections and Jet Resolutions

The influence of hadronization on the quality and reliability of experimental jet reconstruction will be studied for the E, E0, p, D and G jet algorithms, using the JETSET QCD shower plus string hadronization program[13] with parameters optimized to describe global event shapes of hadronic Z^0 decays [12]. The quantities under study are calculated after applying the jet algorithms to the same set of 3000 generated hadronic events. In each case, jets are reconstructed from both the final quarks and gluons at the end of the QCD shower, terminated at a cut-off of $Q_0 = 1$ GeV ("parton level"), and from the particles after hadronization ("hadron level"). For the latter case, all final state particles with lifetimes larger than $3 \cdot 10^{-10}$ s are taken into account, and no simulation of detector acceptance or resolution is applied.

The relative n-jet production rates, R_n, for $n = 2,3,4$ and greater or equal to 5 are shown in Figure 4 as a function of the jet resolution parameter $y_{\rm cut}$. It is evident that not only are the jet composition and the absolute numbers of n-jet events for given values of $y_{\rm cut}$ quite different between the algorithms, but also that the size of the hadronization correction, i.e. the difference between hadrons and partons, varies significantly. The absolute hadronization corrections are smallest in the JADE-E0 and in the new D scheme, while they are largest for the E and G algorithms. The E0 scheme was therefore the most commonly used jet algorithm in experimental QCD studies; however since the D-scheme provides advantages in theoretical calculations [21,20] it may well be the preferred choice in the future.

The magnitude of hadronization effects is further demonstrated in two-dimensional correlation plots (Fig.5), where for each hadronic event the value of y_2 at which its classification is changed from a 3-jet to a 2-jet configuration, calculated at both parton and hadron level, is plotted. Vanishing overall hadronization corrections would lead to an event population which is symmetric around the main diagonal [y_2(hadrons) = y_2(partons)], while a finite jet resolution causes a certain spread around that diagonal. The largest spread is observed for the E algorithm, which also suffers from a sizable systematic shift away from the main diagonal. The E0 and D algorithms are most symmetric with respect to the diagonal, even at very small values of y_2, while the G algorithm shows large asymmetries in the small y_2 region. The overall width of the two-dimensional distributions, perpendicular to the main diagonal and normalized by the average value of y_2, is smallest for the G and p algorithms.

3.3 Testing the Gluon Spin

The distributions of the scaled jet energies within 3-jet events, $x_i = 2E_i^{jet}/E_{cm}$ with $x_1 \geq x_2 \geq x_3$, can discriminate between the hypotheses of scalar (spin = 0) and vector (spin = 1) gluons. Another observable which is sensitive to the gluon spin is the so-called Ellis-Karliner angle θ_{EK} [25], which is the angle between the directions of jets 1 and 2, after they are boosted into the rest frame of the two least energetic jets. Previous studies of these observables already provided experimental evidence for gluons being vector particles, only shortly after gluon jets were observed for the first time [26,2]. It is both mandatory and intuitive to repeat this kind of analysis at LEP, since the larger data statistics, the higher c.m. energies and the better understanding of both the perturbative and the nonperturbative phase of hadron production provide a much increased sensitivity for precize tests of the specific features of QCD.

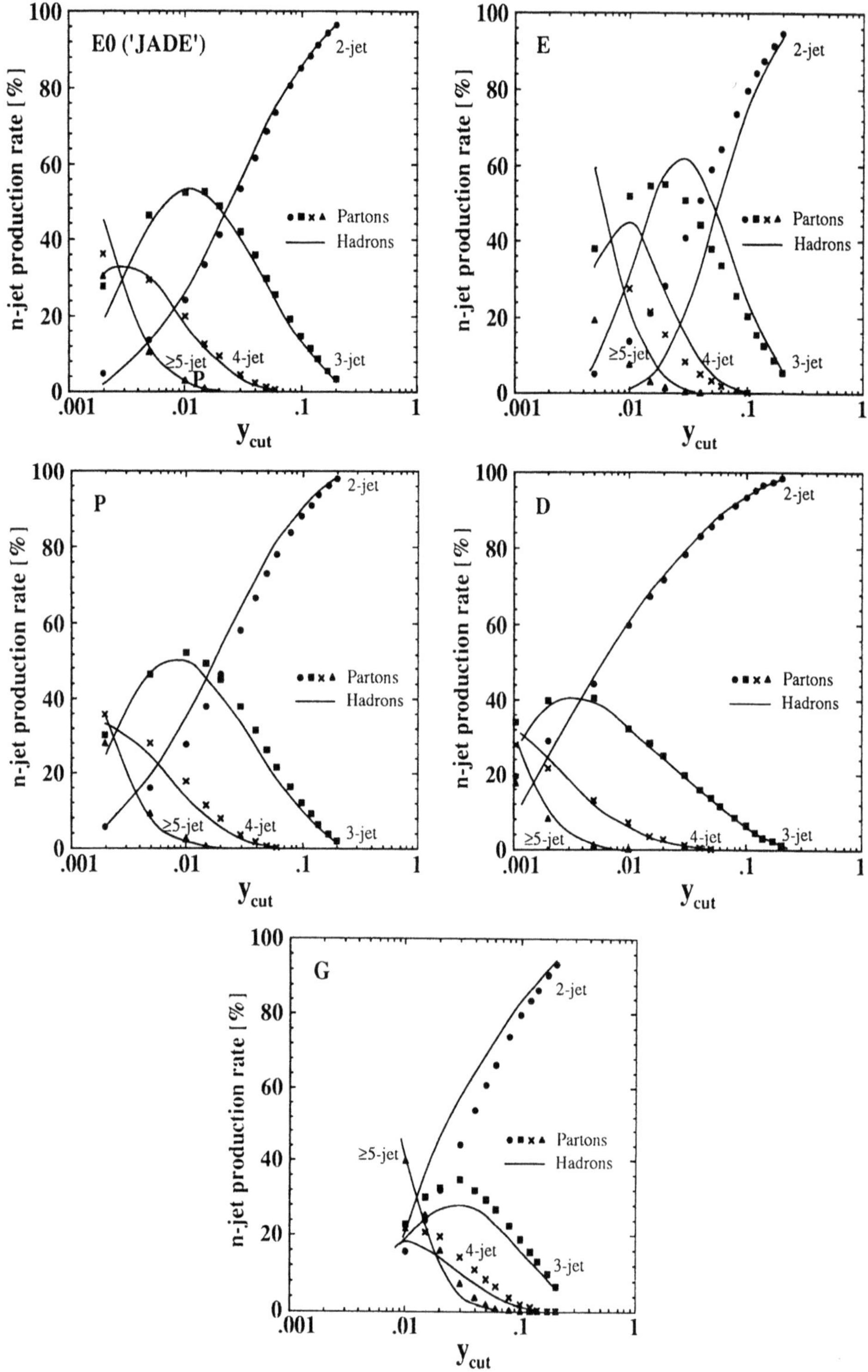

Fig. 4. Jet rates as a function of y_{cut}, determined from JETSET QCD shower model events before (partons) and after (hadrons) the hadronization process.

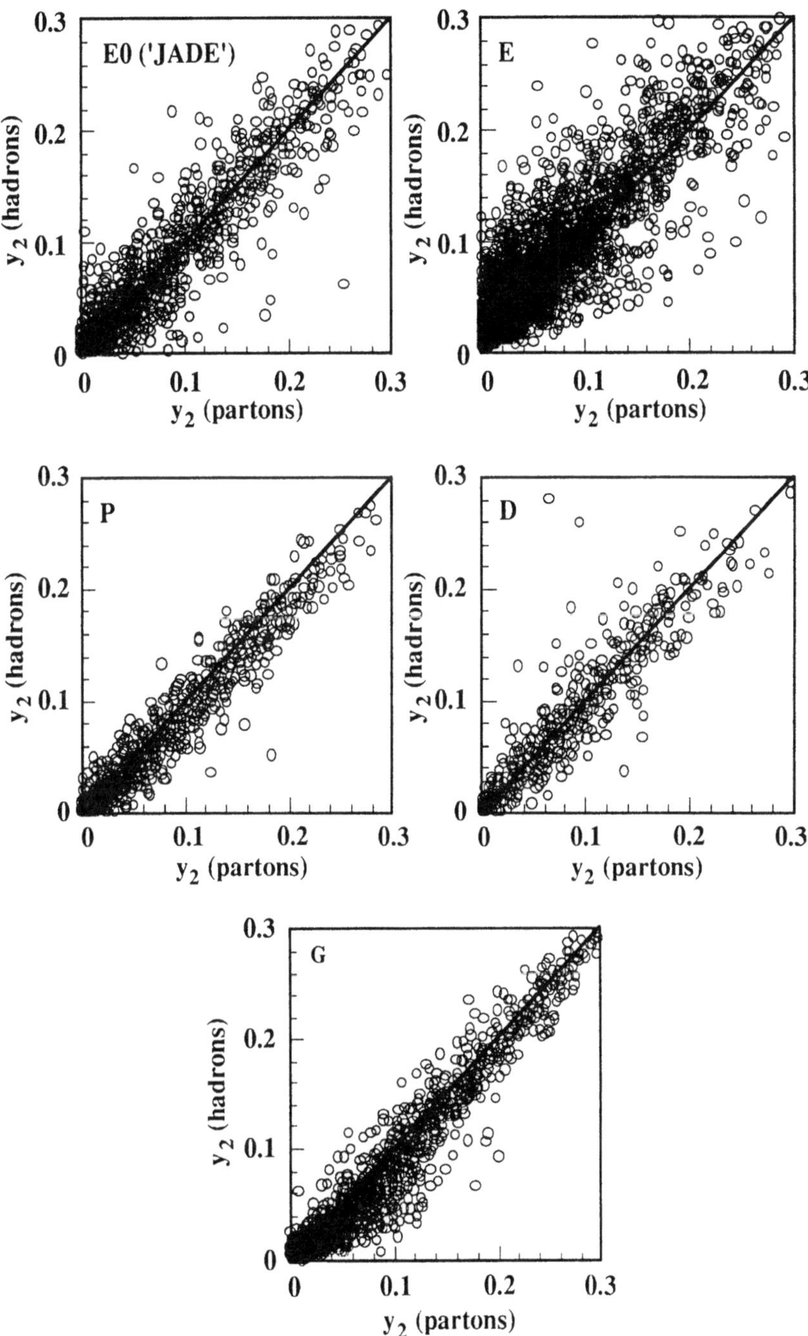

Fig. 5. Distributions of y_2 calculated for JETSET model events before and after the hadronization process.

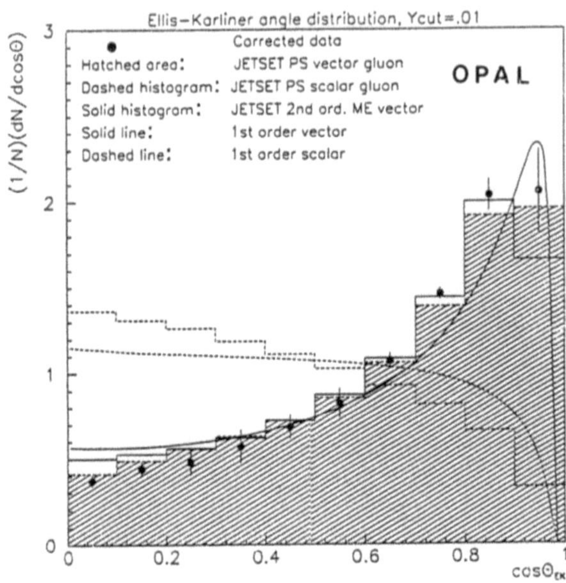

Fig. 6. Distribution of $\cos\theta_{EK}$ from reconstructed 3-jet events, compared to predictions of various models for scalar and for vector gluons.

The distribution of $\cos\theta_{EK}$ measured by OPAL [27] is shown in Fig. 6. The data are compared to various expectations for vector (full lines) and for scalar gluons (dashed lines), predicted by QCD shower models and first and second order analytic QCD calculations. The prominent pole structure for $\cos\theta_{EK} \to 1$ which is predicted by QCD but not by the scalar gluon models is clearly seen in the data. This pole structure could not be seen at lower c.m. energies, where a reliable jet identification for small x_3 was not possible [2].

3.4 Testing the Gluon Self Coupling.

The gluon self coupling is a characteristic feature of the nonabelian nature of QCD. It is predicted that the process $g \to gg$ is the dominant source of 4-jet events in e^+e^- annihilation, besides double gluon bremsstrahlung and gluon splitting into a $q\bar{q}$ pair which also result in 4-jet final states.

OPAL [28] has studied two angles χ_{BZ} [29] and θ^*_{NR} [30,31] defined by the momenta of the jet axes within 4-jet final states. These angles were demonstrated to be sensitive to the difference between QCD and the abelian model, even if quark and gluon jets cannot be identified on an event by event basis but are statistically separated by ordering the jet energies E_j, motivated by the bremsstrahlungs spectrum of gluons, according to $E_1 \geq E_2 \geq E_3 \geq E_4$ [32]. In this case, the difference between QCD and the abelian model is mainly due to the characteristic angular structure of $q\bar{q}q\bar{q}$ events rather than to the triple gluon vertex itself. In leading order (i.e. $O(\alpha_s^2)$) QCD predicts that about 4.7% of all 4-jet events defined with $y_{cut} = 0.01$ are $q\bar{q}q\bar{q}$ final states, while in the alternative abelian gauge theory, due to the absence of the gluon self coupling, this number is 31.3% [32]. Since the relative admixture of such events is definitely given by the gauge structure of the theory, a measurement of these angles is suited to provide evidence for the nonabelian structure of QCD.

The angular distributions determined from about 4000 4-jet events, defined by using the JADE (E0) algorithm and corrected for detector acceptance and for hadronization

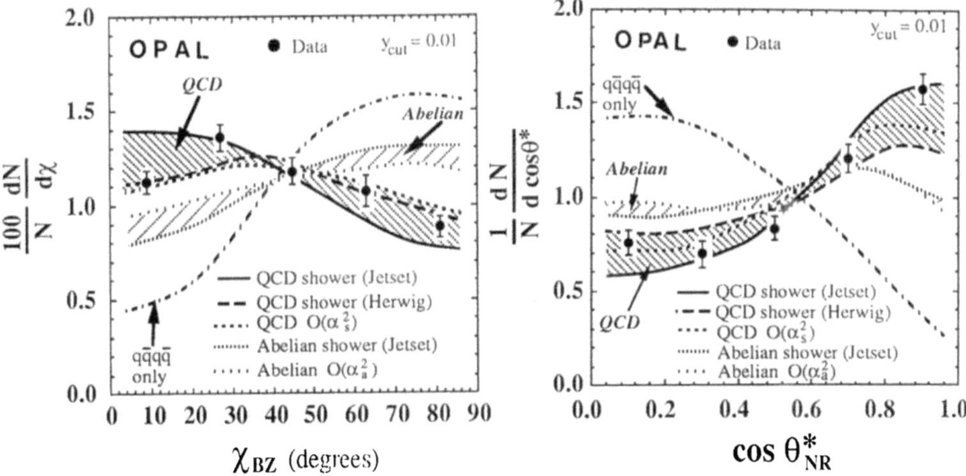

Fig. 7. Measured distributions of χ_{BZ} and $\cos\theta^*_{NR}$, together with the predictions of QCD and abelian vector gluon models.

effects, are shown in Fig. 7. They are compared to various predictions of perturbative QCD and of the abelian theory, both calculated using second order perturbation theory and QCD parton shower models which partly include higher than second order contributions. The differences between these predictions are treated as theoretical uncertainties within the QCD and abelian model predictions. The data are in good agreement with QCD and significantly rule out the abelian hypothesis [28].

3.5 Experimental Evidence for Asymptotic Freedom.

A further direct consequence of the nonabelian nature of QCD is the specific energy dependence of α_s, which is predicted, in leading order QCD, to decrease with increasing energy μ as $\alpha_s \sim 1/ln(\mu)$, c.f. Eq. 2. Direct measurements of α_s from various processes at different energies hardly provided conclusive evidence for the "running" coupling constant, due to the relatively large systematic uncertainties in determinations of α_s. Studies of the energy dependence of 3-jet event production rates, within a given recombination scheme and for a given value of y_{cut}, offer a more direct way to test the energy dependence of α_s [18,4,16].

According to Eq. 1 and for any given choice of f , the energy dependence of R_3 is expected to be entirely determined by the energy dependence of α_s. Previous studies have shown that this assumption is correct, within a hadronization uncertainty of $\pm 2\%$, for jet rates defined in the JADE (E0) scheme at $y_{cut} \geq 0.08$ for e^+e^- centre of mass energies above 25 GeV [16]. In Fig. 8, updates of the measurements of 3-jet event production rates at LEP [19,33,34,35] are compared with similar measurements at lower c.m. energies [18,36,37,38,39]. Also shown are fits of analytic $O(\alpha_s^2)$ QCD calculations [11,40] to the data at $E_{cm} \geq 29$ GeV, using two different scale factors f, of the hypothetical case with an energy independent coupling constant and of the abelian vector theory in complete $O(\alpha_A^2)$ [16]. The results of the fitted parameters and of the corresponding values of χ^2 are listed in Table 2. While an energy independent coupling constant is excluded with a significance of more than 7 standard deviations, the abelian theory with an increasing coupling strength is entirely ruled out. The QCD calculations are in good agreement with the data, almost independent of the renormalization scale chosen; however the

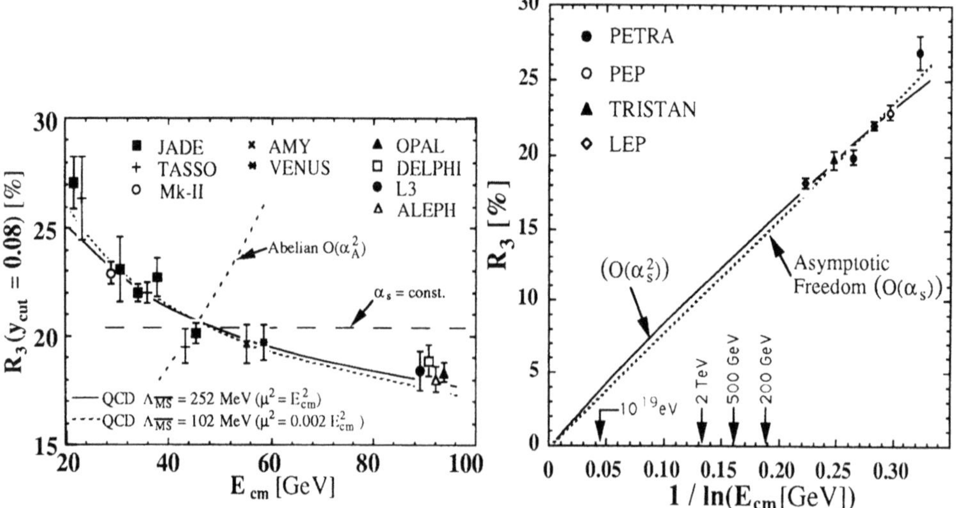

Fig. 8. Energy dependence of 3-jet event rates at $y_{cut} = 0.08$, compared with fits of $O(\alpha_s^2)$ QCD calculations, with the hypothesis of an energy independent α_s and with the abelian theory in $O(\alpha_A^2)$.

Fig. 9. The data as shown in Fig. 8, combined at similar c.m. energies, as a function of $1/ln(E_{cm})$ and compared with the prediction of asymptotic freedom.

Table 2. Fit results of various assumptions about the energy dependence of 3-jet rates, from the data between $E_{cm} = 29$ GeV and 91 GeV shown in Fig. 8. An additional relative systematic point-to-point error of 2% was included in the fits [16].

Theory	fit result	χ^2/d.o.f.
QCD, $f = 0.002$	$\Lambda_{\overline{MS}} = (102 \pm 4)$MeV	10.8 / 12
QCD, $f = 1.0$	$\Lambda_{\overline{MS}} = (252 \pm 11)$MeV	7.4 / 12
α_s = const.	$< R_3 > = 20.4 \pm 0.2$	72.3 / 12
Abelian theory	$\alpha_A(44 GeV) = 0.26$	****

actual values of $\Lambda_{\overline{MS}}$ and of α_s do strongly depend on the scale (see also section 4.1).

Another way to demonstrate that measured jet production rates are in good agreement with asymptotic freedom is to plot R_3 as a function of $1/ln(E_{cm})$, as shown in Fig. 9. The dashed line indicates the leading order QCD prediction ($R_3 \propto \alpha_s \propto \frac{1}{lnE_{cm}}$). The corresponding prediction in $O(\alpha_s^2)$ is also shown, indicating that higher order terms affect the energy dependence of R_3 only slightly. At infinite energy ($1/ln(E_{cm}) \to 0$), R_3 and α_s are expected to vanish; an assumption which is in good agreement with the data.

4. Determinations of $\alpha_s(M_{Z^0})$.

The determination of α_s, which is the only free parameter of the theory, was always one of the key analyses of hadronic final states in e^+e^- annihilation. Comprehensive summaries of α_s measurements before the era of LEP and SLC can be found e.g. in [2,3] (1987) and in [4] (1989). Measurements of $\alpha_s(M_{Z^0})$ with the OPAL detector at LEP will be summarized in the following paragraphs.

4.1 $\alpha_s(M_{Z^0})$ from Jet Production Rates.

The value of $\alpha_s(M_{Z^0})$ from jet production rates is determined in fits of the analytic

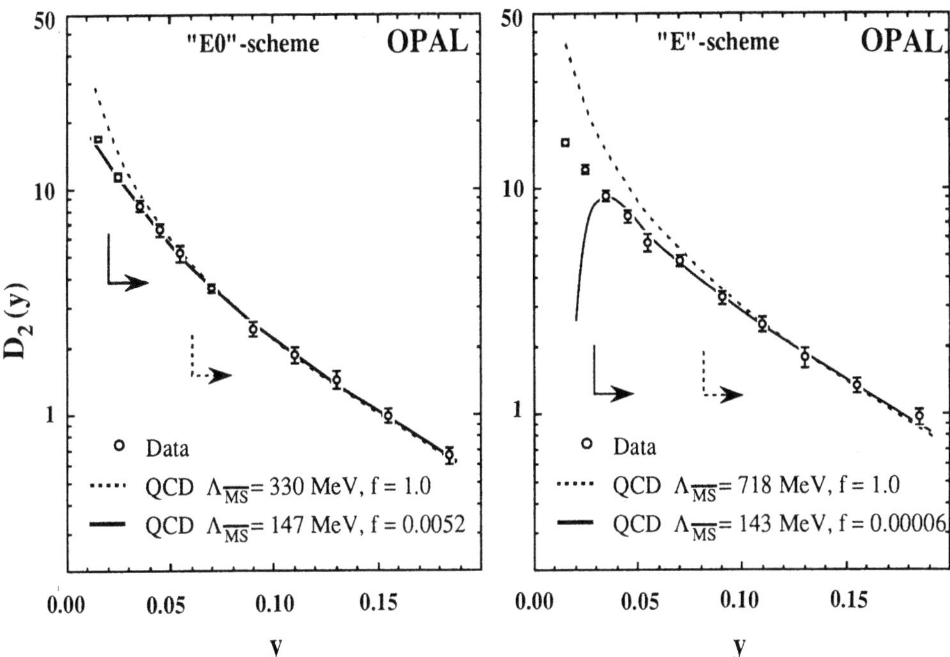

Fig. 10. Measured distributions of $D_2(y)$, corrected for detector acceptance and hadronization effects, together with the corresponding analytic $O(\alpha_s^2)$ QCD calculations.

Table 3. Final results of $\alpha_s(M_{Z^0})$ for different jet recombination schemes.

Scheme	$\alpha_s(M_{Z^0})$	$\Delta\alpha_s$ (exp)	$\Delta\alpha_s$ (had)	$\Delta\alpha_s$ (Q_0)	$\Delta\alpha_s$ (scale)	$\Delta\alpha_s$ tot.
E0	0.118	±0.003	±0.003	±0.003	±0.007	±0.009
E	0.126	±0.003	±0.003	±0.003	±0.013	±0.014
p0	0.118	±0.003	±0.003	±0.005	±0.004	±0.008
p	0.118	±0.003	±0.003	±0.006	±0.003	±0.008

$O(\alpha_s^2)$ QCD calculations to the measured differential $D_2(y)$ distributions ($y \equiv y_{cut}$),

$$D_2(y) = \frac{R_2(y) - R_2(y - \Delta y)}{\Delta y}.\tag{3}$$

D_2 measures the distribution of y_{cut} values for which the classification of events changes from 3-jet to a 2-jet classification [41].

The recombination scheme dependence of jet rates and of α_s, for the E0, E, p0 and p schemes defined in Table 1, was studied in detail [19]. For each recombination scheme, the measured jet rates were corrected for hadronization effects according to the predictions of the JETSET QCD shower model (see Fig. 4). This model provides an excellent description of the measured jet production rates in all the four schemes [19]. The QCD parameter $\Lambda_{\overline{MS}}$ was then determined from fits of the $O(\alpha_s^2)$ QCD calculations corresponding to that scheme [11]. The renormalization scale dependence was separately studied within each scheme, allowing the scale to vary between $\mu^2 = E_{cm}^2$ and the best fit result for μ. The resulting values of $\Lambda_{\overline{MS}}$ (which depend on the choice of μ) are then converted into values of $\alpha_s(M_{Z^0})$ using Eq. 2. The spread of these results due to the variation of f is taken as a systematic uncertainty, which can also be interpreted as an estimate of the influence of the unknown higher order terms if the range of scale variations considered is sufficiently large such that it is likely to cover the hypothetical result to all orders. Uncertainties of the hadronization correction are studied by using the HERWIG model to calculate these corrections, instead of JETSET, and the difference between the results is taken into account in the final error calculation.

The corrected distributions of $D_2(y)$ for the E0 and the E recombination scheme are presented in Figure 10, together with the fit results of $\Lambda_{\overline{MS}}$ for the two choices of μ mentioned above. The arrows indicate the regions of fits, which are determined by the demand that $R_5 < 1\%$ if both $\Lambda_{\overline{MS}}$ and μ are fitted (5-jet events are only predicted in higher than second order QCD), and $R_4 < 1\%$ if $\Lambda_{\overline{MS}}$ is fitted for $\mu = E_{cm}$ (with this scale, data cannot be described in regions where a sizable fraction of 4-jet events is observed). In agreement with previous observations [42,38,41] it is found that $\Lambda_{\overline{MS}}$ largely depends on the choice of μ, and that the fit values of $f = \mu^2/E_{cm}^2$ are in general small. Further conclusions are that the results for f, the overall sensitivity of $\Lambda_{\overline{MS}}$ and $\alpha_s(M_{Z^0})$ on variations of f and the dependence on the parton virtuality Q_0 to which the data are corrected, are different for each recombination scheme, as summarized in Table 3. Within the overall uncertainties, however, the values of $\alpha_s(M_{Z^0})$ from the four different recombination schemes agree with each other, resulting in an overall $\alpha_s(M_{Z^0}) = 0.118 \pm 0.008$ from studies of jet production rates.

4.2 $\alpha_s(M_{Z^0})$ from Hadronic Event Shape Distributions.

OPAL has also determined $\alpha_s(M_{Z^0})$ from a variety of event shape distributions like Thrust T, Oblateness O and the C-parameter [43], from measurements of energy-energy correlations (EEC) and their asymmetry (AEEC) [44] and from planar triple energy correlations [45]. These analyses are performed in a similar way as for $\alpha_s(M_{Z^0})$ from jet production rates described above, as suggested in [11]: the measured distributions are corrected for detector acceptance and resolution and also for hadronization effects as sudied with the JETSET QCD shower model. The QCD parameter $\Lambda_{\overline{MS}}$ is then determined in fits of the corresponding analytical $O(\alpha_s^2)$ calculations to the corrected data distributions. Systematic uncertainties due to the hadronization correction and due to renormalization scale variations (between $f = 1$ and the best fit result of f or - if no best fit of f emerges - the range of f for which the data are suitably well described) are determined and included in the quoted results of $\alpha_s(M_{Z^0})$. The results of these studies are summarized in section 4.4 and in Fig. 11.

4.3 $\alpha_s(M_{Z^0})$ from the Hadronic Partial Width of the Z^0

The cleanest way to determine $\alpha_s(M_{Z^0})$ is a precise measurement of the ratio R of the hadronic and leptonic partial widths of the Z^0,

$$R \equiv \left(\frac{\Gamma_{had}}{\Gamma_{lept}}\right)_{exp} = \left(\frac{\Gamma_{had}}{\Gamma_{lept}}\right)_0 \cdot (1 + \delta_{QCD}) . \tag{4}$$

The QCD correction δ_{QCD}, which is only about 4%, has been calculated to third order $(O(\alpha_s^3))$ perturbation theory [46]. Including quark mass corrections it is of the form [47]

$$\delta_{QCD} = 1.05 \cdot \left(\frac{\alpha_s}{\pi}\right) + 0.9 \cdot \left(\frac{\alpha_s}{\pi}\right)^2 - 13 \cdot \left(\frac{\alpha_s}{\pi}\right)^3 . \tag{5}$$

The standard model expectation for $(\Gamma_{had}/\Gamma_{lept})_0$, without QCD corrections, is 19.97 [48] with only little uncertainty due to the unknown masses of the top quark and of the higgs particle. Basically, the study of R requires only counting of events and does not depend on non-perturbative hadronization effects. However, the precision of $\alpha_s(M_{Z^0})$ from a measurement of R is given by

$$\Delta\alpha_s \approx \frac{\Delta R}{R} \cdot \pi,$$

such that R is required to be known to about 4 per mille to reach $\Delta\alpha_s(M_{Z^0}) = 0.01$.

The value of R measured by OPAL is $R = 20.95 \pm 0.22$ [10]. Together with Eqs. 4 and 5 this gives

$$\begin{aligned}
\alpha_s(M_{Z^0}) &= 0.146 \pm 0.032 \quad (\text{in } O(\alpha_s^3)) \\
\text{or } \alpha_s(M_{Z^0}) &= 0.141 \pm 0.032 \quad (\text{in } O(\alpha_s^2)).
\end{aligned}$$

The uncertainties of these results are currently dominated by the available event statistics and will thus gradually improve in the near future.

The significance of $\alpha_s(M_{Z^0})$ from R can be increased in a combined fit to the hadronic and leptonic Z^0 line shape and asymmetry measurements. With the additional constraint of the W boson mass from $p\bar{p}$ collider experiments, a two-parameter fit of $\alpha_s(M_{Z^0})$ and of the top quark mass m_t results in [10]

$$\begin{aligned}
\alpha_s(M_{Z^0}) &= 0.141^{+0.022}_{-0.020} \quad (\text{in } O(\alpha_s^2)). \\
\text{and } m_t &< 207 \text{ GeV}.
\end{aligned}$$

4.4 Summary of $\alpha_s(M_{Z^0})$.

A compilation of $\alpha_s(M_{Z^0})$ values, obtained from event shapes, jet production rates and from the analysis of Z^0 line shapes and decay asymmetries by OPAL, is given in Fig. 11. The quoted uncertainties include the experimental as well as all the theoretical uncertainties, like hadronization and scale uncertainties as described above, all added in quadrature. Apart from the results obtained in our own analyses, the outcome of a study of our previous measurements of event shape distributions [12], done by Magnoli et al. [49], is also given in Fig. 11. Within the quoted uncertainties, which (with the exception of the line shape measurement) are largely dominated by the theoretical uncertainties, all these results are in good agreement with each other.

The observables, the corresponding theoretical calulcations and their uncertainties are partly correlated with each other. We therefore quote the weighted average of all these results (excluding the number from Magnoli et al.) as our overall final result and assign the smallest uncertainty which is achieved for a single observable as the final error:

$$\alpha_s(M_{Z^0}) = 0.119 \pm 0.008.$$

This is in good agreement with the prediction of $\alpha_s(M_{Z^0}) = 0.11 \pm 0.01$ [50] which was based on a compilation of α_s measurements from various processes at lower c.m. energies, before LEP was turned on. It is also compatible with the average of $\alpha_s(M_{Z^0})$ from the four LEP experiments [51] of $\alpha_s(M_{Z^0}) = 0.120 \pm 0.007$, and with recent analyses of scaling violations in structure functions from deep inelastic scattering which result, if scaled to the Z^0 mass, in $\alpha_s(M_{Z^0}) = 0.109^{+0.007}_{-0.008}$ [52] and in $\alpha_s(M_{Z^0}) = 0.113 \pm 0.005$ [53].

5. Soft Hadron and Gluon Coherence Phenomena

Experimental tests of perturbative QCD do not only cover the domain of jet production and related items which are typically described by fixed order $(O(\alpha_s^2))$ perturbation

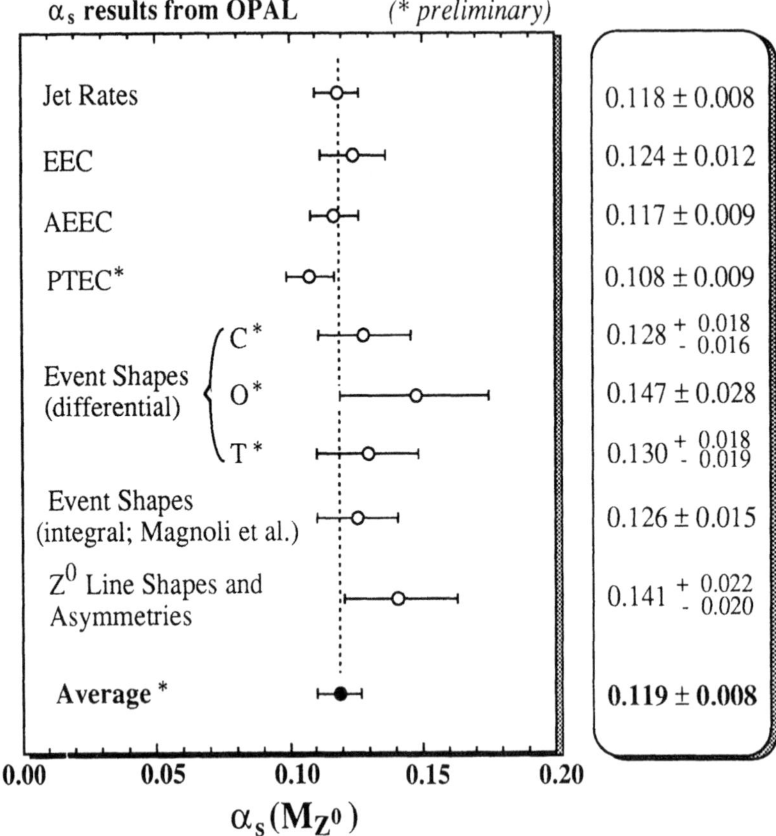

Fig. 11. Summary of measurements of $\alpha_s(M_{Z^0})$ from OPAL.

theory, but also include studies of hadron distributions both inside (intrajet) and in between (interjet) the regions of jets. Quantitative predictions from analytic perturbative calculations exist which are able to describe global features of hadronic systems, such as particle multiplicities, inclusive energy spectra and correlations of particles, distributions of particle flow *etc.* Such calculations are typically carried out in the so called (Modified) Leading Logarithmic Approximation [(M)LLA] of perturbative QCD. With the hypothesis of Local Parton Hadron Duality [LPHD] [54], these calculations are believed to provide reliable predictions for hadron spectra without reliance on any hadronization model. An important ingredient of the perturbative approach to hadron physics (i.e. MLLA plus LPHD; see [55,56,57] for recent reviews) is the destructive interference between soft gluons, which among other effects leads to a suppression of low-energetic partons within jets and to an asymmetric parton flow between the jet axes in $q\bar{q}g$ events.

5.1 Inclusive Particle Spectra

The destructive interference of soft gluons in jets is studied in terms of the observable $\xi_p = ln(1/x_p)$, where x_p is the particle momentum normalized by the beam energy. Measured ξ_p distributions of charged particles [58] and of neutral kaons [59] can be well described by the analytic calculations. In particular, the predicted hump-backed structure and the depletion of hadrons at small x_p are observed and the energy dependence of the peak positions of these spectra - if compared with similar measurements made at lower c.m. energies - are well reproduced by the QCD predictions; see Figs. 12 and

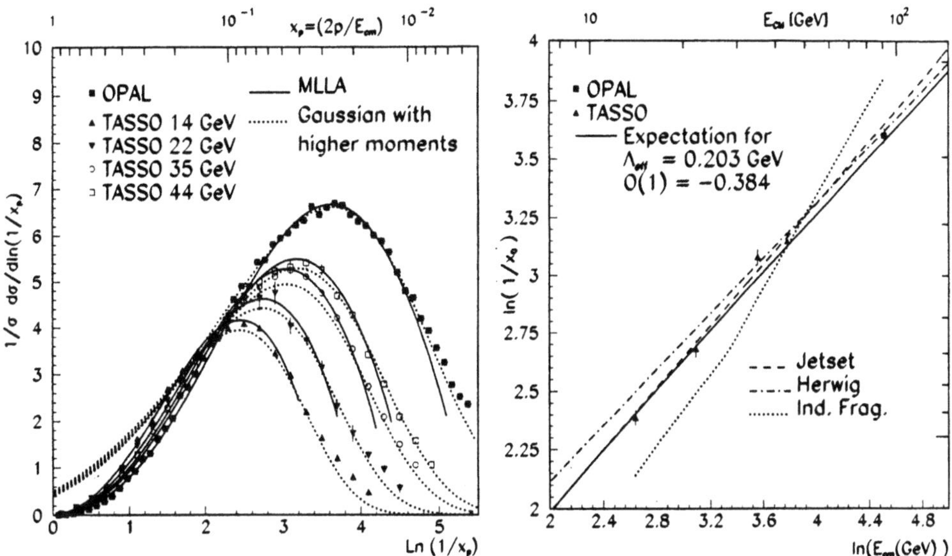

Fig. 12. $\xi_p = ln(1/x_p)$ for charged particles from OPAL, compared to results from lower c.m. energies and to MLLA analytic calculations.

Fig. 13 Peak positions of the measured ξ_p spectra as a function of $\ln(E_{cm})$, compared with MLLA and QCD model calculations.

13. Especially the latter result is interpreted as evidence for the reliabilty of MLLA calculations, for soft gluon destructive interference and for the validity of LPHD.

5.2 Studies of the 'String Effect'

About 10 years ago, the JADE collaboration was the first to report that in $q\bar{q}g$ events less particles are produced between the quark and antiquark jet then between a quark and the gluon jet [60]. This effect can be explained by both the nonperturbative string hadronization picture [61] and by soft gluon coherence effects within analytic QCD calculations [62].

While previous studies of this effect relied on the comparison of data with QCD plus hadronization models, the OPAL collaboration has performed a new and model independent analysis based on 3-jet events with identified quark jets [63]. Three-jet events which are symmetric under transposition of the two least energetic jets and with one of these two jets containing an energetic lepton are selected. The lepton allows to tag the corresponding jet as a (heavy) quark jet, and the non-tagged of the two least energetic jets is thus identified as the gluon jet, with a purity of more than 80%. Since the energies of the low energy quark and gluon jets and the angles ψ_{qg} (between the high energy quark and the gluon jet) and $\psi_{q\bar{q}}$ (between the two quark jets) are the same, this method allows to compare the particle populations between the jets in a model independent way.

For example, Fig. 14 shows the inclusive multiplicity distribution $(1/N)\ dN/d\psi$ with respect to the azimuthal angle ψ in the 3-jet event plane, for an event sample in which $\psi_{qg} \approx \psi_{q\bar{q}} \approx 150^\circ$. The histogram describes the energy flow starting at the high energy quark jet ($\psi = 0$), proceeding via the gluon jet and the low energy quark back to the high energy quark jet (q-g-q), while the points show that same energy flow for comparison but commencing in the opposite sense in the event plane (q-q-g). In the interval labelled (1), which is the region between the higher energy quark jet and either the gluon (q-g) or the low energy quark jet (q-q), a higher particle flow is observed for

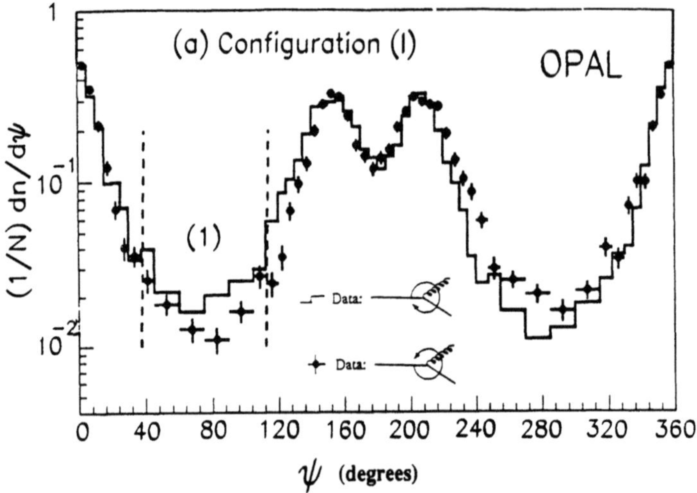

Fig. 14. Particle flow in symmetric 3-jet events.

the q-g than for the q-q case (the identical information - with exchanged symbols - is seen in the region around $\psi \approx 280°$). The conclusion of this and similar studies is that there is a significant population asymmetry in the regions between two quark jets and between a quark and a gluon jet. This can either be explained by the phenomenological string fragmentation picture and/or by gluon coherence effects. Further investigations with the aim to discriminate between these two possibilities are being worked on.

5.3 Difference between Quark and Gluon Jets

Since gluons carry a larger colour charge than quarks, QCD qualitatively predicts that gluon jets are broader and contain more and softer hadrons than quark jets of the same energy. OPAL studied differences between quark and gluon jets [64] in a model independent way, using the same sample of symmetric 3-jet events with a lepton tag as described in section 6.2. Fig. 15 shows the scaled particle energy spectra from the core regions of gluon jets compared to the core region of the lower energy quark jets, for the event sample shown in Fig. 14. Gluon jets are found to yield a softer particle energy spectrum than quark jets. Gluon jets are also observed to be broader than quark jets both in and out of the event plane [64]. It is shown that the results are not biased on the specific selection of heavy quark events.

Little or no difference is observed, however, between the mean value of particle multiplicities for the two jet types: the ratio of average hadron multiplicities in gluon and quark jets is determined to be $1.02 \pm 0.04^{+0.06}_{-0.00}$, in the environment of 3-jet events as described above. Note that the leading order QCD expectation for this ratio to be 9/4 is valid only for pure colour singlet quark or gluon jet systems at asymptotic energies. Finite energy, higher order corrections and the different event environments prohibit the direct comparison of this QCD prediction with the experimental result.

6. Nonperturbative Aspects of Hadron Production.

Local density fluctuations of partciles, also called "intermittency", have been studied in terms of factorial moments as a function of e.g. decreasing rapidity binsize δy. The rapidity region of interest, Y, is devided into M bins of width $\delta y = Y/M$. The normalised

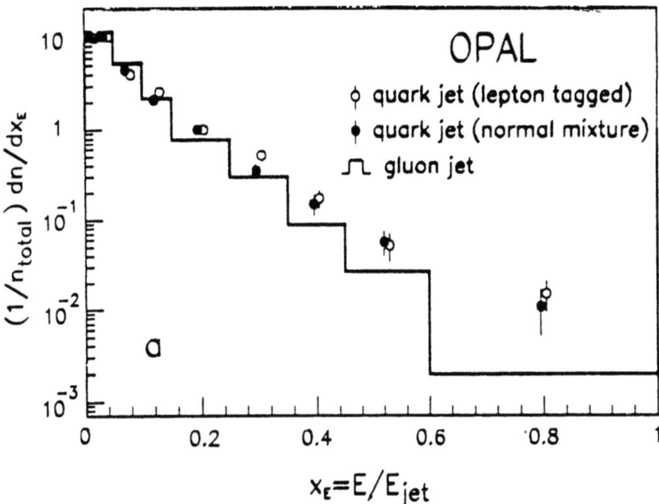

Fig. 15. Particle energy spectra from the cores of quark (symbols) and gluon jets (line).

factorial moment of order j is then defined as

$$F_j(M) = \frac{1}{<\bar{n}_m>^j} < \frac{1}{M} \sum_{m=1}^{M} n_m(n_m - 1)...(n_m - j + 1) >$$

where $\bar{n}_m = \frac{1}{M}\sum_{m=1}^{M} n_m$, n_m is the number of particles in bin m and the angle brackets imply an average over all bins [66]. The moments can also be defined in more than one dimension.

In Fig. 16, the measured 2nd to 5th factorial moments of the rapidity distribution (w.r.t. the sphericity axis) are compared with several QCD plus hadronisation models [67]. The increase of the moments with decreasing bin size can be well described by all the models studied. This statement is also true for factorial moments defined in more than one dimension. Further studies show that the origin of the observed behaviour is largely due to the self similar cascade process of jet evolution, and no unexplained fluctuations are observed.

OPAL also studied correlations between like sign charged particle pairs [68]. Figure 17 shows the ratio of like sign and oppositely charged particle pairs as a function of Q, the four-momentum difference of the pair ($Q^2 = (p_1 - p_2)^2$). Since charged particles are prodominantly pions, the enhancement observed for particle pairs at small Q is explained by Bose-Einstein statistics: identical bosons prefer to occupy the same quantum state. From the measurement, the effective radius of the pion emitting source was found to be $R_0 = 0.93 \pm 0.02 \pm 0.15$ fm which is in agreement with results from e^+e^- experiments at lower c.m. energies.

7. Summary

During the first two years of LEP operation, a large number of significant tests of QCD was performed by OPAL. The data are in excellent agreement with the predictions of perturbative QCD, and no significant deviation from the standard model expectations was so far observed. The results significantly increased the understanding of and confidence in pertubative QCD:

- Global event structures of hadronic final states in e^+e^- annihilations are well

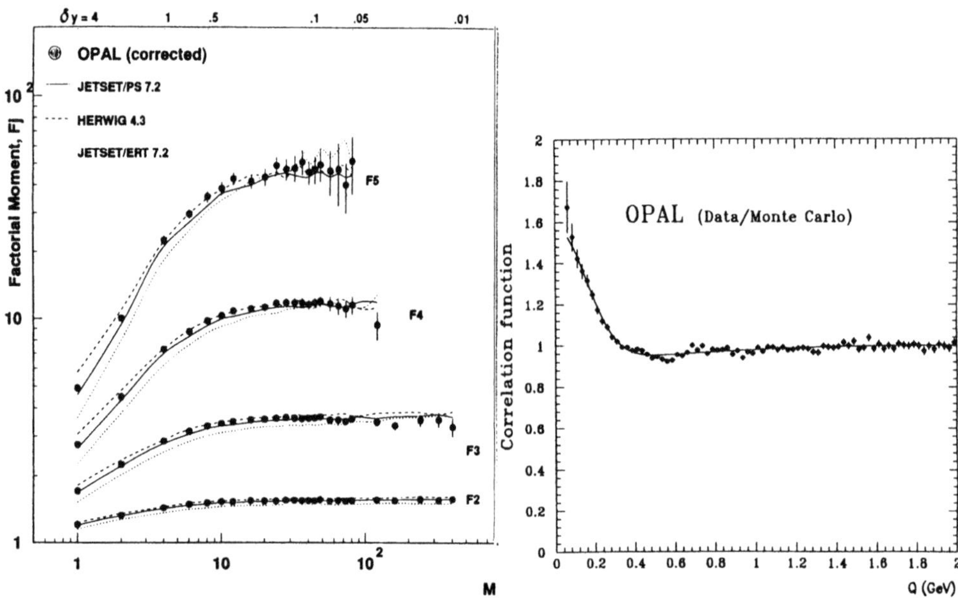

Fig.16. Factorial moments of the rapidity distribution, as a function of decreasing bin size.

Fig.17. Bose-Einstein correlations for like sign charged particle pairs.

described by QCD (shower) models, and their observed energy dependence is understood in terms of QCD scaling violations plus an energy *in*dependent parametrization of hadronization.

- Jet production rates significantly test the validity of QCD and its nonabelian structure: α_s undoubtedly runs as expected from asymptotic freedom.

- Detailed studies of 4-jet final states provide evidence for the gluon self coupling.

- $\alpha_s(M_{Z^0})$ is consistently measured, from a variety of different observables and in $O(\alpha_s^2)$, as $\alpha_s(M_{Z^0}) = 0.119 \pm 0.008$. The error includes all experimental theoretical uncertainties which are accessible so far. This corresponds to $\Lambda_{\overline{MS}}^{(N_f=5)} = 240^{+120}_{-90}$ MeV.

- Inclusive particle spectra are in good agreement with MLLA calculations and with the hypothesis of Local Parton Hadron Duality.

- New methods of quark and gluon jet tagging, together with the large data statistics available at LEP, allowed to study the 'string' effect and differences between quark and gluon jets in a model independent way.

Many studies of similar nature and with compatible results are also avaible from the ALEPH, the DELPHI and the L3 experiments at LEP. A recent and comprehensive summary of "QCD studies at LEP" and the corresponding references can be found in [51].

Future QCD studies at LEP will profit from even larger event statistics and new experimental techniques, like comparisons of final state photon radiation with gluon radiation from quarks [65], studies of QCD with heavy quarks using improved quark tagging methods, and measurements of identified particle spectra. New theoretical input is needed, however, in order to increase the precision on the strong coupling $\alpha_s(M_{Z^0})$, which is currently limited by uncertainties of the unknown higher order contributions to calculations of event shapes and jet production rates. These observables provide the smallest experimental uncertainties in α_s determinations of 3% and less, and in many cases even the hadronization uncertainties, as inferred from hadronization models which provide realistic descriptions of the data, is smaller than the estimates of higher order effects. Resummations of next to leading logarithms to all orders, as recently done for some of the observables [23], may significantly improve this situation; however a full third order calculation, preferentially for jet production rates, is indispensable for a deeper understanding of perturbative QCD and for more precise determinations of α_s.

Acknowledgements: This review would not have been possible without the effort and efficient cooperation of the members of the OPAL collaboration, the support staff of the participating institutions and of the SL Division at CERN.

References

[1] H. Fritzsch, M. Gell-Mann, 16th Intern. Conf. on High Energy Physics, Chicago-Batavia (1972);
H. Fritzsch, M. Gell-Mann and H. Leutwyler, Phys. Lett. B47 (1973) 365;
Gross, F. Wilczek, Phys. Rev. Lett. 30 (1973) 1343;
H.D. Politzer, Phys. Rev. Lett. (1973) 1346.

[2] S.L. Wu, Phys. Rep. 59 (1984) 107.

[3] B. Naroska, Phys. Rep. 148 (1987) 67.

[4] S. Bethke, LBL-28112 (1989).

[5] J.R. Carter, plenary talk at the LP-HEP91 Conference on High Energy Physics, Geneva 1991.

[6] ALEPH Collaboration, D. Decamp et al., Nucl. Instr. Methods A294 (1990) 121.

[7] DELPHI Collaboration, P. Aarnio et al., Nucl. Instr. Methods A303 (1991) 233.

[8] L3 Collaboration, B. Adeva et al., Nucl. Intr. Methods A289 (1990) 35.

[9] OPAL Collaboration, K. Ahmet et al., Nucl. Instr. Methods A305 (1991) 275.

[10] OPAL Collaboration, G. Alexander et al., CERN-PPE/91-67.

[11] Z. Kunszt and P. Nason [conv.] in "Z Physics at LEP 1" (eds. G. Altarelli, R. Kleiss and C. Verzegnassi), CERN 89-08 (1989).

[12] OPAL Collaboration, M.Z. Akrawy et al., Z. Phys. C47 (1990) 505.

[13] T. Sjöstrand, Comp. Phys. Comm. 39 (1986) 347;
T. Sjöstrand and M. Bengtsson, Comp. Phys. Comm. 43 (1987) 367.

[14] G. Marchesini and B.R. Webber, Nucl. Phys. B310 (1988) 461.

[15] U. Petterson, LU TP 88-5 (1988);
L. Lönnblad and U. Petterson, LU TP 88-15 (1988);
L. Lönnblad, LU TP 89-10 (1989).

[16] S. Bethke, Proc. of the Workshop on Jet Physics at LEP and HERA, Durham, Dec 9-15, 1990; CERN-PPE/91-36.

[17] JADE Collaboration, W. Bartel et al., Phys. C33 (1986), 23.

[18] JADE Collaboration, S. Bethke et al., Phys. Lett. B213 (1988), 235.

[19] OPAL Collaboration, M.Z. Akrawy et al., Z. Phys. C49 (1991) 375.

[20] S. Bethke, Z. Kunszt, D.E. Soper, and W.J. Stirling, CERN-TH.6222/91.

[21] S. Catani, these proceedings, and CERN-TH.6281/91.

[22] S. Catani et al., Cavendish-HEP-91/5.

[23] S. Catani et al., Phys. Lett. B263 (1991) 491; CERN-TH.6231/91.

[24] Review of Particle Properties, Phys. Lett B204 (1988), 96.

[25] J. Ellis and I. Karliner, Nucl. Phys. B148 (1979) 141.

[26] TASSO collaboration, R. Brandelik et al., Phys. Lett. B97 (1980) 453;
PLUTO Collaboration, Ch. Berger et al., Phys. Lett. B97 (1980) 459;
CELLO Collaboration, H.J. Behrend et al., Phys. Lett. B110 (1982) 329.

[27] OPAL Collaboration, G. Alexander et al., CERN-PPE/91-97.

[28] OPAL Collaboration, M.Z. Akrawy et al, Z. Phys. C49 (1991) 49.

[29] M. Bengtsson and P. Zerwas, Phys. Lett. B208 (1988) 306.

[30] O. Nachtmann and A. Reiter, Z. Phys. C16 (1982) 45.

[31] M. Bengtsson, Aachen University preprint PITHA 88/12 (1988).

[32] S. Bethke, A. Ricker and P.M. Zerwas, Z. Phys. C49 (1991) 59.

[33] L3 Collaboration, B. Adeva et al., Phys. Lett. B248 (1990) 473.

[34] ALEPH Collaboration, D. Decamp et al., Phys. Lett. B255(1991) 623.

[35] DELPHI Collaboration, P. Abreu et al., Phys. Lett. B247 (1990) 167.

[36] TASSO Collaboration, W. Braunschweig et al., Phys. Lett. B214 (1988) 286.

[37] MARK2 Collaboration, S. Bethke et al., Z. Phys. C43 (1989) 325.

[38] AMY Collaboration, I. Park et al., Phys. Rev. Lett 62 (1989) 1713.

[39] VENUS Collaboration, K. Abe et al., Phys. Lett. B240 (1990) 232.

[40] R.K. Ellis, D.A. Ross and A.E. Terrano, Nucl. Phys, B178 (1981) 421.

[41] OPAL Collaboration, M.Z. Akrawy et al., Phys. Lett. B235 (1990) 389.

[42] S. Bethke, Z. Phys. C43 (1989) 331.

[43] S. Sanghera, OPAL internal note PN033.

[44] OPAL Collaboration, M.Z. Akrawy et al., Phys. Lett. B252 (1990) 159.

[45] G. Azuelos, OPAL internal note PN032.

[46] L.R. Surguladze and M.A. Samuel, Phys. Rev. Lett. 66 (1991) 560;
S.G. Gorishny, A.L. Kataev and S.A. Larin, Phys. Lett. B259 (1991) 144.

[47] T. Hebbeker, Aachen report PITHA 91/08 (revised version).

[48] D. Bardin et al., Program *ZFITTER*, Phys. Lett. B255 (1991) 290.

[49] N. Magnoli, P. Nason and R. Rattazzi, Phys. Lett. B252 (1990) 271.

[50] G. Altarelli in: Ann. Rev. of Nuclear and Particle Science, Vol. 39 (1989) 357.

[51] T. Hebbeker, plenary talk presented at the LP-HEP91 Conference on High Energy Physics, Geneva 1991; Aachen preprint PITHA 91/17.

[52] A.D. Martin, W.J. Stirling and R.G. Roberts, Phys. Lett. B266 (1991) 173.

[53] A. Milsztajn, talk presented at the LP-HEP91 Conference on High Energy Physics, Geneva 1991.

[54] Yu.L. Dokshitzer and S.I Troyan, preprint LNPI-922 (1984); Ya.I. Azimov et al., Z. Phys. C27 (1985) 65; Z. Phys. C31 (1986) 213.

[55] V.A. Khoze, LU TP 90-12.

[56] Yu.L. Dokshitzer, V.A. Khoze and S.I. Troyan, LU TP 91-12.

[57] Yu.L. Dokshitzer, V.A.Khoze and S.I.Troyan, in *Perturbative QCD*, ed. A.H. Mueller, World Scientific, Singapore, 1989, p.241. Yu.L. Dokshitzer, V.A. Khoze, A.H. Mueller and S.I. Troyan, *Basics of Perturbative QCD*, Editions Frontieres, Paris,1991.

[58] OPAL Collaboration, M.Z. Akrawy et al., Phys. Lett B247 (1990) 617.

[59] OPAL Collaboration, G. Alexander et al., Phys. Lett. B264 (1991) 467.

[60] JADE Collaboration, W. Bartel et al., Phys. Lett. B101 (1981) 129; Z. Phys. C21 (1983) 37.

[61] B. Andersson, G. Gustafson and T. Sjöstrand, Z. Phys. C6 (1980) 235; Nucl. Phys. B197 (1982) 45.

[62] V.A. Khoze, L. Lönnblad, Phys. Lett. B241 (1990) 123.

[63] OPAL Collaboration, M.Z. Akrawy et al., Phys. Lett. B261 (1991) 334.

[64] OPAL Collaboration, G. Alexander et al., Phys. Lett. B265 (1991) 462.

[65] OPAL Collaboration, M.Z. Akrawy et al., Phys. Lett. B246 (1990) 285.

[66] A. Bialas and R. Peschanski, Nucl. Phys. B273 (1986) 703; Nucl. Phys. B308 (1988) 857.

[67] OPAL Collaboration, M.Z. Akrawy et al., Phys. Lett. B262 (1991) 351.

[68] OPAL Collaboration, P.D. Acton et al., Phys. Lett. B267 (1991) 143.

QCD AT (..ALMOST..) 2 TEV

Michelangelo L. Mangano (presented by)

The CDF Collaboration[1]
Istituto Nazionale di Fisica Nucleare, Scuola Normale Superiore
and Dipartimento di Fisica, Pisa, Italy

INTRODUCTION

As pointed out during the presentation of the Eloisatron project [1], one of the main milestones for the realisation of the 200 TeV Collider is the 10% Eloisatron. What I will present here today is a selection of achievements obtained at the $\sim$ 1% Eloisatron energy, namely results on QCD from the CDF Collaboration at the 1.8 TeV Fermilab $\bar{p}p$ Collider. Since this Workshop is dedicated to tests of our understanding of QCD and our ability to extrapolate its behaviour to the highest energies available in the future, I will mostly dwell on the comparisons between data and QCD predictions rather than on a detailed description of the experimental analysis and systematics. For these I will refer the reader to the CDF publications quoted in the references.

Most of the results presented here are based on data collected during the 1988-89 run, when approximately 4.4 pb^{-1} were gathered. I will discuss the following items:

1. 1-jet inclusive E_T distributions;

2. di-jet invariant mass spectrum;

3. 3-jet topologies and initial/final state coherence effects;

4. properties of events with large total E_T;

5. jet fragmentation properties;

6. direct photon production;

7. W and Z p_T distributions.

The data will be compared to either parton level calculations or to shower Monte-Carlo programs, which provide complementary tools for the description of hadronic phenomena. In each case I will indicate the significance of the measurement as a test of QCD or as a possible tool to improve our understanding of the hadronic structure.

---

[1]CDF Collaborating Institutions: ANL - Brandeis - Univ. of Chicago - Fermilab - INFN, Frascati - Harvard - University of Illinois - KEK - LBL - University of Pennsilvania - INFN, University and Scuola Normal Superiore, Pisa - Purdue - Rockefeller - Rutgers - Texas A&M - Tsukuba - Tufts - University of Wisconsin .

EXPERIMENTAL SYSTEMATICS AND DATA SELECTION

Trigger

The CDF detector has been described in full detail elsewhere[2]. For the measurements presented here the following set of triggers were used:

- Single jet online triggers, requiring the presence of at least one energy cluster in the calorimeter with transverse energy greater than, respectively, 20, 40 and 60 GeV. The 20 and 40 GeV triggers were prescaled. These triggers were used for the one- and two-jet analyses.

- For the three-jet and large-E_T analyses the trigger required a total transverse energy in the calorimeter larger than 120 GeV.

- Photons were collected using a prescaled E_T trigger ($E_T > 10$ GeV) and a high E_T trigger ($E_T > 23$ GeV). The photon clusters were required to have a small fraction of hadronic energy ($E_T^{had}/E_T^{em} < 1/8$) and to be isolated: the extra energy within a cone of radius $R = (\Delta\phi^2 + \Delta\eta^2)^{1/2} = 0.7$ centered around the photon was required to be less than 15% of the neutral cluster energy.

- Electrons were triggered on by requiring a cluster in the central electromagnetic (EM) calorimeter with $E_T > 12$ GeV, a track in the central drift chamber (CTC) with $p_T > 6$ GeV pointing toward the EM cluster and $E_T^{had}/E_T^{em} < 1/8$.

- The muon trigger required hits in the muon chamber to match a track from the CTC with $p_T > 6$ GeV.

The additional off-line selection criteria used in the analyses can be found in the references quoted in the following.

Jet Definition

At the leading order in QCD (LO), jet production is described by the scattering of two partons from the beam hadrons and results in final states with two back-to-back partons which we associate to the jets. Jets so defined have no size, and the relative cross-sections are independent of the clustering procedure which experiments are forced to introduce because of the finite size of the detector. Since unavoidably the final state partons will emit radiation during their evolution towards a stable hadronic configuration, the energy carried by the parton will eventually spread over a finite volume. This results in a cross-section which depends on the specific definition of jet used in the measurement, contrarily to the LO result. At the next-to-leading order in QCD (NLO) the initial or final state partons are allowed to radiate one gluon, and the sharing of energy between the resulting three partons results in a QCD prediction which is dependent on the jet definition.

The effect of the final state radiation is clearly visible in figure 1, where we show the average energy flow around the jet axis as a function of the azimuthal distance ϕ from the axis itself. The plot is generated using clean two-jet events, namely excluding events where a third hard jet is irradiated. The peaks at $\phi = 0°$ and $180°$ correspond to the jets, with wide tails due to emission of radiation, while the energy plateau at $\phi = 90°$ can be interpreted as a measure of the average energy density due to the evolution of the leftovers of the struck hadrons (the *underlying event*, UE). It is customary to define the jet energy as the amount of energy contained within a cone with $(\Delta\phi^2 + \Delta\eta^2)^{1/2} < \Delta R$, with the energy from the UE subtracted away. The UE energy is subtracted because its effect is not present in the QCD calculations we will try to compare the data with. This procedure is consistent provided the two sources of radiation – final state bremsstrahlung and UE – are independent, so that we can assume the latter to contribute with a constant overall shift in the total energy of the jet.

A comparison between the energy density detected at 90° from the jet axis in QCD events and the energy density observed in minimum bias events shows the former to be larger by a factor of two, suggesting that in fact half of the energy plateau is due to large angle soft final state radiation. This is also confirmed in part by theoretical calculations[3]. In the jet measurements presented here, unless otherwise stated, the full plateau is taken as UE event and is subtracted from the energy of the jet. Since this is in any case a constant term, the relative effect on the measurement of the *true* jet-energy becomes negligible for hard enough jets – typically for $E_T > 100$ GeV.

If we intend to compare the measurement with the LO prediction we should try to collect all of the energy of the primordial parton, and therefore we will add also the so called *out of cone* energy (see fig. 1). However, in order to match as closely as possible the definition of jets which can be implemented using the recent higher order calculations of jet production[4] and to provide measurements which could one day be compared to NLO predictions once these calculations will be available, it has now become convention[5] NOT to apply the out-of-cone corrections, these being included in the theoretical systematic error in the case of comparisons with LO calculations. According to this convention the energy and position of the jet are defined by the following energy-weighted averages:

$$E_T = \sum_{R_i < \Delta R} E_{T_i} \quad , \quad \eta = \frac{1}{E_T} \sum_{R_i < \Delta R} E_{T_i}\eta_i \quad , \quad \phi = \frac{1}{E_T} \sum_{R_i < \Delta R} E_{T_i}\phi_i, \tag{1}$$

where i indicates the i-th calorimeter cell in the case of the experiment or the i-th parton in the case of the theoretical calculation.

Calorimeter Response

The measured energy of a jet is the result of the response of the calorimeter to each single particle contained in the jet. Therefore the jet energy response function:

$$F(E_{jet}^{true}, E_{jet}^{obs}) \tag{2}$$

is the convolution of the jet fragmentation function with the single particle detector response, accounting for possible energy losses in underinstrumented regions. F represents the probability the the true energy of the jet E_{jet}^{true} will be detected as E_{jet}^{obs}. The observed spectrum will therefore be given by:

$$\frac{d\sigma}{dE_T^{obs}} = \int \frac{d\sigma}{dE_T} F(E_T, E_T^{obs}) dE_T \tag{3}$$

which results into a smearing of the true distribution. Since both the fragmentation function and the single particle detector response can be measured (the latter both on the test beam

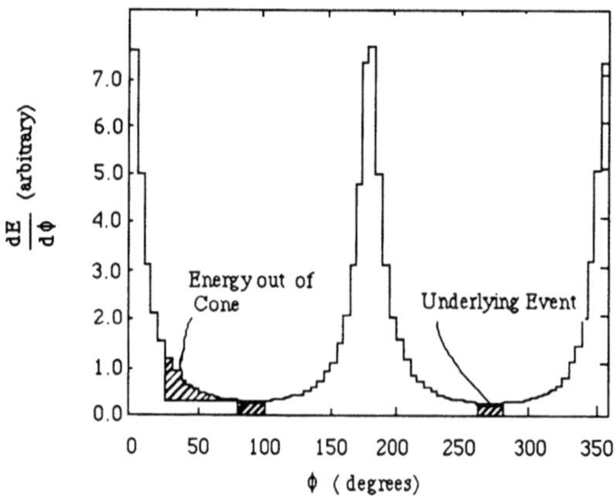

Figure 1. Azimuthal energy profile in di-jet events.

and in situ by matching the measured momentum of single isolated charged tracks with the calorimeter signal), it is possible to explicitly evaluate using a Monte Carlo the function F and *unsmear* the observed distribution to obtain the true one to be compared to the theoretical calculations. The systematic errors, primarily due to uncertainties in the fragmentation tuning of the Monte Carlo and in the low-energy detector response, result in a transverse energy scale uncertainty varying between 4 and 6% (the smaller value applying to the higher energy jets). This reflects itself into an uncertainty in the determination of the cross-section for the single inclusive jet E_T distribution of about 22% (constant above E_T=80 GeV).

RESULTS ON JET PHYSICS

As was pointed out by J. Stirling's in his talk[6], one of the main uncertainties in predicting the behaviour of cross-sections at very high energy is the poor knowledge we have of the gluon structure functions even at current energies. Contrarily to the quark structure functions, which are measured directly in deep inelastic scattering experiments and which are rather well known, gluon structure functions have up to now only been extracted indirectly by studying the scaling violation in the evolution of quarks. Jet production in high energy hadronic collisions is mostly due to the scattering of gluons and therefore precise measurements of jet production in these experiments might lead to a better understanding of the distribution of gluons inside the nucleon. The recent completion of the NLO prediction for the jet inclusive E_T cross-section reduces the theoretical uncertainties intrinsic in the LO calculation (dependence on the jet definition and on the factorization scale Q^2) and opens the possibility to test gluon structure function parametrizations in a direct and precise way. The extension of these calculations to exclusive quantities (such as the invariant mass of the di-jet systems or di-jet correlations), together with large samples of jets with energies varying over a wide range will hopefully lead to global constraints on the gluon structure functions which will reduce their uncertainties. In the following section I will present some attempts in this direction made by CDF.

Jet Inclusive E_T Distribution[7]

For the inclusive cross-section measurement we used jets in the central pseudo-rapidity region, $0.1 < |\eta| < 0.7$. Fig. 2a shows the differential cross-section as a function of E_T for a cone size of 0.7, compared to the NLO QCD calculation[4]. As mentioned earlier the data were corrected for the detector response, the UE energy was subtracted and no out-of-cone corrections were applied. The theory curve was calculated using MRSB structure functions[8] and factorization scale $Q^2 = E_T^2$. The normalization is absolute. Fig. 2b shows the residuals on linear scale – $(data - theory)/theory$ – where the set MRSB was used as reference and normalized to the data. The dashed lines correspond to the 22% energy-independent systematic error, while the solid lines correspond to the residuals of different sets of input structure functions[9, 10]. Allowing for a floating absolute normalization, the confidence levels for the various sets of parton distributions (PDF's) are given in the following table (systematic errors included):

PDF	Conf. Level (%)	relative normalization
MRSB	19	1.001
HMRSE	< 1	1.143
MTS	49	1.265
MTB	56	1.328

The agreement between data and theory is very good, at least for some sets of PDF's, and notice that it spans several orders of magnitude in cross-section. Notice also that HMRSE, based on obsolete data from EMC which have now been corrected, has the worse agreement with data. One is tempted to hope that the jet E_T spectrum has become such a sensitive probe of the hadronic structure that it can select good PDF's from "bad" ones!

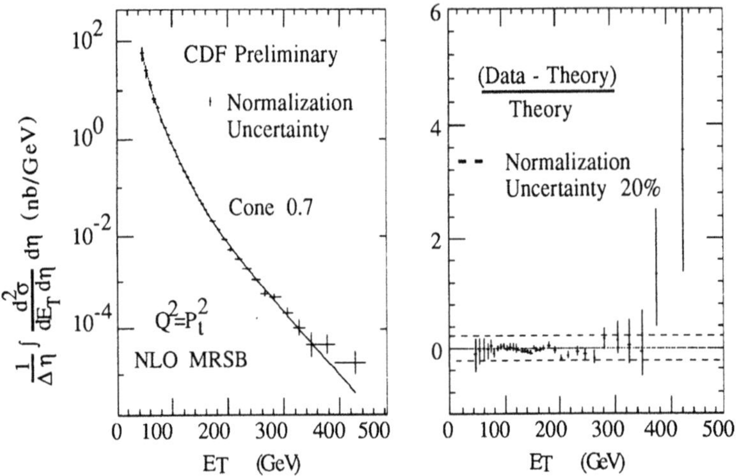

Figure 2. The inclusive jet cross-section.

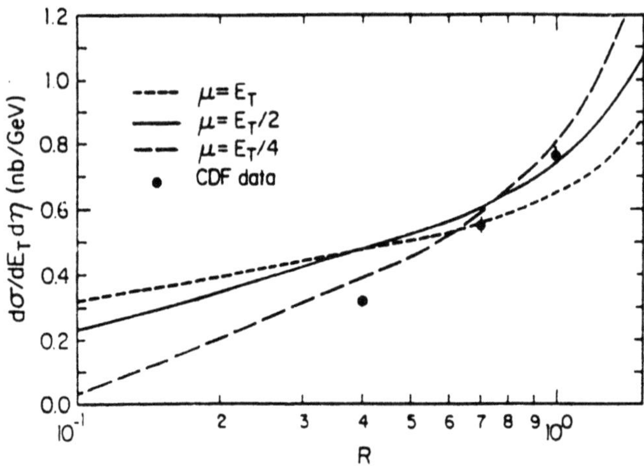

Figure 3. Cone size dependence of the jet cross-section at 100 GeV.

In order to rely on the QCD NLO calculations now available for the jet E_T spectrum, we should however check that these calculations properly describe the dependence of the cross-section on the jet definition, and in particular with the jet size R. In fig. 3 I show the jet size dependence of the differential cross-section for 100 GeV jets. Three different choices of factorization scale were chosen for the theoretical calculation[11]. A fit to the data of the form:

$$\frac{d\sigma}{dE_T} = A + B\log(R) \tag{4}$$

gives the following result:

$$A = 0.79 \pm 0.02\,nb/GeV \quad , \quad B = 0.49 \pm 0.03\,nb/GeV, \tag{5}$$

with a slope evaluated at $R = 0.7$ to be $0.70 \pm 0.05\,nb/GeV$. The theory slope at $R = 0.7$ is $0.5 \pm 0.2\,nb/GeV$, with the uncertainty coming from the range of factorization scales shown in the figure. This result is not inconsistent with the data, but clearly indicates a need to choose a relatively small value of Q^2 to better describe the behaviour at small values of R. This is reminiscent of similar problems encountered in the study of jet multiplicities in e^+e^-

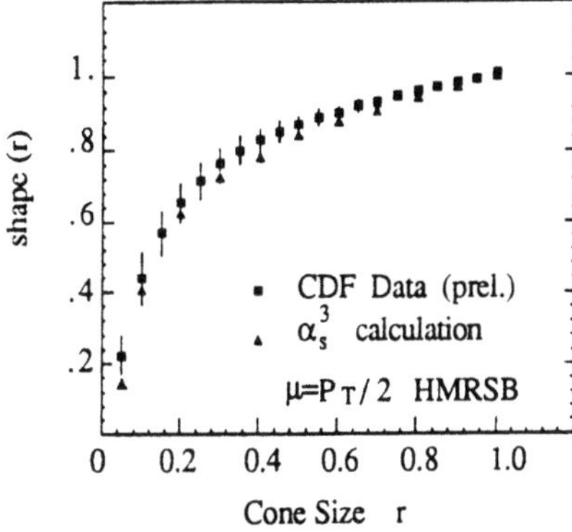

Figure 4. Integrated jet energy profile.

collisions[12] when the jet invariant mass cutoff is chosen too small ($y_{cut} \to 0$). Even in this case if one wants to properly match the data to a QCD calculation at a fixed order in perturbation theory one is forced to choose a value of renormalization scale which is much smaller than the natural energy scale of the collision. In e^+e^- the solution is likely to come from the resummation of the leading logarithms in y_{cut} which appear at any order in perturbation theory and are large when y_{cut} is small[2]. Physically this corresponds to accounting for the suppression of jets with large energy and small mass, suppression due to the large probability that multiple gluon radiation will spread the energy of the jet and increase its mass. Similarly one would expect that here the resummation of terms of order $\alpha_s(Q^2) * \log(1/R)$ relative to the leading order result will give a better agreement between QCD and data in the small R region. Waiting for this important improvement to be performed, it still remains true that the prediction corresponding to $R = 0.7$ is rather stable under changes in Q^2, and might therefore provide a good reference point for QCD tests.

It is interesting to check how the NLO calculation describes another important inclusive feature of jets, namely their transverse energy profile. The idea behind this measurement is that even though several gluons will be emitted during the jet evolution, only those emitted with the largest energy and at the largest angle from the jet parent parton will contribute significantly to a variable such as the energy-weighted shape:

$$shape(r) \;=\; \frac{\int_0^r \rho(r')dr'}{\int_0^R \rho(r')dr'}, \tag{6}$$

where $\rho(r)dr$ is the transverse energy flowing through the annulus of radius r and width dr around the jet axis and R is the overall size of the jet. Since one does not expect to have too often more than one such gluon at large angle and large energy, a calculation at the NLO which describes the emission of only one gluon from the final state partons might be sufficient to describe the data. A comparison between CDF data and theory[11] is shown in fig. 4. In order to reduce the systematic errors due to the non-linearity of the calorimeter, the analysis was performed defining $\rho(r)$ as the density of charged momentum within the jet – as measured by the CTC – rather than the calorimetric energy; the latter was however still used to determine the jet axis. Jets are central ($0.1 < |\eta| < 0.7$) and with $95 < E_T < 120$ GeV. The main systematic error, shown in the plot, comes from the determination of the jet axis, which varies depending on the size of the reference cone. The agreement between data and theory is rather good, and gives us even more confidence in the use of the NLO QCD calculation to perform tests of QCD.

---

[2]As an example of this, see the treatment of the thrust distribution in the $T \to 1$ region in [13].

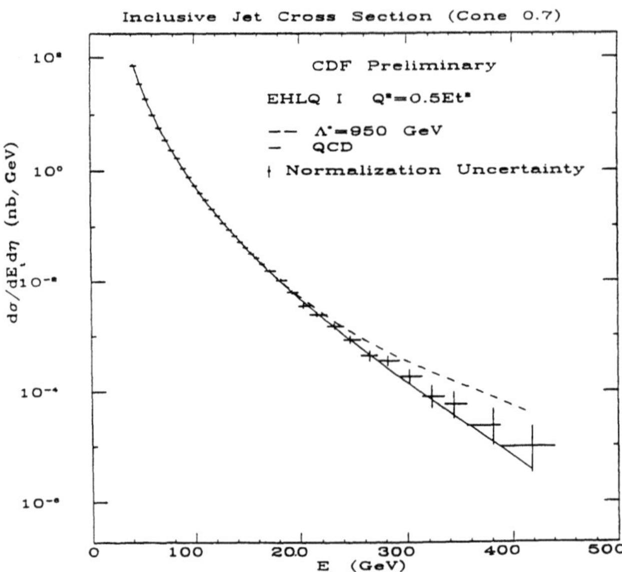

Figure 5. The inclusive jet cross-section (solid line) compared to what is expected from quark compositness with $\Lambda = 950 GeV$.

Having acquired confidence that our QCD tools are reliable, we can then go ahead and try to use the high energy component of the inclusive E_T spectrum to probe the short distance structure of partons. A departure at high energy from the QCD prediction, normalized to the observed behaviour of the cross-section at small values of E_T, might indicate the presence of anomalous parton interactions. A typical example of such departure is the presence of four fermion couplings within the quark Lagrangian:

$$\Delta \mathcal{L} = \frac{f}{\Lambda^2} \bar{\psi} \gamma_\mu \psi \; \bar{\psi} \gamma^\mu \psi, \tag{7}$$

where Λ is a variable with the dimension of a mass introduced to parametrize the scale at which some possible new interactions (*e.g.* due to quarks being composite) show up. It is customary to put the adimensional coupling constant f to be $f = 4\pi$, in such a way that $f/4\pi$, the equivalent of $\alpha = e^2/4\pi$, will be equal to 1. In fig. 5 we show CDF data compared to the theoretical prediction based on a model with $\Lambda = 950$GeV. Using its data CDF can now rule out the presence of such interactions with $\Lambda < 1400$ GeV at 95% confidence level[7]. Notice that this is three times larger than the energy of the most energetic jet observed, the reason being that the quartic interaction given above gives cross-sections which rise very quickly with energy, similarly to the Fermi interactions. With additional statistics we should expect to be able to increase the sensitivity to compositeness scales well above the center of mass energy of the $p\bar{p}$ system. Extrapolating to Eloisatron energies, one therefore expects that quark structure will one day be explored down to scales even larger than 200 TeV!

Di-jet mass distribution

Additional tests of QCD involve the study of di-jet events. We measured the di-jet invariant mass spectrum. The di-jet mass is defined as $M_{jj}^2 = (P_1 + P_2)^2$, where P_1 and P_2 are the quadrimomenta of the two leading jets, defined by:

$$E = \sum_i \epsilon_i \quad , \quad \vec{P} = \sum_i \epsilon_i \hat{n}_i , \tag{8}$$

where ϵ_i is the energy deposited in the i-th calorimeter tower contained within the jet, the position of the tower being defined by the versor $\hat{n}_i$. Since a jet is usually formed by many

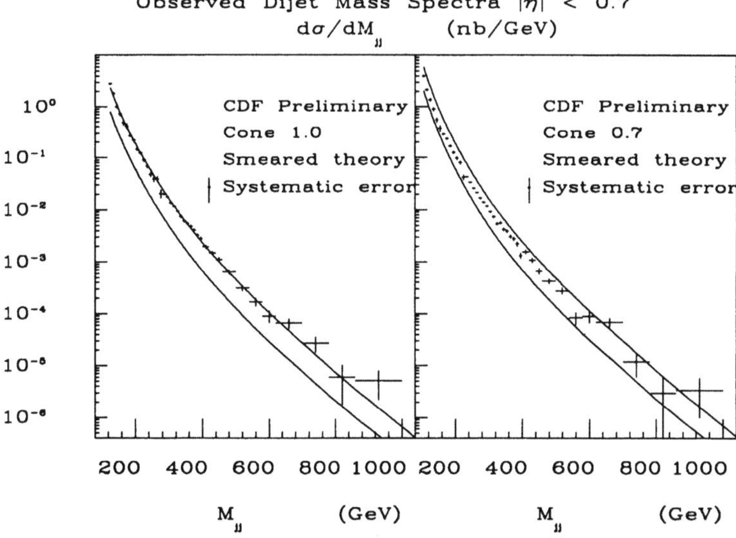

Figure 6. Di-jet mass spectra.

towers, jet quadrimomenta defined in this way will be massive and their mass will enter in the definition of the di-jet mass itself. The resulting spectrum obtained for two different values of the jet cone size ($R=0.7$ and 1.0) is shown in fig. 6. The comparison is made against the LO QCD prediction, since the full NLO calculation has become available only very recently[14] and a systematic comparison has not been carried out as yet. In the figure we show the envelope of the theoretical curves obtained by varying the factorization scale in the interval $E_T^2/2 < Q^2 < 2E_T^2$ (E_T being the transverse energy of the jet) and using the following sets of PDF's: DFLM[15], DO[16], EHLQ[17], HMRS[9] and MT[10]. In the analysis jets from the central region were used ($|\eta| < 0.7$), no out-of-cone nor UE corrections were applied and instead of unsmearing the data by deconvoluting the calorimeter response function from the observed spectrum the choice was made to smear the theory curves using the detector response as a function of M_{jj} as given by the Monte Carlo[18]. To test the shape of the distributions we also normalized the theoretical predictions to the data by fitting a global normalization factor, taking into account the systematic uncertainties. The corresponding confidence levels are shown in the following Table:

| $Q^2/E_T^2=$ | 1 | 2 | 0.5 | 1 | 2 | 0.5 |
PDF	CONE 1.0			CONE 0.7		
DFLM 190	48	46	47	1	1	< 1
DFLM 260	54	52	54	1	1	2
DFLM 360	50	52	53	2	2	2
DO1	51	51	47	2	1	< 1
DO2	49	48	48	2	2	2
EHLQ1	40	38	40	< 1	< 1	< 1
EHLQ2	24	21	25	< 1	< 1	< 1
HMRSB	46	48	47	2	1	1
HMRSE	46	46	39	3	4	4
MT 155	57	58	54	3	3	2
MT 187	56	56	56	2	2	1
MT 191	66	65	64	5	6	6
MT 212	62	61	59	3	4	3

As we see most of the sets of PDF's fit the data rather well for $R = 1$, while fits are very poor for the smaller cone. This is most likely due to the effect of radiation outside the cone – which is less in the $R = 1$ case – and the effect will hopefully be reproduced by the full NLO calculation. It is common lore that the effect of the higher order corrections can be estimated by varying the scale Q^2 chosen for the LO calculation. Notice that in the case of M_{jj} the χ^2 seem to be independent of the Q^2 scale chosen; this is because at the LO changing scale turns out to result in a simple overall change of normalization. However this should not be interpreted as a sign that the NLO will just induce a scale change with respect to the LO prediction. In fact, in order to bring the LO prediction for the cone 0.7 in agreement with data, the NLO will have to change the shape of the LO distribution as well, while hopefully leaving the cone 1.0 shape invariant as this already reproduces well the data. In conclusion, one should not think of the NLO effects as an overall K-factor independent of energy and jet definition, but as subtle changes both in normalization and in shape, changes which will usually depend on the jet definition.

Three-jet topologies[19]

It is well known[20] that the shapes of the partonic cross-sections for QCD $2 \to 2$ processes, such as those responsible for di-jet events, are to good approximation independent of the initial state partons:

$$\frac{d\sigma}{dt}(gg \to jj) \simeq \frac{9}{4}\frac{d\sigma}{dt}(qg \to jj) \simeq \left(\frac{9}{4}\right)^2 \frac{d\sigma}{dt}(q\bar{q} \to jj). \tag{9}$$

Therefore by measuring correlations in di-jet events it is only possible to extract information on the following combination of partonic distributions:

$$F(x, Q^2) = g(x, Q^2) + \frac{4}{9} \sum_f \left[q_f(x, Q^2) + \bar{q}_f(x, Q^2) \right]. \tag{10}$$

This combination of structure functions is known as *effective structure function*. This interesting result is not true anymore for three-jet events: in this case QCD predicts that the shape of various correlation distributions does depend on whether the initial state contains quarks or gluons. Using the large statistics available at CDF we can therefore check the predictions of QCD, improving on the previous analysis performed at the CERN $S\bar{p}pS$ Collider[21, 22].

Using the UA1 conventions[21] we parametrize a three-jet event as shown in fig. 7. The event is boosted to the center of mass frame of the colliding partons (labelled 1 and 2), and the two planes indicated are determined one by the beam axis and the direction of the leading jet, the other by the direction of the leading jet and the direction of the third jet. We define the energy fractions of the three jets to be $x_i = 2E_i/M_{3j}$, in such a way that $x_3 > x_4 > x_5$

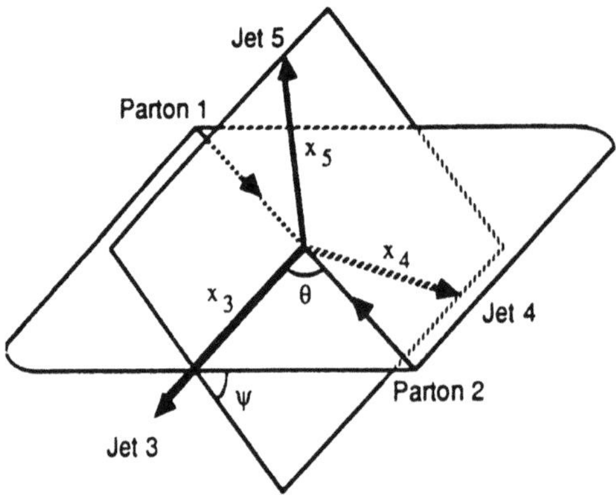

Figure 7. Parametrization of a three-jet event

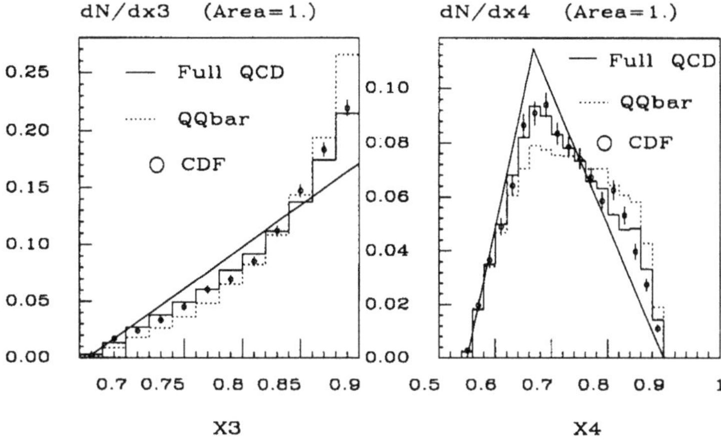

Figure 8. Jet energy fractions in 3-jet events

and $\sum_{i=3}^{5} x_i = 2$. The events are selected by requiring at least three calorimetric clusters with $E_T > 15$ GeV, well separated from each other ($\Delta R \geq 0.85$) and with $|\eta| < 3.5$. In addition M_{3j} (defined as the three jet invariant mass) is required to be larger than 250 GeV, to eliminate trigger biases. To avoid singular regions in the matrix element calculation and to reduce the acceptance systematics, we finally add the following constraints: $x_3 < 0.9$, $|\cos\theta| < 0.6$ and $30^o < |\psi| < 150^o$. The QCD distributions have been obtained generating parton events with the $O(\alpha_s^3)$ tree level matrix elements[23], fragmenting the partons into jets and simulating the detector response. The same constraints applied to the data are applied to the simulated events.

Some of the resulting distributions[19] are shown in fig. 8 and 9, where a comparison between the shapes of data, the full QCD result and the QCD prediction for $q\bar{q}$ initial states only are displayed. The distributions show an excellent agreement with the full QCD prediction and in particular the angular correlations show the expected enhanced peaking in the forward/backward regions due to the more singular collinear emission spectrum of the dominating initial-state gluons. These distributions can be used in a combined fit where the fraction of events generated by the $q\bar{q}$ initial state is a free parameter. The best value for the $q\bar{q}$ fraction thus obtained is a small number – $3\%^{+12\%}_{-3\%}$ –, compatible with the theoretical prediction of $11\%\pm4\%$. As for the absolute rate after the cuts discussed above, we measure

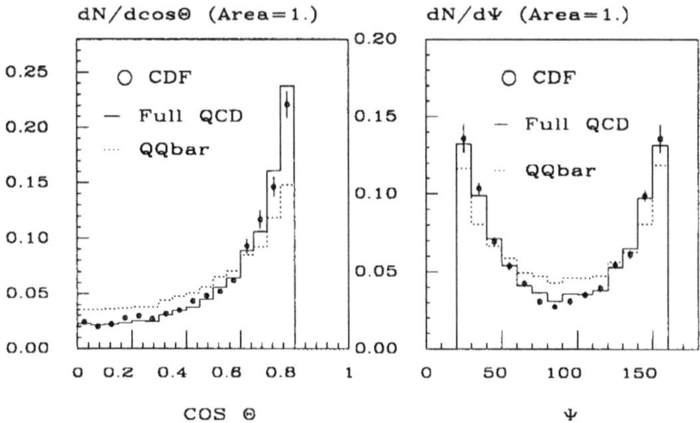

Figure 9. Angular correlations in 3-jet events

1.2±0.4 nb, to be compared to the QCD result of 1.0nb, with an uncertainty of 50% due to structure functions and factorization scale choice.

Initial/final State Coherence Effects
‾‾‾‾‾‾‾‾‾‾‾‾‾‾‾‾‾‾‾‾‾‾‾‾‾‾‾‾‾‾‾‾‾‾‾

Even though parton-level matrix elements calculations are a very important tool to perform QCD calculations, a full fledged parton-shower QCD Monte Carlo is often fundamental in order to describe the complete features of an event. QCD Monte Carlo programs usually contain only the LO expression for a given hard process, and additional partons are emitted during the cascade evolution driven by the soft and collinear emission probabilities[24]. This scheme, leading to a Markov-like evolution, can only partly account for the interference effects which are inherent in any quantum process: the probability that an additional parton be emitted at some point of the cascade only depends in the MC on the momenta of the partons next to it in the shower, while in principle we should expect more general correlations. In particular we know that the structure of the color flows within the shower determines kinematical constraints on the the structure of the emitted radiation. Since a MC which only contains the $2 \rightarrow 2$ hard scattering processes can only describe the presence of a third jet through the emission of a hard gluon, it is important to verify that the algorithm implemented in the MC reproduces the correlations between this additional jet and the rest of the event, showing that the interference effects are correctly accounted for.

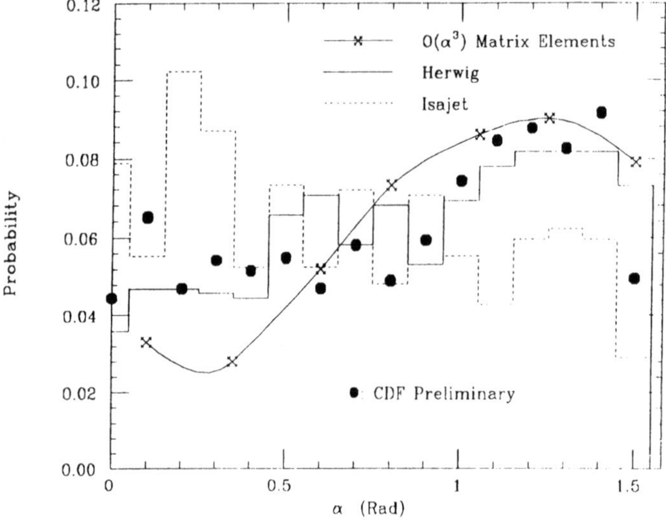

Figure 10. Initial/final state correlations in 3-jet events

A MC which does contain in part the effects of quantum interference and the color coherence is Herwig[25]. It is interesting to compare its predictions with those of a MC with independent fragmentation such as Isajet[26] and with the results of a parton level calculation. To this purpose, we will consider three-jet events and will study the following variable:

$$\alpha = \arctan \frac{\Delta\eta}{\Delta\phi}, \tag{11}$$

where we define $\Delta\eta$ and $\Delta\phi$ as the difference in pseudo-rapidity and in azimuth between the second and third most energetic jets in the event (taken in absolute value). $\alpha=90°$ would correspond to an event with the 2nd and 3rd jets lying coplanar to the beam axis, while $\alpha=0$ corresponds to an acoplanar event. One would expect that the color coherence between initial and final state hard partons will enhance the emission of the third hard jet in the plane of the event, therefore favoring the $\alpha=90°$ configurations. In fig. 10 I show the distribution in α as measured in CDF data[27] and as estimated using Herwig (solid line), Isajet (dashed line) and

227

the 3-jet tree-level matrix elements. The results are preliminary, and more detailed studies are under way; nevertheless it is quite obvious that while the parton level calculation and the coherent MC do reproduce the qualitative behaviour of the data, the incoherent emission MC predicts a significantly different shape. An interesting result[28] is obtained if one generates using Herwig a set of events where initial and final state are not color connected (such as $q\bar{q} \to q\bar{q}$): in this case the shape is more similar to that given by Isajet, indicating that probably the key element in the shape of this distribution are in fact the color coherence effects taking place between initial and final state.

Properties of Events with Large $\sum E_T$

The large energy available at the Fermilab Collider allows for very spectacular events, with very high total transverse energy. After rejecting cosmic rays and beam-halo collisions requiring the out-of-time energy deposition in the central hadronic calorimeter to be small, or requiring a small missing transverse energy ($\not{E}_T/\sqrt{\sum E_T} \leq 6$ GeV$^{1/2}$), and after rejecting events with multiple vertices, we are left with 279 events with $\sum E_T > 400$ GeV. Several distributions were studied[29] and compared to the predictions of the Herwig MC. We show the jet multiplicity distribution in fig. 11, the multi-jet mass spectra in fig. 12 and the E_T flow around the jet axis in azimuth and pseudorapidity for different ranges of jet-E_T in fig. 13. The points correspond to the CDF data and the curves correspond to the Herwig simulation. The different curves in fig. 13 correspond to different Herwig runs performed using two different versions of the detector simulation MC; the spread of the four curves indicates the statistical significance of the MC sample and the systematics related to differences between different tunings of the detector simulation.

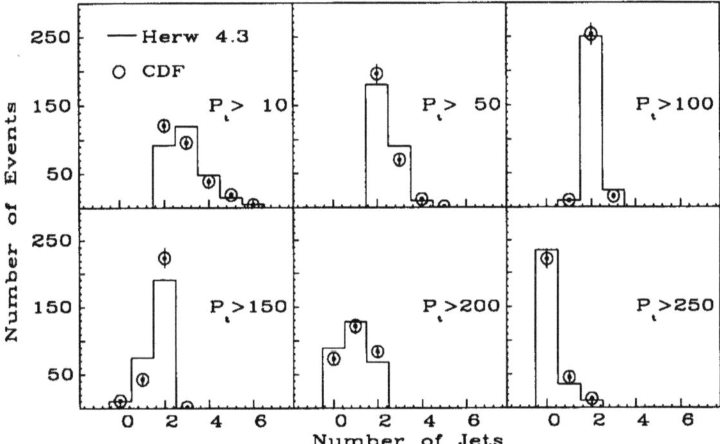

Figure 11. Jet multiplicity in large $\sum E_T$ events (CDF prelim)

Jets are clustered using a 0.7 cone size, $|\eta| < 2$ and unless otherwise indicated in the plots, $E_T > 10$ GeV. The agreement with Herwig is excellent: multi-jet events are reproduced correctly both in rate and in their mass and p_T distribution; the energy flow around the jet axis is also correctly reproduced both in the ϕ and η direction, and the evolution of the jet width with increasing jet-E_T is properly described. The theoretical description of the jet shape provided by the MC is complementary to that given by the NLO QCD calculation (see previous Section) and it is a very important result that both agree with the data.

Jet Fragmentation

It has by now become common belief that the fragmentation properties of jets can be properly described at high energy by shower MC based on the Leading Log Approximation to

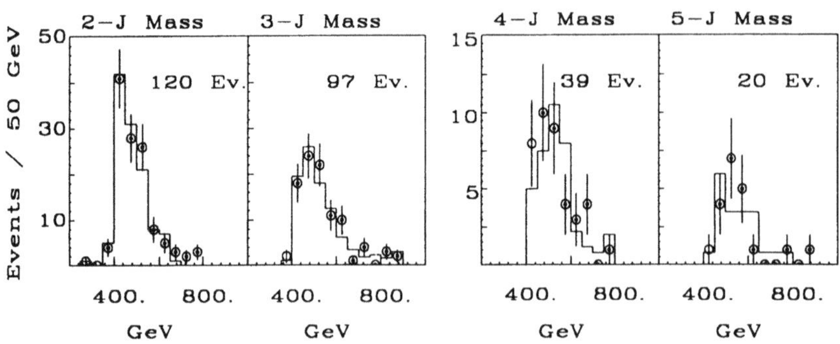

Figure 12. Multi-jet mass distributions (CDF prelim)

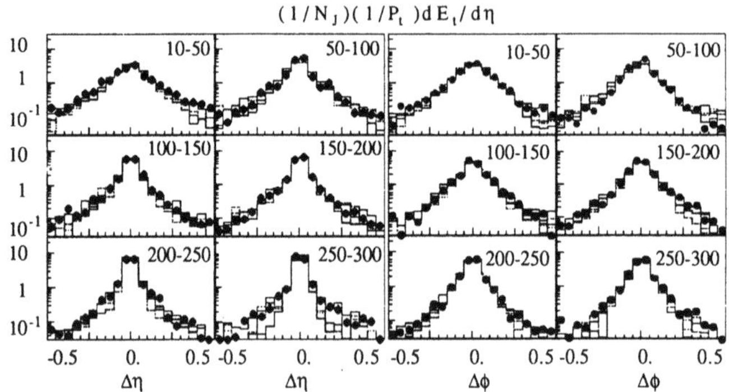

Figure 13. Transverse energy flow around the jet axis (CDF prelim)

perturbative QCD. Partons from the hard scattering are evolved from the large Q^2 scale of the hard process down to small virtualities, of the order of the GeV. Simple parametrizations of the hadron-formation phase can be included at the end of the perturbative evolution[25, 30] and tuned at some fixed energy and for some fixed process (such as jet production in e^+e^- collisions). These non-perturbative effects are *universal*, namely they do not depend on the hard process which originated the jets. All of the differences in the final state of independent processes are accounted for by the perturbative QCD evolution. On the basis of this, one expects for example that gluon jets will have softer fragmentation than quark jets and that average multiplicities will increase with energy. The variety of phenomena which in hadronic collisions give rise to jets is such that several independent measurements of fragmentation can be performed, providing unique consistency checks of QCD. For example one can measure the scaling violation effects in the fragmentation functions by looking at jets with different transverse energies, or one can compare the fragmentation of gluon jets (which dominate the standard QCD production of jets) with the fragmentation of quark jets (which can be found either by constraining the kinematics of a di-jet QCD event or by selecting a sample of jets recoiling against a prompt photon). All of these tests are under way at CDF, as well as attempts to separate mixed samples of quark and gluon jets using their difference as obtained from QCD as the input for global likelihood fits[31] or neural networks[32].

In fig. 14 I show the inclusive fragmentation function of charged particles as measured in CDF using data from the 1987 data taking[33]. The momentum fraction z is defined as $z = P_{||}/|\mathbf{P}_{jet}|$, jets are defined by the cluster algorithm with cone size 1.0 and were constrained to $0.1 < |\eta| < 0.7$. Furthermore only di-jet events with invariant mass in the range 80-200 GeV were used. Out-of-cone and UE corrections were performed on the jet energies, as well as jet-by-jet corrections based on the spectrum of charged tracks within the jet. The data

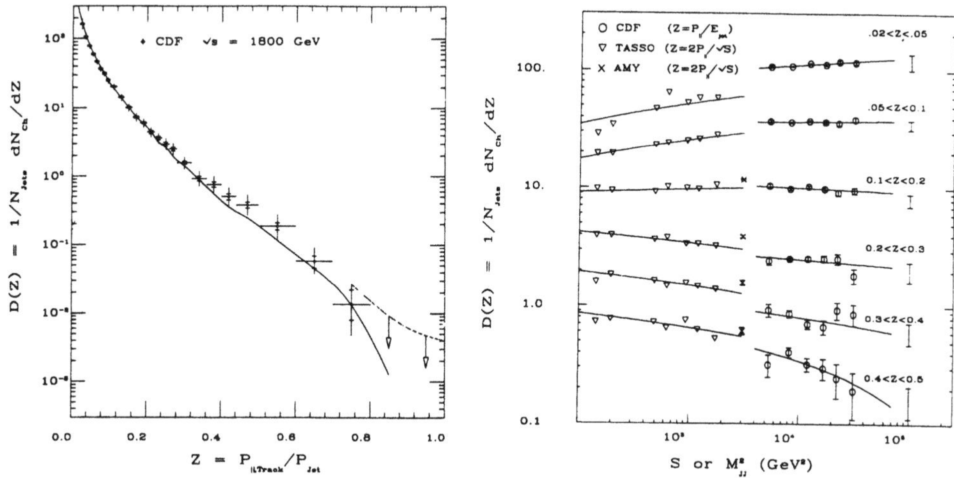

Figure 14. Jet fragmentation function and its Q^2 evolution

are compared to the prediction of Herwig (version 3.2), which seems to reproduce data rather well. The dashed line for $z > 0.7$ represents an upper limit to the data.

As a measurement of the scaling violations I also show in fig. 14 the fragmentation function for different bins in z as a function of energy. CDF data are presented together with lower energy data from e^+e^- colliders[34]. The trend of scaling violation indicated by the e^+e^- data is confirmed, and in addition a finite offset is present between CDF and e^+e^- curves, as expected because of the different content of quarks and gluons in the two jet samples.

P_T DISTRIBUTIONS OF ELECTROWEAK VECTOR BOSONS

Photon p_T distribution

The measurement of the prompt photon production in hadronic collisions is important both as a test of QCD and as a tool to learn more about the structure of the proton. First of all there exist NLO calculations of the inclusive production cross-sections[35]: this means that the pure partonic cross-section is supposedly known with good accuracy and a precise measurement of the production rate could provide a test of the structure functions. Selecting particular kinematical configurations one becomes sensitive to relatively small values of x and probe the small-x behaviour of gluon distributions. The experimental input on prompt photon events could be used together with information coming from jet measurements or Drell-Yan production to further constrain the set of acceptable gluon parametrizations. Furthermore, since most of the jets recoiling against a γ are expected to be produced by quarks – contrarily to the *average* jet in pure QCD events – one can study the differences predicted by perturbative QCD in the fragmentation properties of quark and gluon jets[31].

In CDF we have two methods to separate direct photons from π^0 decays[36]. The first uses a shower profile measurement with a wire chamber (CES) located at shower maximum (5 radiation lengths) inside the EM calorimeter. A single γ showering in the EM calorimeter has a narrow profile, similar to that of a test beam electron, contrarily to the broader profile of a pair of collimated γ's such as those coming from a π^0 decay. A χ^2 fit to the EM shower profile can therefore be used to separate prompt γ's from decay γ's. In fig. 15 I show a comparison between the χ^2 distribution for isolated γ's – tagged as η decay products – and for γ's from a π^0 decay – with the π^0 tagged as product of a $\rho \to \pi^0\pi^\pm$ decay. The data are superimposed onto the detector simulation, showing that this provides a good description of the detector response. For any given bin in E_T the χ^2 q distribution of γ's is then fitted as a superposition of prompt and decay γ's; from the fit we obtain the rate of prompt γ's. This technique allows a good separation between signal and background up to $E_T \sim 35$ GeV, because for larger energies the two photons from π^0 decay are too collimated.

The other technique extracts the direct γ flux by counting the number of conversions produced just outside the tracking chambers and detected by the central drift tubes in front of the coil. The amount of conversion material is about 18% of radiation length. The probability for a single γ to convert is $10\pm2\%$ and roughly twice as much for the γ's from a π^0 decay. The preliminary CDF data are shown in fig. 16, together with the NLO calculation by Aurenche et al.[35]. At small Pt, where the shower profile technique gives a more precise measurement, data seem to be above the theoretical prediction, evaluated using several independent sets of structure functions. Still it is unclear whether this is due to an improper estimate of higher order contributions to the photon fragmentation function[37] or whether the origin of the discrepancy is in our poor knowledge of the gluon structure functions[38]. The better detection efficiency and the larger statistics we expect to enjoy during the coming run will hopefully help clarifying this important issue.

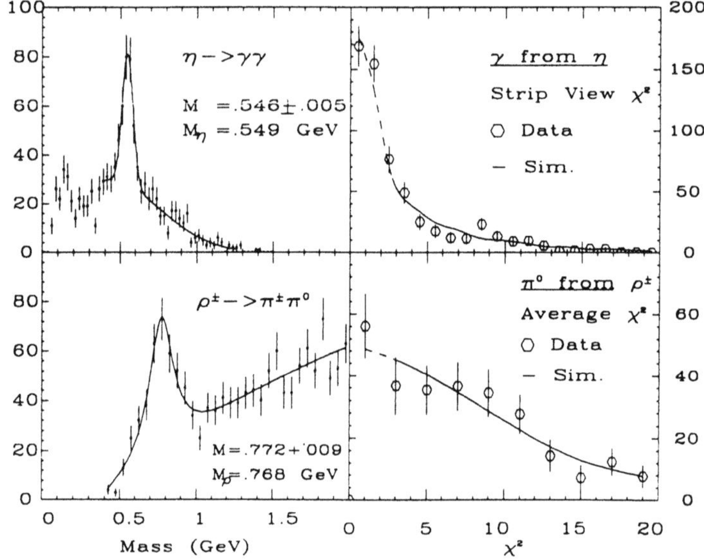

Figure 15. χ^2 distributions for isolated and π^0 photons (CDF prelim.)

W and Z p_T distributions

The full $O(\alpha_s^2)$ calculation of the inclusive p_T distribution for the W and the Z has been recently completed[39]. Depending on the p_T range observed, this process is mostly due to either $q\bar{q}$ or qg initial states, and therefore a measurement of the distribution throughout the full range of p_T accessible provides once again an independent test of the parametrizations of parton distributions. The study of the p_T in the region close to 0 – which is possible in the case of the Z where the momentum resolution is very good thanks to a precise reconstruction of the energies of the decay electrons in the EM calorimeter or of the momentum of the decay muons in the tracking system – is important because in this region the dominant effect comes from the emission of multiple soft gluons. Techniques exist to resum these contributions and precise comparisons between data and theory will help theorists gaining more confidence in these tools. Finally, a comparison of the data with the QCD prediction can furthermore indicate whether there are any anomalous sources of gauge boson production at large p_T, possible signals of new physics. The W-p_T analysis was performed using W decays into electrons[40]. In addition to the trigger selection – described in a previous Section – electrons were defined by requirements of isolation, shower/track matching, $E/P < 1.5$, shower profile χ^2. Furthermore only electrons with $p_T > 20$ GeV and falling within the fiducial region of the calorimeter were kept, and events with missing E_T smaller than 20 GeV were rejected. After eliminating events consistent with γ conversions or Z decays, a sample of 2496 events was selected.

The missing E_T measured by the detector was corrected for cracks and non-linearities before being identified as the p_T of the neutrino. The corrections were checked using a sample of $Z \rightarrow ee$ events. Finally the energy resolution function was used to unsmear the p_T spectrum, similarly to what was done for the inclusive jet-E_T analysis. The fully corrected distribution is shown on the left in fig. 17. The integrated cross-section for $p_T > 50$ GeV is $423\pm58(\text{stat})\pm108(\text{syst})$ pb, consistent with the theoretical prediction[39] of 428 ± 64 pb.

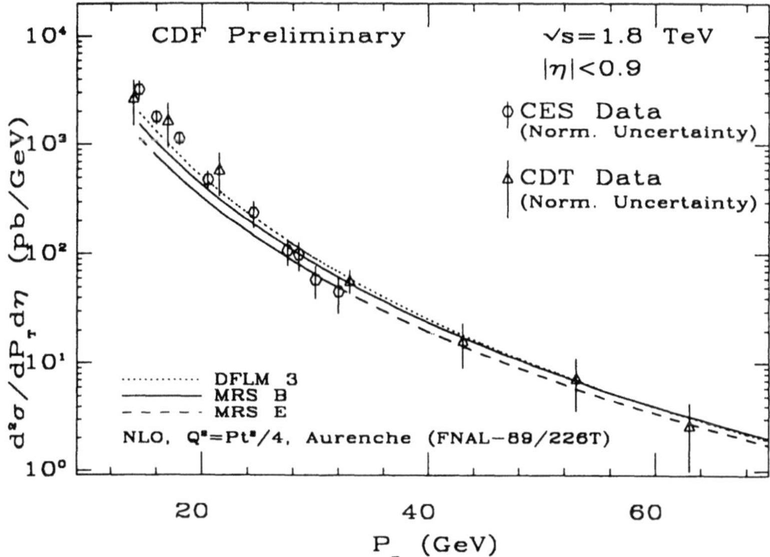

Figure 16. Photon P_T spectrum

The Z analysis was performed using both electron and muon decays (238 and 106 events respectively in the final plots). The p_T spectrum is shown on the right in fig. 17, with a blow-up of the small-p_T region. We measure $\langle p_T \rangle = 11.3\pm1$ GeV. In the high-p_T region we find $\sigma(p_T > 45 GeV) = (214 \pm 59)pb$, in agreement with the QCD prediction of 205 ± 31 pb.

Both distributions for the W and for the Z agree with the theoretical predictions, even though a better statistics is auspicated in order to use these measurements as stringent tests of QCD. The tails of the distributions show no deviation from what expected from the Standard Model.

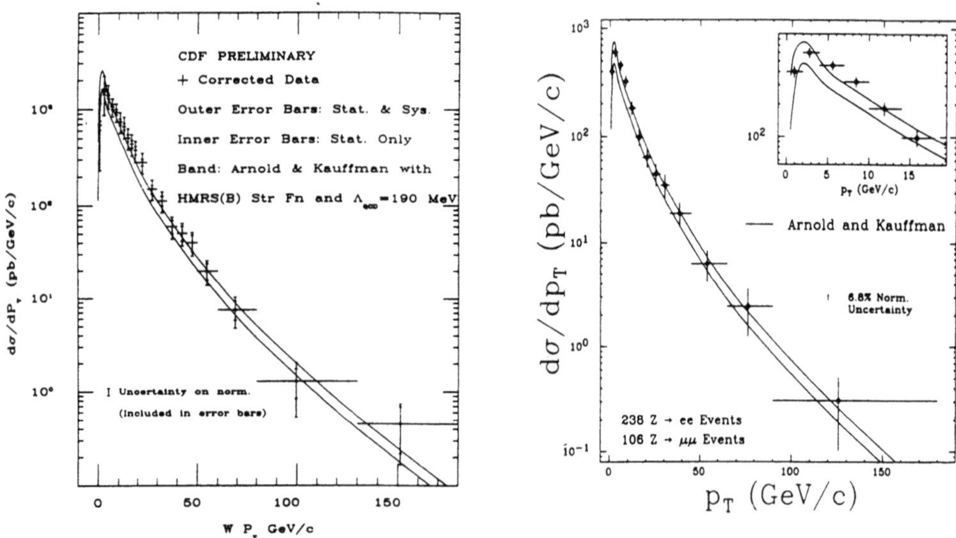

Figure 17. W and Z P_T spectrum

CONCLUSIONS

We presented results on QCD measurements from the 1988-89 data taking at CDF. A comparison between data and theoretical expectations shows an impressive *quantitative* agreement for a large set of inclusive variables (jet-E_T, W- and Z-p_T, jet shapes, ...) and over a wide range of cross-section rates (up to 7 orders of magnitude for the jet cross-sections).

We also find a good description of multi-jet topologies and correlations in terms of both parton-level and coherent shower Monte Carlo calculations. QCD Monte Carlo calculations, furthermore, provide a good description of the fragmentation properties of jets and provide the basis for the understanding of scaling violations in jet fragmentation functions and possible tools for the separation of quark and gluon jets.

Acknowledgements: I wish to thank Luisa Cifarelli and Yuri Dokshitzer for the kind invitation and for the lively scientific environment provided in Erice. In addition I would like to thank my friends and colleagues in CDF for their help and for the stimulating discussions, and in particular P. Giannetti, M.Dell'Orso and J. Huth for the help provided in the preparation of this talk.

References

[1] A. Zichichi, in these Proceedings.

[2] CDF Collaboration, *Nucl. Instr. and Meth.* **A267**:249,257,272,280,301,315,330 (1988); *Nucl. Instr. and Meth.* **A268**:24,33,41,46,50,75,92 (1988).

[3] G. Marchesini and B. Webber, *Phys. Rev.* **D38**:3419 (1988).

[4] R.K. Ellis and J. Sexton, *Nucl. Phys.* **B269**:445 (1986);
F. Aversa, P. Chiappetta, M. Greco and J.Ph. Guillet, *Nucl. Phys.* **B327**:105 (1989),*Z. Phys.* **C46**:253 (1990);
F. Aversa, P. Chiappetta, L. Gonsalez, M. Greco and J.Ph. Guillet, CERN-TH-5710/90;

S. Ellis, Z. Kunszt and D. Soper, *Phys. Rev.* **D40**:2188 (1989), *Phys. Rev. Lett.* **64**:2121 (1990).

[5] J. Huth et al., Fermilab-CONF-90/249-E, to appear in the Proceedings of the Summer Study on High Energy Physics (1990), Snowmass, Co.

[6] J.W. Stirling, in these Proceedings.

[7] F. Abe et al. (CDF Collab.), *Phys. Rev. Lett.* **62**:613 (1989);
F. Abe et al. (CDF Collab.), FERMILAB-PUB-91/231-E, submitted to Phys. Rev. Lett.

[8] A. Martin, R. Roberts and J. Stirling, *Phys. Rev.* **D37**:1161 (1988).

[9] A. Harriman, A. Martin, R. Roberts and J. Stirling, *Phys. Rev.* **42**:798 (1990).

[10] J. Morfin and W.K. Tung, Fermilab-PUB-90/74, IIT-PUB-90/11 (1990), to appear in Z. Phys. C.

[11] Courtesy of S. Ellis, private communication to the CDF Collaboration.

[12] S. Bethke, Invited talk given at the Workshop on Jet Physics at LEP and HERA, Durham (UK), Dec. 9-15, 1990, CERN-PPE/91-36; see also the present Proceedings.

[13] S. Catani, L. Trentadue, G. Turnock and B. Webber, *Phys. Lett.* **263B**:491 (1991).

[14] Z. Kunszt and D. Soper, preprint OITS475, ETH-TH/91/29 (1991).

[15] M. Diemoz, F. Ferroni, E. Longo and G. Martinelli, *Z. Phys.* **C39**:21 (1988).

[16] D.W. Duke and J.F. Owens, *Phys. Rev.* **D30**:49 (1984).

[17] E. Eichten, I. Hinchliffe, K. Lane and C. Quigg, *Rev. Mod. Phys.* **56**:579 (1984).

[18] M. Dell'Orso et al., CDF Internal Note 1056 (1989).

[19] F. Abe et al. (CDF Collab.), Fermilab-PUB-91/232-E, submitted to Phys. Rev. D.

[20] F. Halzen and P. Hoyer, *Phys. Lett.* **130B**:326 (1983);
B.L Combridge, C.J. Maxwell, *Nucl. Phys.* **B239**:429 (1984).

[21] G. Arnison et al. (UA1 Collab.), *Phys. Lett.* **158B**:494 (1985).

[22] J. Appel et al. (UA2 Collab.), *Z. Phys.* **C30**:341 (1986).

[23] F. A. Berends, R. Kleiss, P. De Causmaeker, R. Gastmans and T.T. Wu, *Phys. Lett.* **103B**:124 (1981);
T. Gottschalk and D. Sivers, *Phys. Rev.* **D21**:102 (1980) ;
Z. Kunszt and E. Pietarinen, *Nucl. Phys.* **B164**:45 (1980).

[24] B.R. Webber, Ann. Rev. of Nucl. Sci. **36** (1986) 253.

[25] G. Marchesini and B.R. Webber, *Nucl. Phys.* **B310**:461 (1988).

[26] F. Paige and S.D. Protopopescu, Brookhaven report BNL-38034 (1986).

[27] M. Dell'Orso, Proceedings of the 8th Topical Workshop on Proton-Antiproton Collider Physics, 1-5 Sept. 1989, Castiglione della Pescaia (Italy), eds. G. Bellettini and A. Scribano, p.177.

[28] E. Meschi, private communication.

[29] E. Buckley, S. Geer and C. Wendt, CDF/DOC/JET/CDFR/1381 (1991); F. Abe et al (CDF Collab.), to be submitted to *Phys. Rev. D*.

[30] B. Andersson, G. Gustafson and T. Sjöstrand, *Z. Phys.* **C6**:235 (1980); *Nucl. Phys.* **B197**:45 (1982).

[31] F. Abe et al. (CDF Collab.), CDF/ANAL/JET/CDFR/1572. to be submitted to *Phys. Rev. D*.

[32] B. Denby et al., work in progress.

[33] F. Abe et al. (CDF Collab.), *Phys. Rev. Lett.* **65**:968 (1990).

[34] M. Althoff et al. (TASSO Collab.), *Z. Phys.* **C22**:307 (1984).

[35] P. Aurenche, R. Baier, M. Fontannaz and D. Shiff, *Nucl. Phys.* **B297**:661 (1988);
A.P. Contogouris, S. Papadopoulos and D. Atwood, Mc Gill Univ. Report (1989);
H. Baer, J. Ohnemus and J.F. Owens, *Phys. Rev.* **42**:61 (1990).

[36] R. Harris (CDF Collab.), Proceedings of the Workshop on Hadron Structure Functions
and Parton Distributions, Eds. D.F. Geesaman et al. World Scientific, Singapore, 1990,
p. 278.

[37] E.L. Berger and J. Qiu, ANL-HEP-PR-90-104, ITP-SB-90-81 (December 1990).

[38] P. Aurenche, R. Baier, M. Fontannaz, J.F. Owens and M. Werlen, *Phys. Rev.* **39**:3275
(1989).

[39] P.B. Arnold and M.H. Reno, *Nucl. Phys.* **B319**:37 (1989);
P.B. Arnold and R.P. Kaufmann, *Nucl. Phys.* **B349**:381 (1991).

[40] F. Abe et al. (CDF Collab.), *Phys. Rev. Lett.* **66**:2951 (1991).

[41] F. Abe et al. (CDF Collab.), Fermilab PUB-91/199-E, to appear in *Phys. Rev. Lett.*

QCD CORRECTIONS TO Z PAIR HADRONIC PRODUCTION

Barbara Mele

INFN, Sezione di Roma, Italy

1. INTRODUCTION

The detailed knowledge of the production of two Z vector bosons at future hadron colliders is a crucial problem if one considers the search for Higgs bosons.

In proton-proton collisions a heavy Higgs , with mass larger than twice $M_{W,Z}$, is produced, through gluon [1] or vector boson fusion [2,3], mostly accompanied by two forward-backward peaking jets that usually escape detection. Hence what one observes experimentally are just the Higgs decay products, that is a couple of W or Z with an invariant mass about m_H [4]. The detection of the Higgs decay into W's is particularly difficult due to the overwhelming background from $t\bar{t}$ strong-interaction production with the subsequent top decay $t \to Wb$. On the other hand a very clear signal arises from the Higgs decay into a pair of Z's [5]. In a considerable fraction of the events one observes four charged leptons from which the two Z kinematical features can be easily recostructed. This channel is almost free from strong-interaction background. The production of four charged leptons not coming from two Z's have been studied and can be reduced by means of appropriate kinematical cuts [6]. The most dangerous and irreducible background comes then from the ZZ continuum production $pp \to ZZ + X$.

In Fig.1 [7] the four-charged-leptons invariant mass distribution relative to the Higgs case for $m_H = 600$ GeV and $m_H = 800$ GeV is compared with the ZZ continuum production at LHC. Various kinematical cuts are applied in order to enhance the signal to background ratio. The ZZ continuum curve includes the QCD leading-order contribution from $q\bar{q} \to ZZ$ plus the order α_S^2 channel $gg \to ZZ$ proceeding through a quark loop [8]. Although important, this channel is not dominant over the $q\bar{q}$ annihilation process at the LHC and SSC energies. One should

QCD at 200 TeV, Edited by L. Cifarelli
and Y. Dokshitzer, Plenum Press, New York, 1992

should also stress that $gg \to ZZ$ is only part of the order α_S^2 correction to the ZZ pair production in pp collisions. All the other subprocesses contributing to the QCD corrections to this order should be included in order to know accurately the relevance of the process $gg \to ZZ$.

From Fig.1 one can see that an accurate prediction for the process $pp \to ZZ + X$ is crucial in order to disentangle the hadronic production of a heavy Higgs.

Until recently QCD radiative corrections to ZZ production were known only for the tree-level production of high transverse momentum Z pairs. The cross

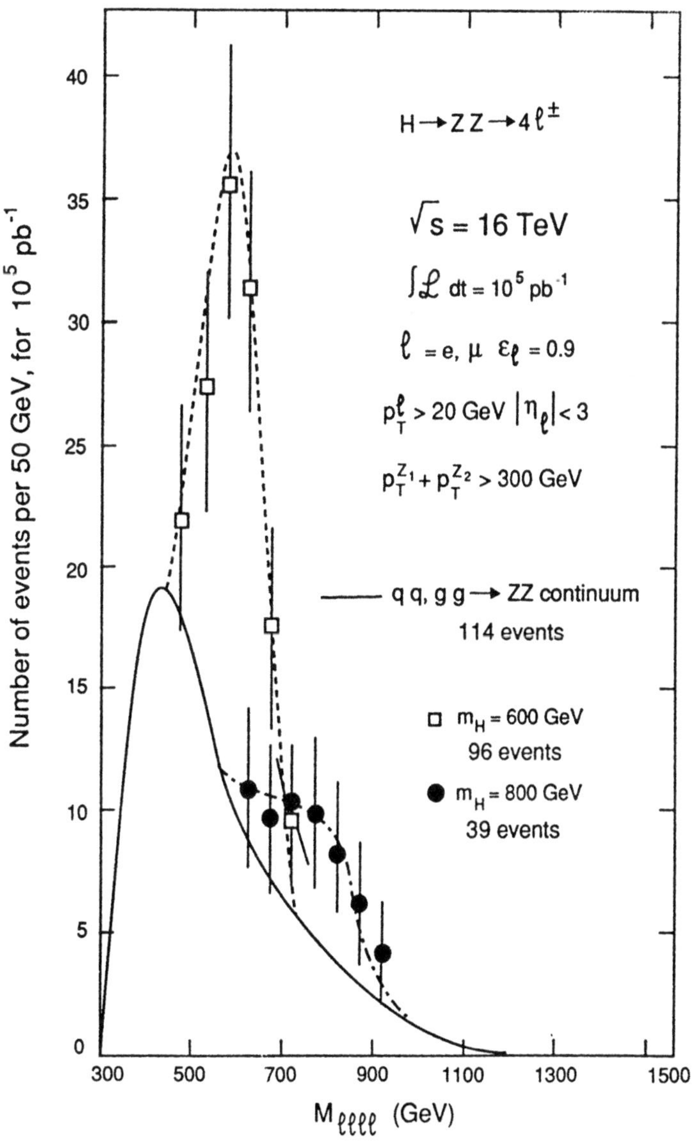

Fig. 1. ZZ background to heavy Higgs production.

section for the process $q\bar{q} \to ZZ$ was calculated in the Born approximation in ref. [9]. In ref. [10], vector boson pair production accompanied by a large transverse momentum jet was studied in leading order QCD. For the ZZ case this process arises at the parton level through the channels $q\bar{q} \to ZZg$, $qg \to ZZq$ which can give an important background for Higgs production at large $p_\perp$ [11]. Also ZZ pair production accompanied by two large transverse momentum jets, $pp \to ZZJJ + X$, arising in second order QCD, was studied in refs. [12,13]. In ref. [13] the purely electroweak contribution to this process, including Higgs boson exchange, was analysed as well. All these calculations are based on tree-level amplitudes and do not include virtual corrections.

Here we present the results of a complete calculation (including both real and virtual gluon corrections) of the first order QCD corrections to the Z pair production cross section at hadron colliders [14]. The same calculation has been performed in ref. [15] that appeared shortly after our paper. Besides a better determination of background for new physics, this calculation will provide a further way to test perturbative QCD predictions.

2. SOME COMMENTS ON THE CALCULATION

The order α_S $pp \to ZZ + X$ cross section can be obtained by properly convoluting proton structure functions accurate to next-to-leading order with the order α_S short-distance cross sections σ_{ij}, where i,j stand for the different incoming parton types $q\bar{q}$, gq, $g\bar{q}$. The diagrams involved in the process $q\bar{q} \to ZZ + X$ up to the order α_S are shown in figs. 2, 3, 4 and those for the process $gq(g\bar{q}) \to ZZ + X$ are shown in fig. 5. We have not included self-energy corrections on the external fermion lines, since they vanish in dimensional regularization, which is the scheme we adopted. The cross section to order α_S is given by the square of the Born term, plus the interference of the Born with the virtual graphs, plus the square of the sum of the real graphs

$$\sigma_{ij} = \sigma_{ij}^{(b)} + \sigma_{ij}^{(v)} + \sigma_{ij}^{(r)}. \tag{2.1}$$

The calculation is rather lengthy and complicated. Here we make just some comments on the general strategy we adopted and refer for the details to the more detailed description in ref. [14].

One of the most delicate point is the treatment of ultraviolet, collinear and infrared divergences that appear in intermediate steps of the calculation. An appropriate regularization scheme must be introduced. The dimentional regularization presents the advantages of simplifying calculations and yielding a result directly in the $\overline{MS}$ scheme [16]. On the other hand, the presence of the γ_5 matrix in the $Zq\bar{q}$ vertex $-i\gamma^\mu(g_V - g_A\gamma_5)$ would in general prevent a straightforward use of this reg-

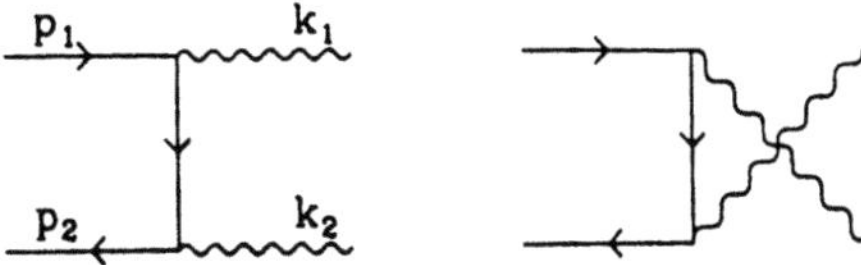

Fig. 2.   Born diagrams for the hadronic production of a pair of Z bosons.

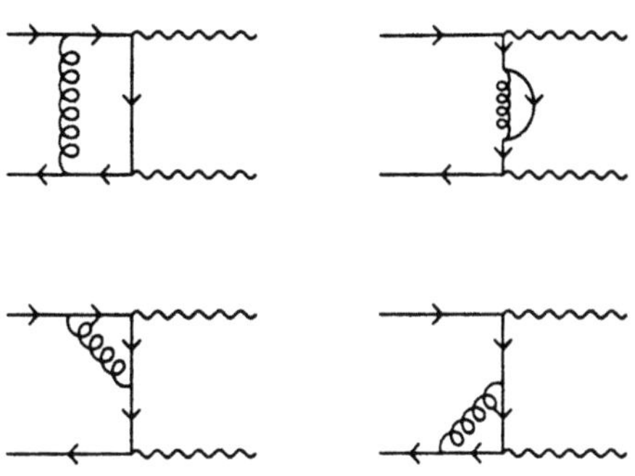

+ crossed diagrams

Fig. 3.   Virtual diagrams.

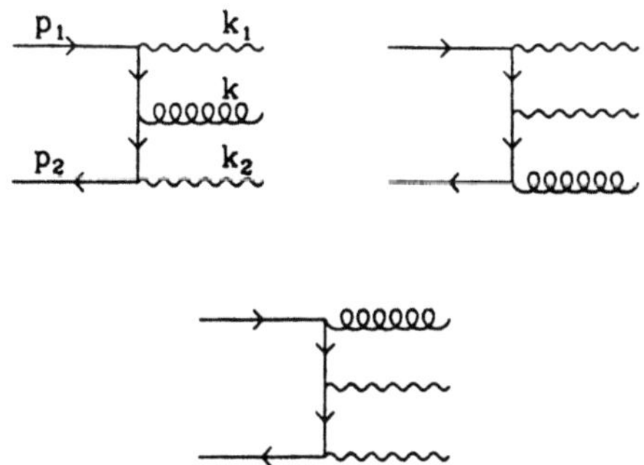

+ crossed diagrams

Fig. 4.   Real diagrams.

ularization scheme due to the problem of the γ_5 definition in $n \neq 4$ dimentions. As a matter of fact, in our particular case one can show that the γ_5 problem can be avoided completely and dimentional regularization can be safely used. Indeed, one can consider another renormalization scheme, that eventually can be related to the $\overline{MS}$ scheme through definite trasformation rules. Since Z pair production at the order α_S involves neither non-abelian QCD vertices nor divergent fermionic loops, one can regularize all divergences preserving vector and axial current conservation by introducing the following substitution in the gluon propagator

$$\frac{1}{l^2} \rightarrow \frac{1}{l^2 - \lambda^2} - \frac{1}{l^2 - \Lambda^2} \tag{2.2}$$

with Λ large and λ small. As a consequence, Λ regulates the ultraviolet divergences and λ the collinear and soft ones. Ultraviolet divergences cancel between wave function renormalization factors, vertex corrections and self-energy corrections. Therefore Λ disappears from the result. Logarithms of λ do instead remain in the inclusive cross section, and they collect themselves into collinear factors convoluted with the Born cross section. The separation of these collinear factors would yield a cross section which can be related to the $\overline{MS}$ cross section through a well defined transformation of scheme.

In the scheme proposed above γ_5 anticommutes with all the other Dirac matrices. Hence, one can easily show that, in the order α_S matrix element, all terms containing an even power of the axial coupling g_A can be collected as $g_V^4 + g_A^4 + 6g_A^2 g_V^2$ times a common factor. The contribution of terms with an odd power of g_A is finite. In fact, all divergences must be factorizable in terms of the Born cross section, and there are no terms with an odd number of g_A in the Born term. Furthermore, in the case of the production of two identical vector bosons this contribution is zero, because it must be proportional to the antisymmetric tensor contracted with a set of independent momenta in the process, while it should also be symmetric with respect to the momenta of the two vector bosons. It is therefore sufficient to calculate our process setting $g_A = 0$ at the beginning, and then replacing g_V^4 with $g_V^4 + g_A^4 + 6g_V^2 g_A^2$ at the end of the calculation. In this case, since we only have vector couplings around, we can use directly dimensional regularization.

At this point one can proceed with the evaluation of the virtual and real contributions to the order α_S cross section. Due to the extension of the calculation, we used both Reduce and Macsyma for the algebric manipulation of the matrix elements in order to check the results.

The interference of the virtual graphs, shown in Fig.3, with the Born graphs gives the virtual contribution, that we treated in the following way. We carried out all the traces and contraction of colour and spin indices. The scalar products appearing in the numerator of the resulting expression were then decomposed in terms of linear combinations of denominators. All these algebric steps were performed by computer. At the end of this procedure one is left with a small number of virtual integrals of bosonic type. We evaluated these integrals by hand . The complete result for their analytic expressions and for the virtual matrix element can be found in ref. [14].

For the real part, we evaluated the square of the graphs in Figs.4 and 5. In the contribution from diagrams in Fig.4, soft and collinear divergences appear as single poles in the gluon kinematical variables, while double poles, that appear in individual diagrams, cancel in the total. In ref. [14] a set of gluon variables is chosen that is particularly convenient to single out and regularize the different divergent terms. An analogous procedure has been applied to treat the final-quark divergences in Fig. 5. At the end, collinear divergences (whose form can also be predicted on the basis of the factorization theorem [17]) are subtracted and reabsorbed into hadronic structure functions, while soft divergences cancel by adding the real contribution to the virtual one. As a result of this procedure, some residual dependence upon the renormalization scale μ is left in the subtracted parton cross section $\hat{\sigma}_{ij}$.

The final analytic result for the matrix element and further details on the calculation can be found in ref. [14]. In the following section, we will give some approximated formula for the total partonic cross section that can be useful for most of the practical applications.

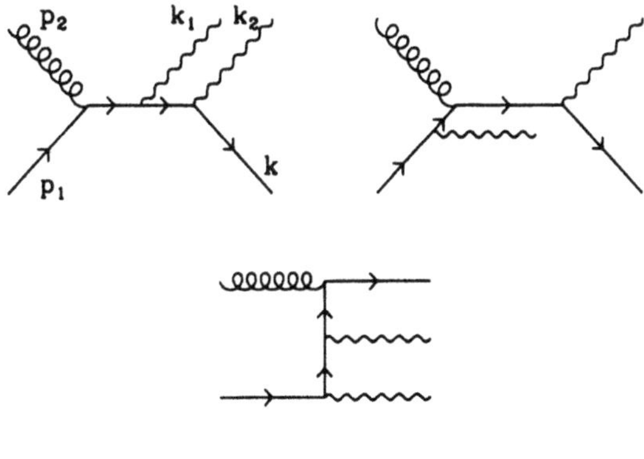

+ crossed diagrams

Fig. 5.   Quark-gluon graphs.

3. TOTAL PARTON CROSS SECTIONS

The physical cross section for the process $pp \to ZZ + X$ is given by

$$\sigma_{H_1 H_2}(p_1, p_2) = \sum_{kl} \int \hat{\sigma}_{kl}(x_1 p_1, x_2 p_2) f_{kH_1}(x_1) f_{lH_2}(x_2) dx_1 dx_2 \qquad (3.1)$$

where $f_{kH_{1(2)}}$ is the structure function of the parton k in the hadron $H_{1(2)}$ and $\hat{\sigma}_{kl}$ are the total short-distance cross sections . As a consequence of the subtraction of collinear divergencies, both $\hat{\sigma}_{kl}$ and structure functions become dependent upon the scale μ. In particular, the expressions for $\hat{\sigma}_{kl}$ depend on the subtraction scheme adopted. In ref. [14] we give the explicit form of $\hat{\sigma}_{kl}$ in the $\overline{\text{MS}}$ scheme. If we want to use our results in conjunction with structure function sets defined in a different scheme (like the DFLM[21] sets) we must correct our short-distance cross section in order to account for the redefinition. In ref. [14], therefore, we give also the transformation functions needed to go from the $\overline{\text{MS}}$ scheme to a scheme in which the deep inelastic structure function F_2 receives no corrections to order α_S[19,20], which we will call from now on DIS scheme. These transformation functions must be used in conjunction with the structure functions of ref. [21].

We proceed now to discuss the final result for the total parton cross sections. The total cross sections for the processes $q\bar{q} \to ZZ + X$, $qg \to ZZ + X$ can be written in the following way:

$$\hat{\sigma}_{ij}(s, m^2, \mu^2) = \frac{g_V^4 + g_A^4 + 6g_V^2 g_A^2}{16\pi N_C m^2} h_{ij}\left(\rho, \frac{\mu^2}{m^2}\right), \qquad (3.2)$$

where $\rho = 4m^2/s$ and μ is the renormalization scale. The indices i and j refer to the type of the colliding partons. Since the functions $h_{ij}(\rho, \mu^2/m^2)$ depend on the choice of the renormalization scheme; we will give our results both in the $\overline{\text{MS}}$ and in the DIS scheme.

The quantities $h_{ij}(\rho, \mu^2/m^2)$ have a perturbative expansion in α_S:

$$h_{ij}\left(\rho, \frac{\mu^2}{m^2}\right) = h_{ij}^{(0)}(\rho) + \frac{\alpha_S}{4\pi}\left[h_{ij}^{(1)}(\rho) + \hat{h}_{ij}^{(1)}(\rho) \log \frac{\mu^2}{m^2}\right] + \mathcal{O}(\alpha_S^2). \qquad (3.3)$$

From the Born cross section one easily obtains

$$h_{q\bar{q}}^{(0)}(\rho) = \rho \left[\frac{1 + \rho^2/4}{1 - \rho/2} \log \frac{1 + \beta}{1 - \beta} - \beta\right], \qquad (3.4)$$

where $\beta = \sqrt{1 - \rho}$, while $h_{qg}^{(0)}(\rho) = 0$ since there is no $\mathcal{O}(\alpha_S^0)$ contribution to the qg process. The function $h_{q\bar{q}}^{(0)}(\rho)$ is plotted in fig. 6.

In eq. (3.3), the dependence on the renormalization scale μ has been shown explicitly. The coefficient functions $\hat{h}_{ij}^{(1)}$ can be expressed in term of the lowest order functions $h_{ij}^{(0)}$ and the Altarelli-Parisi splitting functions [18]:

$$\hat{h}_{ij}^{(1)}(\rho) = -2\sum_k \int_\rho^1 P_{ki}(x)h_{kj}^{(0)}\left(\frac{\rho}{x}\right) dx - 2\sum_k \int_\rho^1 P_{kj}(x)h_{ik}^{(0)}\left(\frac{\rho}{x}\right) dx. \tag{3.5}$$

The above equation is obtained by differentiating both sides of eq. (3.1) with respect to $\log\mu^2$. The left-hand side of eq. (3.1) is μ-independent, and the derivative of the structure functions are given by the Altarelli-Parisi equation. In our case we get

$$\hat{h}_{q\bar{q}}^{(1)}(\rho) = -4\int_\rho^1 P_{qq}(x)h_{q\bar{q}}^{(0)}\left(\frac{\rho}{x}\right) dx$$

$$\hat{h}_{qg}^{(1)}(\rho) = -2\int_\rho^1 P_{qg}(x)h_{q\bar{q}}^{(0)}\left(\frac{\rho}{x}\right) dx \tag{3.6}$$

where

$$P_{qq}(x) = C_F\left[\frac{1+x^2}{(1-x)_\rho} + \left(\frac{3}{2} + 4\log\beta\right)\delta(1-x)\right]$$

$$P_{qg}(x) = T_F\left[x^2 + (1-x)^2\right] \tag{3.7}$$

are the relevant Altarelli-Parisi kernels. The distribution in round brackets in $P_{qq}(x)$ is defined according to the prescription

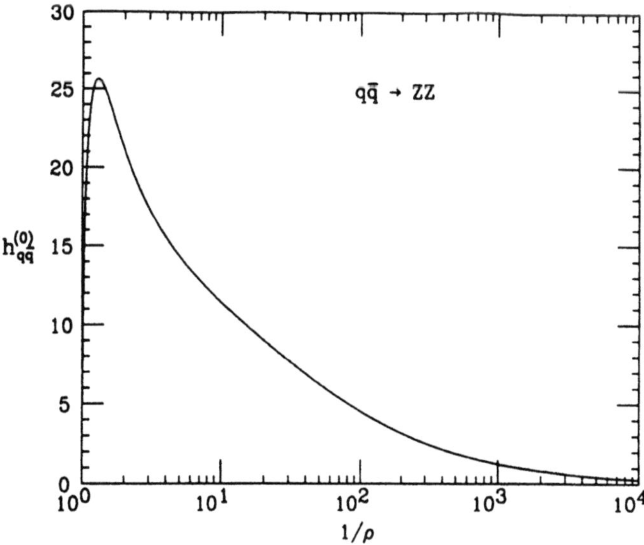

Fig. 6.   The function $h_{q\bar{q}}^{(0)}(\rho)$.

$$\int_\rho^1 h(x)\left(\frac{1}{1-x}\right)_\rho dx = \int_\rho^1 \frac{h(x)-h(1)}{1-x}dx \tag{3.8}$$

where $h(x)$ is any sufficiently regular test function.

We do not have analytic espressions for $h_{ij}^{(1)}(\rho)$ and $\hat{h}_{ij}^{(1)}(\rho)$. Instead, we will provide for them a fit that agrees to better than 1% with the results obtained by numerical integration of the formulae for the matrix elements in ref. [14].

For the $q\bar{q}$ process, the behaviour of $h_{q\bar{q}}^{(1)}(\rho)$ and $\hat{h}_{q\bar{q}}^{(1)}(\rho)$ near the threshold of ZZ production, $s \sim 4m^2$ or $\rho \sim 1$, is dominated by the contribution of the soft-real and virtual terms. We find

$$
\begin{aligned}
h_{q\bar{q}}^{(1)\overline{\mathrm{MS}}}(\rho) &= \left\{ C_F\left[(6+16\log\beta)\log\frac{s}{m^2} + 32\log^2\beta - \frac{4}{3}\pi^2\right]\right.\\
&\quad + a_0 + a_1\rho - a_2\log\beta - a_3\rho\log\beta - a_4\log\rho\\
&\quad \left. - a_5\rho\log\rho + a_6\log^2\rho + a_7\rho\log^2\rho\right\} h_{q\bar{q}}^{(0)}(\rho)
\end{aligned}
\tag{3.9}
$$

$$
\begin{aligned}
h_{q\bar{q}}^{(1)\mathrm{DIS}}(\rho) &= \left\{ C_F\left[(6+16\log\beta)\log\frac{s}{m^2} + 16\log^2\beta + 12\log\beta + 18\right]\right.\\
&\quad + b_0 + b_1\rho - b_2\log\beta - b_3\rho\log\beta - b_4\log\rho\\
&\quad \left. - b_5\rho\log\rho + b_6\log^2\rho + b_7\rho\log^2\rho\right\} h_{q\bar{q}}^{(0)}(\rho)
\end{aligned}
\tag{3.10}
$$

$$
\begin{aligned}
\hat{h}_{q\bar{q}}^{(1)}(\rho) &= \left\{ -C_F(6+16\log\beta)\right.\\
&\quad + c_0 + c_1\rho - c_2\log\beta - c_3\rho\log\beta - c_4\log\rho\\
&\quad \left. - c_5\rho\log\rho + c_6\log^2\rho + c_7\rho\log^2\rho\right\} h_{q\bar{q}}^{(0)}(\rho),
\end{aligned}
\tag{3.11}
$$

with the coefficients a_i, b_i and c_i given in Table I. The functions $h_{q\bar{q}}^{(1)\overline{\mathrm{MS}}}(\rho)$, $h_{q\bar{q}}^{(1)\mathrm{DIS}}(\rho)$ and $\hat{h}_{q\bar{q}}^{(1)}(\rho)$ are plotted in fig. 7.

In the qg case our result is

$$
\begin{aligned}
h_{qg}^{(1)\overline{\mathrm{MS}}}(\rho) &= \left[d_0 + d_1\rho - d_2\log\beta - d_3\rho\log\beta - d_4\log\rho\right.\\
&\quad \left. - d_5\rho\log\rho + d_6\log^2\rho\right]\rho\beta\log\rho
\end{aligned}
\tag{3.12}
$$

$$
\begin{aligned}
h_{qg}^{(1)\mathrm{DIS}}(\rho) &= \left[e_0 + e_1\rho - e_2\log\beta - e_3\rho\log\beta - e_4\log\rho\right.\\
&\quad \left. - e_5\rho\log\rho + e_6\log^2\rho\right]\rho\beta\log\rho
\end{aligned}
\tag{3.13}
$$

$$
\begin{aligned}
\hat{h}_{qg}^{(1)}(\rho) &= \left[f_0 + f_1\rho - f_2\log\beta - f_3\rho\log\beta - f_4\log\rho\right.\\
&\quad \left. - f_5\rho\log\rho + f_6\log^2\rho\right]\rho\beta\log\rho,
\end{aligned}
\tag{3.14}
$$

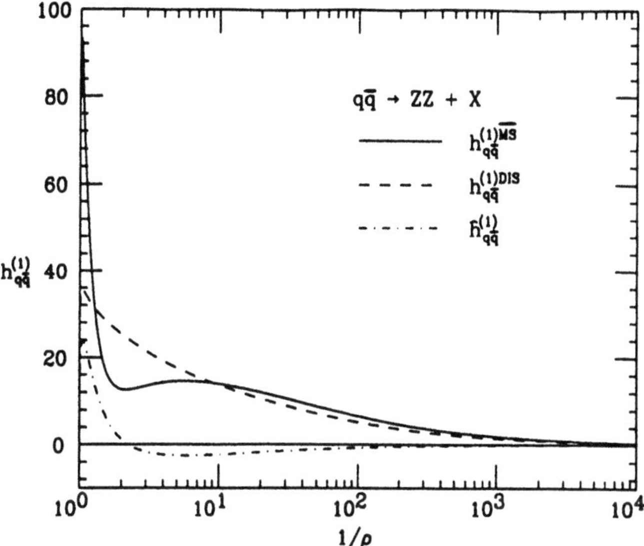

Fig. 7.   The functions $h_{q\bar{q}}^{(1)\overline{\mathrm{MS}}}(\rho)$, $h_{q\bar{q}}^{(1)\mathrm{DIS}}(\rho)$ and $\hat{h}_{q\bar{q}}^{(1)}(\rho)$.

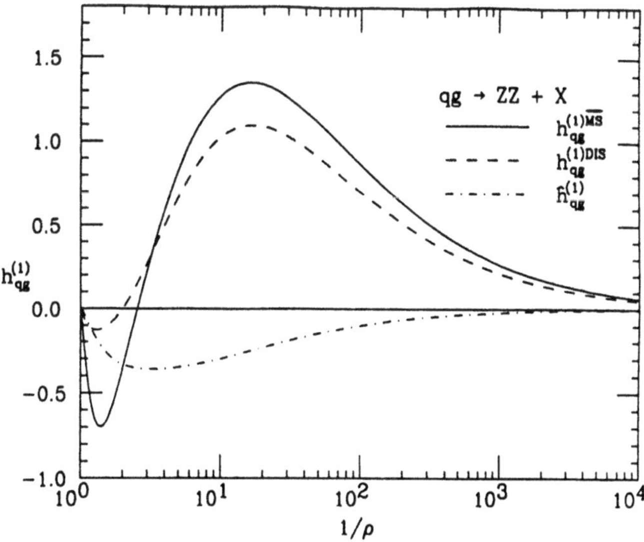

Fig. 8.   The functions $h_{qg}^{(1)\overline{\mathrm{MS}}}(\rho)$, $h_{qg}^{(1)\mathrm{DIS}}(\rho)$ and $\hat{h}_{qg}^{(1)}(\rho)$.

Table I. Coefficients for the fit of $h_{q\bar{q}}^{(1)\overline{\text{MS}}}(\rho)$, $h_{q\bar{q}}^{(1)\text{DIS}}(\rho)$ and $\hat{h}_{q\bar{q}}^{(1)}(\rho)$.

j	a_j	b_j	c_j
0	30.2	21.075	-0.16875
1	-2.56	-5.5530	6.9135
2	-67.5	-29.430	-2.4114
3	93.0	42.245	2.3582
4	-6.96	-16.597	-1.2961
5	17.9	9.5026	-3.5857
6	4.82	4.2457	-0.092729
7	0	-12.941	5.2871

Table II. Coefficients for the fit of $h_{qg}^{(1)\overline{\text{MS}}}(\rho)$, $h_{qg}^{(1)\text{DIS}}(\rho)$ and $\hat{h}_{qg}^{(1)}(\rho)$.

j	d_j	e_j	f_j
0	-2.431	-2.957	0.6104
1	6.096	-0.8835	2.366
2	-12.46	3.134	-2.778
3	22.93	2.552	2.665
4	-0.4829	-0.1864	0.2999
5	5.339	6.566	-0.4115
6	-0.6835	-0.5562	-0.01072

with d_i, e_i and f_i given in Table II. The functions $h_{qg}^{(1)\overline{\text{MS}}}(\rho)$, $h_{qg}^{(1)\text{DIS}}(\rho)$ and $\hat{h}_{qg}^{(1)}(\rho)$
are plotted in fig.8.

4. $pp \to ZZ + X$ TOTAL CROSS SECTION

As anticipated in the previous section, the total cross section for $pp \to ZZ + X$
is obtained by convoluting parton cross sections with proton structure function
through the eq. (3.1) . In order to give a consistent result at the next-to-leading
level, we must use sets of structure functions which are accurate to next-to-leading
order. We have chosen as reference set the DFLM[21] set with $\Lambda_4 = 260$ MeV. We
use the two-loop expression for α_S. In our calculation, we set $M_Z = 91.18$ GeV,
$\sin^2 \theta_W = 0.228$, $\alpha_{\text{em}} = 1/128$ and $g^2 = 4\pi\alpha_{\text{em}}/\sin^2 \theta_W$.

Table III. Cross sections (in pb) for Z pair production at various colliders, and for different sets of structure functions. The scale μ is chosen equal to 90 GeV, and the errors quoted correspond to $\mu = 45$ and $\mu = 180$ GeV. Both the Born and the full (Born $+\ \mathcal{O}(\alpha_S)$) cross sections are shown.

		DFLM160	DFLM260	DFLM360	HMRSE	HMRSB
$p\bar{p}$	Born	$0.0551^{+0.01}_{-0.007}$	$0.0475^{+0.01}_{-0.007}$	$0.0417^{+0.01}_{-0.007}$	$0.0682^{+0.01}_{-0.009}$	$0.0571^{+0.01}_{-0.008}$
$\sqrt{S} = 0.63$ TeV	Full	$0.0693^{+0.005}_{-0.004}$	$0.0610^{+0.005}_{-0.004}$	$0.0542^{+0.005}_{-0.004}$	$0.0889^{+0.005}_{-0.005}$	$0.0766^{+0.005}_{-0.005}$
$p\bar{p}$	Born	$0.925^{+0.07}_{-0.06}$	$0.869^{+0.07}_{-0.06}$	$0.820^{+0.07}_{-0.06}$	$1.02^{+0.07}_{-0.04}$	$0.972^{+0.03}_{-0.07}$
$\sqrt{S} = 1.8$ TeV	Full	$1.17^{+0.04}_{-0.04}$	$1.12^{+0.05}_{-0.04}$	$1.07^{+0.06}_{-0.05}$	$1.25^{+0.04}_{-0.04}$	$1.22^{+0.05}_{-0.04}$
pp	Born	$3.00^{+0.07}_{-0.1}$	$3.02^{+0.05}_{-0.1}$	$2.99^{+0.02}_{-0.07}$	$3.01^{+0.1}_{-0.2}$	$3.29^{+0.1}_{-0.1}$
$\sqrt{S} = 6$ TeV	Full	$3.97^{+0.08}_{-0.06}$	$4.06^{+0.1}_{-0.1}$	$4.06^{+0.1}_{-0.1}$	$3.80^{+0.06}_{-0.03}$	$4.21^{+0.09}_{-0.05}$
pp	Born	$9.8^{+0.9}_{-1.1}$	10.7^{+1}_{-1}	11.4^{+1}_{-1}	$8.13^{+0.9}_{-1}$	9.75^{+1}_{-1}
$\sqrt{S} = 16$ TeV	Full	$13.7^{+0.04}_{-0.08}$	$15.2^{+0.01}_{-0.02}$	$16.3^{+0.1}_{-0.1}$	$11.0^{+0.1}_{-0.1}$	$13.2^{+0.1}_{-0.08}$
pp	Born	24.2^{+4}_{-4}	29.2^{+4}_{-5}	34.0^{+5}_{-5}	17.7^{+4}_{-3}	23.5^{+4}_{-4}
$\sqrt{S} = 40$ TeV	Full	35.8^{+1}_{-1}	43.6^{+1}_{-1}	$50.8^{+0.7}_{-1}$	$25.6^{+0.9}_{-1}$	33.9^{+1}_{-1}
pp	Born	48.6^{+12}_{-11}	64.6^{+14}_{-14}	82.2^{+17}_{-18}	36.8^{+10}_{-9}	53.2^{+13}_{-13}
$\sqrt{S} = 100$ TeV	Full	77.4^{+3}_{-5}	103^{+3}_{-5}	131^{+3}_{-6}	56.3^{+3}_{-5}	81.3^{+3}_{-6}
pp	Born	70.1^{+21}_{-19}	100^{+28}_{-26}	136^{+35}_{-34}	61.7^{+20}_{-17}	95.0^{+29}_{-26}
$\sqrt{S} = 200$ TeV	Full	120^{+7}_{-10}	172^{+7}_{-12}	232^{+7}_{-13}	98.5^{+8}_{-11}	150^{+12}_{-16}

We start by considering the scale dependence of our final result. In Figs. 9 and 10 we plot the total cross section as a function of the scale μ for $\sqrt{S} = 2$ TeV and 16 TeV respectively. Both the leading and next-to-leading results are shown. We can see that in both cases the next-to-leading result is less sensitive to scale changes, as expected. At LHC energy we observe that the scale dependence of the next-to-leading result is particularly mild.

In the following we will quote cross sections together with their variation coming from taking $\mu = 45, 90, 180$ GeV, keeping however in mind that this cannot always be considered as a good estimate of the theoretical error due to $\mathcal{O}(\alpha_S^2)$ and higher order corrections.

In Fig. 11 we plot the cross section for Z pair production versus energy. We see that radiative corrections are positive and of the order of 20% to 40% in the energy range considered.

In Table III we give cross sections for Z pair production for energies of interest, using the three sets of DFLM structure functions with $\Lambda_4 = 160, 260$ and 360 MeV, and the two sets of structure functions from ref. [22]. We see that most of the un-

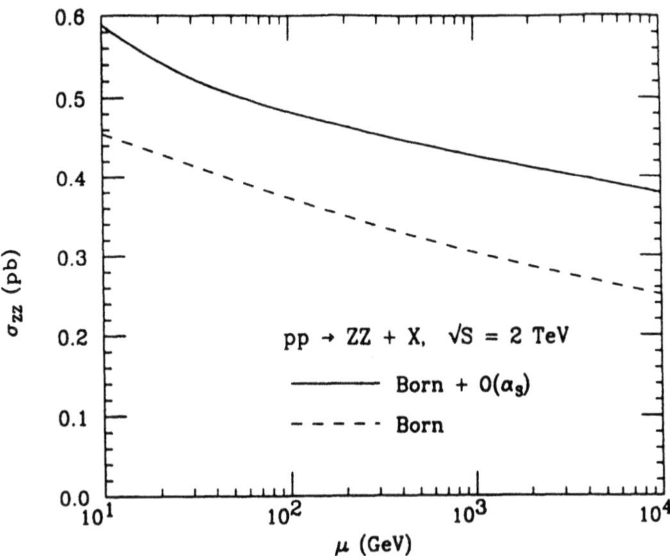

Fig. 9.   Total cross section for Z pair production in *pp* collisions at $\sqrt{S} = 2$ TeV, as a function of the subtraction scale μ. The structure functions are the DFLM set with $\Lambda_4 = 260$ MeV.

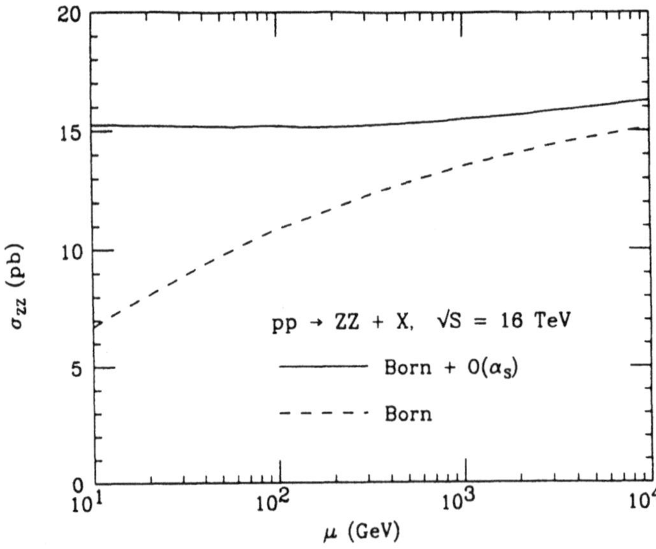

Fig. 10.   Same as in Fig. 9, for $\sqrt{S} = 16$ TeV.

certainty in the cross section comes from structure function dependence. The most extreme case is always given by the HMRSE set, which is based upon EMC data. If we exclude this set, we find that at LHC energy the cross section ranges between 13.1 and 16.4 picobarns, while at the SSC it goes from 32.9 to 51.5 picobarns.

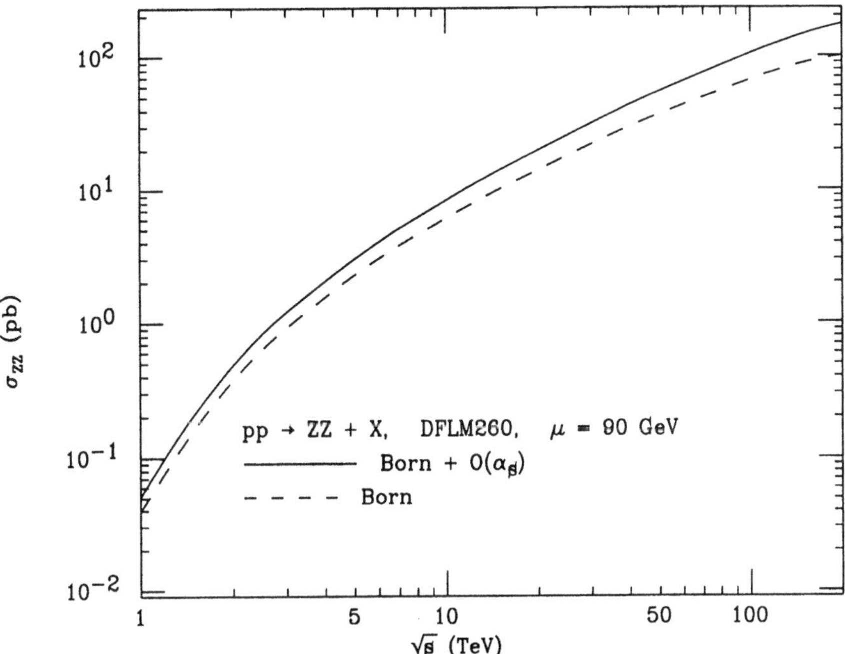

Fig. 11.   Total cross section for Z pair production in pp collisions as a function of the center of mass energy.

ACKNOWLEDGEMENTS

I would like to thank the Theoretical Physics Department of the University of Geneva for its kind hospitality.

REFERENCES

[1]   H. Georgi and S. Weinberg, *Phys. Rev.* **D17** (1978) 275.

[2]   S. Petcov and D. R. T. Jones, *Phys. Lett.* **84B** (1979) 440.

[3]   R. N. Cahn and S. Dawson, *Phys. Lett.* **136B** (1984) 196.

[4]   B. W. Lee, C. Quigg and H. B. Thacker, *Phys. Rev.* **D16** (1977) 1519.

[5]   See for example:
Proc. of the *Large Hadron Collider Workshop*,
Aachen, 4-9 October 1990 , G.Jarlskog and D.Rein eds.,
Report CERN 90-10, ECFA 90-133.

[6]   A. Nisati in ref.[5], vol.II, p.492.

[7]   D. Froidevaux in ref.[5], vol.II, p.444.

[8]   E. W. N. Glover and J. J. van der Bij, *Nucl. Phys.* **B321** (1989) 561.

[9]   R. W. Brown and K. O. Mikaelian, *Phys. Rev.* **D19** (1979) 922.

[10]   U. Baur, E. W. N. Glover and J. J. van der Bij,
Nucl. Phys. **B318** (1989) 106.

[11]   R. N. Cahn, S. D. Ellis, R. K. Kleiss and W. J. Stirling,
Phys. Rev. **D35** (1987) 1626 ;
R. K. Ellis, I. Hinchliffe, M. Soldate and J. J. van der Bij,
Nucl. Phys. **B297** (1988) 211 ;
I. Hinchliffe and S. F. Novaes, *Phys. Rev.* **D38** (1988) 3475.

[12]   V. Barger, T. Han, J. Ohnemus and D. Zeppenfeld,
Phys. Rev. **D41** (1990) 2782.

[13]   U. Baur and E. W. N. Glover, *Nucl. Phys.* **B347** (1990) 12.

[14]   B. Mele, P. Nason and G. Ridolfi, *Nucl. Phys.* **B357** (1991) 409.

[15]   J. Ohnemus and J. F. Owens, *Phys. Rev.* **D43** (1991) 3626.

[16]   W. A. Bardeen *et al.*, *Phys. Rev.* **D18** (1978) 3998.

[17]   J. C. Collins and D. E. Soper,
Ann. Rev. Nucl. Part. Sci. **37**(1987)383 and references therein.

[18]   G. Altarelli and G. Parisi, *Nucl. Phys.* **B126** (1977) 298.

[19]   G. Altarelli, R. K. Ellis and G. Martinelli, *Nucl. Phys.* **B157** (1979) 461.

[20]   W. Furmanski and R. Petronzio, *Zeit. Phys.* **C11** (1982) 293.

[21]   M. Diemoz, F. Ferroni, E. Longo and G. Martinelli,
Zeit. Phys. **C39** (1988) 21 ;
J. V. Allaby *et al.*, *Phys. Lett.* **197B** (1987) 281.

[22]   P. N. Harriman, A. D. Martin, R. G. Roberts and W. J. Stirling,
Phys. Rev. **D42** (1990) 798.

MONTE CARLO EVENT GENERATION FOR

FUTURE SUPER COLLIDERS

Leif Lönnblad

DESY-THEORIE
Notkestraße 85
D-2000 Hamburg 52, Germany
E-mail: lonnblad@apollo3.desy.de, LONNBLAD@DESYVAX.bitnet

Abstract

The importance of Monte Carlo event generation for future super colliders is discussed and a brief overview of available programs is given. To emphasize the need for different models of the hadronization process, an example of a measurement is presented where fragmentation effects 'destroys' an effect predicted by perturbative QCD also at very high energies. Finally the use of modern programming techniques and languages in Monte Carlo event generation is discussed and MC++ - an event generator toolkit for the C++ programming language is presented.

Introduction

Monte Carlo Event Generators (MCEGs) have become invaluable tools in todays high energy physics experiments. They are used to 'translate' theoretical predictions into experimental observables and to make reasonable estimations of systematical errors and backgrounds to these. To do this it is very important that the MCEGs implement reasonable phenomenological models of the non-perturbative fragmentation process of which we, so far, have very poor theoretical understanding.

Overview

If you take a look at the three most commonly used MCEGs for hadron-hadron collisions (HERWIG [1], ISAJET [2] and PYTHIA [3], see table 1) you will see that although they are very similar in structure, they differ very much in the implementation of the different phases in the event generating chain. The strategy of all these programs is to first choose a hard process and the interacting partons, using matrix elements and structure functions. Then an initial state (or space-like) parton shower is applied where the interacting partons are

QCD at 200 TeV, Edited by L. Cifarelli
and Y. Dokshitzer, Plenum Press, New York, 1992

Table 1. Example of MCEGs for hadron-hadron colliders.

author(s)	program	structure functions	hard processes	initial state PS	final state PS	hadron- ization
Webber Marchesini	HERWIG	"any"	"any"	coherent	coherent	cluster
Paige Protopopescu	ISAJET	"any"	"any"	non-coherent	non-coherent	indep-endent
Sjöstand Bengtsson	PYTHIA	"any"	"any"	coherent	coherent	string

evolved 'backwards' in an Altarelli-Parisi kind of way, and the resulting partons are allowed to bremsstrahl in a final-state (or time-like) parton shower before they are transformed into hadrons according to some fragmentation model. Finally these hadrons are allowed to decay.

The largest differences are found in the non-perturbative hadronization phase, where three completely different models are used. PYTHIA uses the Lund String Fragmentation Model [4] as it is implemented in JETSET [5], whereas HERWIG uses a cluster fragmentation model and ISAJET a model where all partons fragment independently. On the other hand the hard interaction is treated more or less in the same manér and the difference here is mostly a question of which matrix elements that are actually implemented. Also the three programs are easily interfaced to the PDFLIB [6] structure function library, so in that respect the programs are equivalent.

In the perturbative parton evolution ISAJET uses a non-coherent shower while both HERWIG and PYTHIA uses coherent shower models. The detail implementation of coherence differs however in HERWIG and PYTHIA so one cannot expect to obtain the same results on parton level in these two programs.

Apart from this the three programs also differs in the treatment of heavy hadron decays and the way they implement 'underlying events'. The programs have also completely different user interfaces, and although they are built in a modular fashion it is extremely difficult to mix modules from different programs and e.g. use the non-coherent shower of ISAJET together with the cluster fragmentation in HERWIG.

All of these three program can easily be used to simulate events at very high energies. E.g. for simulating events at the Eloisatron, just ask for a $100TeV$ proton on a $100TeV$ proton, select the process you are interested of and run. There are however some things to keep in mind:

The default in all programs is to have the four-momentum of all particles in single precision. This means e.g. that when measuring the invariant mass of a jet with an energy of $E \approx 100TeV$, the accuracy is completely lost $\delta M/M \approx 1$ for $M^2 \approx 100GeV$. The workaround is of course to use double precision variables which unfortunately can make the program very slow.

It is also advisable to ask the authors of the programs about 'hidden traps'. One such trap has been presented earlier in this workshop, namely the fact that some structure functions are only defined for $x > 10^{-5}$ [7].

Fragmentation Effects at 'asymptotic' Energies

When extrapolating theoretical predictions to very high (e.g. Eloisatron-) energies, it is important to check what influence the non-perturbative fragmentation process may have. As mentioned before we here have only phenomenological knowledge. As the models used in the three programs above all reproduces data from e.g. LEP satisfactorally [8], it is not possible to say that one is more 'correct' than the other. Therefore it is possible to use MCEGs with different fragmentation models to estimate the the non-perturbative effects at high energies.

The simplest fragmentation 'model' is the concept of Local Parton Hadron Duality (LPHD) [9] based on the idea that at 'asymptotically' high energies, the hadrons will follow the directions of the partons produced in the perturbative phase. An example of the success of LPHD is the measurement of the distribution of the hadrons momentum fraction at LEP [10] which is very well reproduced by the Modified Leading Logarithmic Approximation (MLLA) [11] assuming LPHD. This does however not mean that $91GeV$ is in the asymptotically high energy region and that we can forget about fragmentation effects at even higher energies. As an example of a measurement where fragmentation effects may influence the predictions from MLLA it is instructive to look at jet asymmetries.

It is a prediction of the MLLA that a quark jet in a typical three-jet event in an e^+e^- collision is asymmetric in the sense that if you look at at the number of particles in a cone around the quark, there will be more particles on the gluon jet side than on the anti-quark jet side. If N_g and N_q is the number of particles on the gluon and anti-quark side respectively, the asymmetry can be written as [12]

$$A(\theta) = \frac{N_g - N_q}{N_g + N_q} = \theta g(\alpha_s(E\theta/\Lambda))f(\theta_{qg}, \theta_{q\bar{q}}) \tag{1}$$

where

$$g(\alpha_s) = \frac{2}{\pi}\sqrt{6\alpha_s/\pi} - 11\frac{37\alpha_s}{108\pi^2} + \mathcal{O}(\alpha^{3/2}) \tag{2}$$

and

$$f(\theta_{qg}, \theta_{q\bar{q}}) = \frac{N_c}{2C_F} \cot\frac{\theta_{qg}}{2} + \frac{1}{2N_cC_F} \cot\frac{\theta_{q\bar{q}}}{2} \tag{3}$$

for a cone of opening angle θ at a center of mass energy E and the geometry defined by the angles between the jets, θ_{qg} and $\theta_{q\bar{q}}$. The main effect here is that the asymmetry goes to zero linearly as $\theta \to 0$.

To estimate the fragmentation effects on this prediction I made a Monte Carlo simulation using the colour dipole shower model [13] as it is implemented in Ariadne [14] for the perturbative phase and the Lund string fragmentation Model [4] as implemented in JETSET for the hadronization phase. I also made an attempt to study the fragmentation effects separately by applying string fragmentation directly on three parton configurations produced by a first order matrix element program in JETSET.

The results are presented in figures 1 and 2 for a specific jet geometry with $\theta_{qg} \approx 60°$ and $\theta_{q\bar{q}} \approx 130°$ for $91GeV$ and $100TeV$ center of mass energy respectively.

The first thing to note is that Ariadne indeed reproduces the prediction of the MLLA on the parton level (figures 1a and 2a). In figures 1b and 2b we see that also string fragmentation by itself gives asymmetric quark-jets, but the behavior at small θ is very different, the asymmetry increases slightly for very small θ after a slow decrease for moderately low θ.

Comparing this with figures 1c and 2c where the result produced by adding string fragmentation on the partonic states produced with Ariadne is presented, it seems that the fragmentation effects in some sense actually dominates the measurement even at an energy as high as $100TeV$.

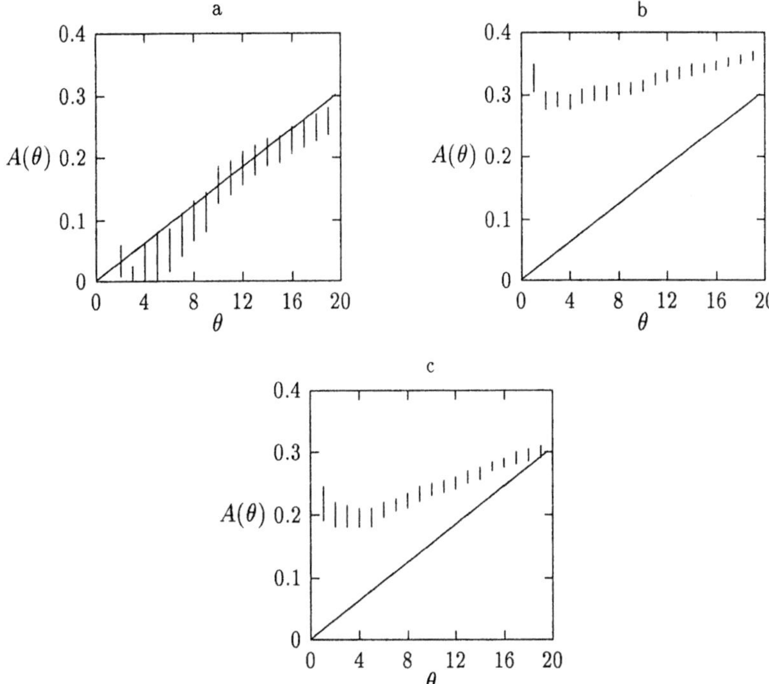

Figure 1. Asymmetries at 91GeV. Error bars shows predictions from (a) partons produced by Ariadne, (b) hadrons produced by JETSET with 1:st order matrix element and (c) hadrons produced by Ariadne+JETSET. The full line is the prediction from MLLA

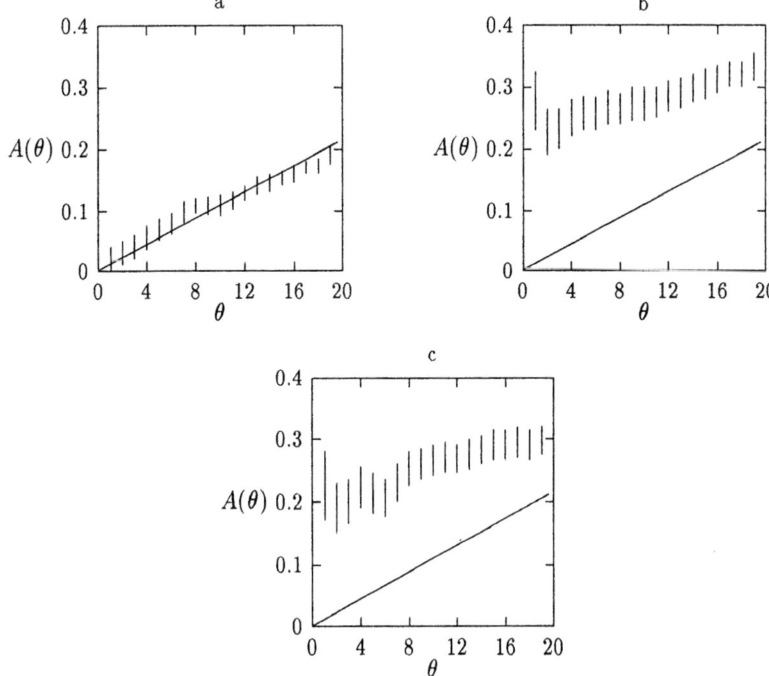

Figure 2. Asymmetries at 100TeV. Error bars shows predictions from (a) partons produced by Ariadne, (b) hadrons produced by JETSET with 1:st order matrix element and (c) hadrons produced by Ariadne+JETSET. The full line is the prediction from MLLA

Future MCEGs for Future Super Colliders

After having advocated the need for good Monte Carlo programs for event simulation at future super colliders it seems fit to discuss the tools available today for designing such programs.

When designing and building new colliders and detectors, large sums of money is spent on the latest high-tech equipment. From sofisticated CAD/CAM computers to the most modern super conducting magnets. But if you look at todays MCEGs, they are all written in FORTRAN and equipped with 'stone-age' user interfaces. This despite the fact that also computer science have made some progress since 1978 (when the FORTRAN-77 standard was introduced).

From my own experience of writing programs [14], I would say that FORTRAN-77 is **not** a suitable programming language. I would estimate that $\approx 90\%$ of the time it takes to write one is spent on things like getting array indexing and memory management right and only 10% is spent on the actual physics problem. This is clearly not a satisfactory situation.

The question is which language to use instead. C? Pascal? ADA? FORTRAN-90? I think the main thing to remember is that no programming language is specially designed for writing MCEGs. There exist however a few languages which one is able to 'customize' so that a 'dialect' can be specially designed for the purpose. One such language is called C++ [15] and recently a project was started by the theory departments in Lund and at SLAC, to write a special toolkit for event generation based on C++, called MC++ [16].

C++ implements one of the latest programming techniques - the concept of Object-Oriented Programming (OOP). Contrary to the conventional procedural languages where subroutines are called to act on fundamental variables (in FORTRAN typically stored in large common blocks) OOP is based on the concept of Objects which are generalized variables which communicate with each other by sending and receiving messages.

This enables you to define classes of objects in close correspondence with the physical objects of interest. Ie. you can define a 'particle' object which understands and can respond to messages like: "What is your charge?", "Boost!" and "Where is you mother?". In addition all memory management is handled by the operating system through the concept of pointers.

The MC++ project was started with a number of goals in mind. Some of them were:

- The code should in a transparent way mirror the different parts of the event generating chain.

- The code should be modular, so that a model describing only a small part of the event generating chain is easily added.

- The code should provide utilities to facilitate the development of new 'modules'.

- The code should be easily interfaced to a GUI.

- The user should easily be able to add, remove, and change decay channels, branching ration and decay methods etc. of particles as well as adjust their mass and lifetimes etc.

- The user should just as easily be able to change between different models for e.g. partonic showers and fragmentation.

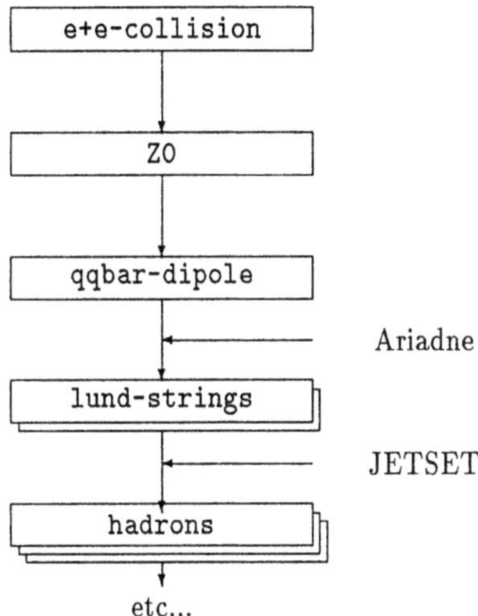

Figure 3. A schematic picture of an e^+e^- event produced with MC++.

The result was a structure where the whole event generating chain is defined in terms of (generalized) particles decaying into (generalized) particles. The particles are implemented as a class of objects in C++. Generalized particles, such as a 'Lund QCD string', are realized through the concept of inheritance and are 'sub classes' of the particle class.

To make a simulation of an e^+e^--event at LEP the procedure would be to first define all particles and how they decay (typically by reading a set-up file). Then to ask the operating system to create the generalized particle 'e^+e^--collision' and send it a message telling it to decay itself and its children recursively as in figure 3

It should be noted that it is not necessary to rewrite all old FORTRAN Monte Carlo programs to fit them into the MC++ toolkit. To begin with, the strategy would be to break them down into modules which can be interfaced to MC++ 'decayer' objects. How far the modularization should go is up to author. One could imagine breaking up e.g. PYTHIA into one module for the initial state radiation and one for the hard sub process etc. One could also implement PYTHIA as one large module which simply decays a 'p-p-collision' into 'hadrons'.

The MC++ Event Generator Toolkit exists today only in a preliminary form, but the first version is planned to be ready in the spring of 1992. [17]

References

[1] G. Marchesini and B.R. Webber, *Nucl. Phys.* **B310**, 461 (1988);
I. G. Knowles, *Nucl. Phys.* **B310**, 571 (1988).

[2] F.E. Paige, B.S. Protopopescu, Proceedings of the Snowmass Summer Study 1986 (QCD184:S7:1986) 320.

[3] H-U. Bengtsson and T. Sjöstrand, *Comp. Phys. Comm.* **46** (1987) 43.

[4] B. Andersson and G. Gustafson, *Z. Phys.* **C3** (1980) 223;
B. Andersson, G. Gustafson, G. Ingelman, T. Sjöstrand, *Phys. Rep.* **97** (1983) 31.

[5] T. Sjöstrand, JETSET 7.3 program and manual see e.g. B. Bambah et. al., QCD Generators for LEP, CERN-TH.5466/89,
T. Sjöstrand, *Computer Phys. Comm.* **39** (1986) 347;
T. Sjöstrand and M. Bengtsson, *Computer Phys. Comm.* **43** (1987) 367.

[6] H. Plothow-Besch, "Parton Density Functions", Proceedings of the 3rd Workshop on Detector and Event Simulation in High Energy Physics, Amsterdam, 8-12 April 1991.

[7] J. Stirling, Talk presented at this workshop.

[8] OPAL collaboration, M.Z. Akrawy et. al., *Phys. Lett.* **B246** (1990) 285.

[9] Ya. I. Azimov, Yu. L. Dokshitzer, S. I. Troyan, V. A. Khoze, Z. Phys. **C27** (1985) 65.

[10] OPAL collaboration, M.Z. Akrawy et.al., *Phys. Lett.* **B247**(1990) 617.

[11] Yu.L. Dokshitzer, V.A. Khoze, A.H. Mueller, S.I. Troyan, *Rev. Mod. Phys.* **60** (1988) 373;
Yu.L. Dokshitzer, V.A. Khoze, S.I. Troyan, Perturbative QCD (ed: A. Mueller), World Scientific, Singapore, (1989) 241;
see also Yu.L. Dokshitzer, V.A. Khoze, A.H. Mueller, S.I. Troyan, "Basics of Perturbative QCD", Editions Frontieres, Paris 1991.

[12] Yu.L. Dokshitzer, V.A. Khoze, S.I. Troyan, *Yad. Fiz.* **50** 808, and private communication.

[13] G. Gustafson, *Phys. Lett.* **B175** (1986) 453,
G. Gustafson, U. Pettersson, *Nucl. Phys.* **B306** (1988) 746.

[14] L. Lönnblad, "ARIADNE 3 - a Monte Carlo for QCD cascades in the Colour Dipole Formulation", Lund Preprint LU TP 89-10 (1989).

[15] See, for example, B. Stroustrup, 'The C++ Programming Language', Addison-Wesley 1987, ISBN 0-201-12078-X.

[16] R. Blankenbecler and L. Lönnblad, "Particle Production and Decays in an Object Oriented Formulation", Lund Preprint LU-TP 91-19 and SLAC preprint SLAC-PUB 5648, to be published in *Particle World*.

[17] L. Lönnblad, A. Nilsson, "The MC++ Event Generator toolkit - version 0", Lund Preprint LU TP 91-35, DESY 91-158, December 1991, to be published in *Comp. Phys. Comm.*.

DIPOLE FORMALISM AND PROPERTIES OF QCD CASCADES

Gösta Gustafson

Department of Theoretical Physics
University of Lund
Sölvegatan 14 A, S-223 62 Lund
Sweden

I. Introduction

Hadron production in e^+e^--annihilation is usually described by two phases, a perturbative phase in terms of quarks and gluons, and a nonperturbative phase, where the energy of these partons is transformed into hadrons. The description of the perturbative phase as a parton cascade has very successfully reproduced the experimental data from LEP. At high energies the general properties of the hadronic state are mainly determined by this perturbative cascade.

It is thus essential to be able to visualize and to check the properties of the QCD cascade. Here the dipole formalism has been very useful. The feature with jets within jets within jets implies a kind of fractal structure. Scaling violations in QCD imply that α_s should run with $k_\perp^2$, and thus also the fractal or anomalous dimensions run with $k_\perp^2$.

The relation partons–hadrons is essential for comparisons with experiments. It is thus important to have variables defined on the parton states, which are closely related to the final state hadrons, as well as to have observables on the hadronic states related to the partons.

The results presented in this talk are mainly obtained in collaborations with colleagues in Lund, B. Andersson, P. Dahlqvist, L. Lönnblad, A. Nilsson, U. Pettersson and C. Sjögren. The outline of the talk is as follows:

II. Dipole formalism
III. Relation partons-hadrons
IV. Jet multiplicities
V. Fractal structures
VI. Conclusions

II. Dipole formalism

A high energy $q\bar{q}$-system radiates gluons according to the dipole formula

$$dn = \frac{3\alpha_s}{4\pi^2} \frac{dk_\perp^2}{k_\perp^2} dy \, d\varphi \qquad\qquad (1)$$

Here the phase space available is given by the relation

$$|y| \lesssim \tfrac{1}{2}\ln(s/k_\perp^2) \qquad\qquad (2)$$

which corresponds to the triangular region in a $y - \ln k_\perp^2$ diagram as shown in fig. 1a. The rapidity range Δy, available for a fixed value of $k_\perp$, is given by $\ln(s/k_\perp^2)$.

If two gluons are emitted, then the distribution of the hardest gluon is described by eq. (1), while the distribution of the second, softer gluon corresponds to two dipoles, one stretched between the quark and the first gluon, and the second between this gluon and the antiquark [1].

This procedure can be generalized [2] so that the distribution of a third, still softer gluon corresponds to three dipoles. With n gluons the emission of gluon number $n+1$ is given by a chain of $n+1$ dipoles. We note that the dipoles connect the gluons in the same way as the string in the Lund fragmentation model.

Each dipole radiates independently. The emission distribution is given by the expression $dk_\perp^2 / k_\perp^2 \cdot dy$ in the rest frame of each dipole. In this way the angular ordering from soft gluon coherence is automatically taken into account.

The dipole formalism is suitable both for Monte Carlo simulations (such a program called ARIADNE is developed by U. Pettersson and L. Lönnblad [3]) and for analytic calculations, corresponding to the double leading log (DLLA) or modified leading log (MLLA) approximations.

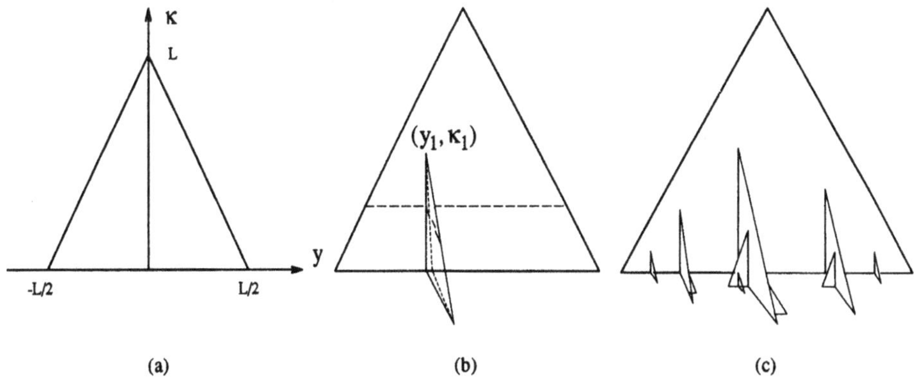

Fig 1. (a) The phase space available for a gluon emitted by a high energy $q\bar{q}$ system is a triangular region in the $y - \kappa$ plane ($\kappa = \ln k_\perp^2/\Lambda^2$; $L = \ln s/\Lambda^2$).
(b) If one gluon is emitted at (y_1, κ_1) the phase space for a second (softer) gluon is represented by the area of this folded surface.
(c) Each emitted gluon increases the phase space for the softer gluons. The total gluonic phase space can be described by this multifaceted surface. The length of the baseline corresponds to the quantity $\lambda(L)$.

Let us study the phase space for the second gluon in case two gluons are emitted. If g_1 is the first gluon and the masses of the qg_1 and $g_1\bar{q}$ systems are $\sqrt{s_{12}}$ and $\sqrt{s_{23}}$ respectively, then the transverse momentum $k_{\perp 1}$ and rapidity y_1 of this gluon are given by the relations

$$
\begin{aligned}
s \cdot k_{\perp 1}^2 &= s_{12} \cdot s_{23} \\
y_1 &= \tfrac{1}{2} \ln \frac{s_{23}}{s_{12}}
\end{aligned}
\tag{3}
$$

For fixed $k_{\perp 1}^2$ the rapidity range Δy available for the second gluon with transverse momentum $k_{\perp 2}$ (measured in the restframe of the parent dipole) is given by

$$
\begin{aligned}
\Delta y &= \ln(s_{12} / k_{\perp 2}^2) + \ln(s_{23} / k_{\perp 2}^2) \\
&= \ln(s) + \ln(k_{\perp 1}^2) - 2\ln(k_{\perp 2}^2)
\end{aligned}
\tag{4}
$$

This corresponds to the dashed line in fig. 1b. The phase space available for the second gluon thus corresponds to the folded surface in fig. 1b, with the constraint $k_{\perp 2}^2 < k_{\perp 1}^2$, as the first gluon is assumed to be the hardest one.

With many gluons the gluonic phase space can be represented by the multifaceted surface in fig. 1c. Each gluon adds a fold to the surface, which increases the phase space for softer gluons.

We note that in this process the recoils are neglected, as is normal in the leading log approximation. Recoil effects and kinematical constraints for hard gluons can be taken into account in a Monte Carlo simulation program, and we note that there are often large quantitative differences between the MC and the analytic calculations.

III. Relation partons-hadrons

Hadron multiplicity

In string fragmentation (or a longitudinal phase space model) the hadronic multiplicity for a simple $q\bar{q}$-system is proportional to $\ln(s / m_0^2)$. Here the parameter m_0 is of the order of one or a few hadron masses. For a $q\bar{q}g$-system, we obtain in the Lund string fragmentation model a bent string with two straight segments. The energy in the segments is $\sqrt{s_{12}}$ and $\sqrt{s_{23}}$, where $\sqrt{s_{ij}} = (k_i + k_j)^2$ and k_1, k_2 and k_3 are the momenta of the q, g and $\bar{q}$ respectively. Thus the average hadron multiplicity, n, is given by the relation

$$
\langle n \rangle \sim \ln(s_{12} / m_0^2) + \ln(s_{23} / m_0^2) \approx \ln(s / m_0^2) + \ln(k_\perp^2 / m_0^2)
\tag{5}
$$

Here $k_\perp$ is the transverse momentum of the gluon (cf eq. 3). We also note that this expression is equal to the length of the baseline of the surface in fig. 1b. For a multigluon state we find in the same way

$$
\langle n \rangle \sim \sum \ln(s_{i,i+1} / m_0^2) \approx \ln(s / m_0^2) + \sum \ln(k_{\perp i}^2 / m_0^2) \equiv \lambda
\tag{6}
$$

where the first sum goes over all dipole masses and the second over all gluon transverse momenta (measured in the rest frame of the parent dipole). This expression, which we call λ, is an "effective rapidity range". It is given by the length of the baseline in fig. 1c, or more precisely the baseline obtained if we cut the surface in fig. 1c at $\ln k_\perp^2 = \ln m_0^2$.

It is possible to calculate the distribution $P(\lambda,s)$ in λ for fixed s [4]. We introduce the Laplace transform $\mathcal{P}$ and use the following notation

$$
\begin{aligned}
&\mathcal{P}(\beta,L) = \int d\lambda e^{-\beta\lambda} P(\lambda,L) \\
&L = \ln(s/\Lambda^2) \; ; \; L_0 = \ln(m_0^2/\Lambda^2)
\end{aligned}
\tag{7}
$$

It is then possible to derive the following differential equation

$$
\frac{d^2}{dL^2}(\ln \mathcal{P}(\beta,L)) = \frac{\alpha_0}{L}(\mathcal{P}(\beta,L)-1)
\tag{8}
$$

where the constant α_0 is defined from the running coupling α_s by the relation

$$
\alpha_0 / L \equiv 3\alpha_s / 2\pi
\tag{9}
$$

We also get the boundary conditions

$$
\begin{aligned}
&\mathcal{P}(\beta,L=L_0) = 1 \\
&\frac{d}{dL}\mathcal{P}(\beta,L=L_0) = -\beta
\end{aligned}
\tag{10}
$$

From these relations, simple expressions are obtained for the first few moments. Thus we find for the average value and the width of the λ-distribution

$$
\overline{\lambda} \underset{s \text{ large}}{\sim} \frac{(\ln s)^{1/4}}{\alpha_0^{3/4}} \exp(2\sqrt{\alpha_0 \ln s})
$$

$$
\overline{\lambda^2} - \overline{\lambda}^2 \approx \tfrac{1}{3}\overline{\lambda}^2
\tag{11}
$$

Similar results have been obtained for the gluon multiplicity distribution [5, 6]. The results in eq. (11) correspond to the so-called double leading log approximation. It is also possible to generalize to the modified leading log approximation (MLLA). If the numerator in the Altarelli-Parisi splitting function is taken into account, it corresponds to a suppression of the phase space at the ends of the rapidity range, see fig. 2. The differential equation (8) is then changed into the following equation

$$
\frac{d^2}{dL^2}(\ln \mathcal{P}) = \frac{\alpha_0}{L}\left(\mathcal{P} - \frac{11}{6}\frac{d\mathcal{P}}{dL} - 1\right)
\tag{12}
$$

with the asymptotic solution

$$
\overline{\lambda} \sim (\ln s)^{1/4 - 11\alpha_s/12} \exp(2\sqrt{\alpha_0 \ln s})
\tag{13}
$$

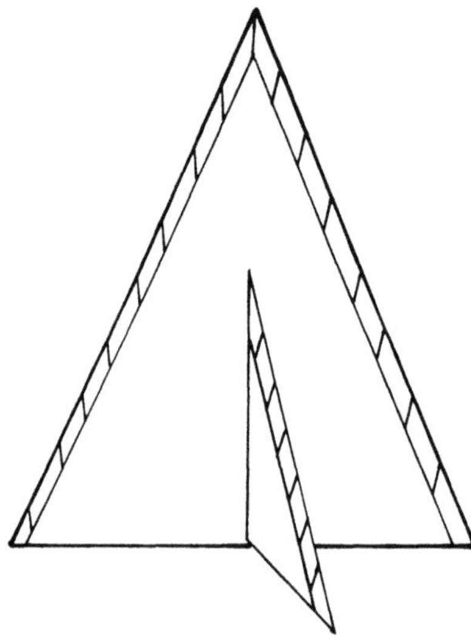

Fig. 2. The phase space in the modified leading log approximation is suppressed at the ends of the rapidity range.

The width of the λ-distribution now becomes more narrow, in better agreement with MC calculations and experiments.

As will be further discussed below, the quantity λ can be generalized in an infrared stable way such that soft gluons give small but positive contributions [7]. This implies that no $k_\perp$-cutoff is needed. (With the expression in eq. (6) a negative contribution would be obtained for gluons with $k_\perp < m_0$.)

We see that for a fixed energy W the perturbative cascade gives a parton state with definite parton momenta, and thus a definite value of λ. The soft hadronization mechanism then gives a certain hadronic state with a hadron multiplicity n. If the hadronization is described by the Lund string fragmentation model, we find that the distribution in n depends only on λ and not on W [7]. Thus for a high energy event with few gluons and a low energy event with many gluons, but with the same λ, we have the same effective string length and the same hadron multiplicity distribution.

Momentum distribution

For a $q\bar{q}$-system the hadrons are evenly distributed in rapidity, which means that their energy-momentum four-vectors, when plotted after each other, are distributed around a hyperbola as seen in fig. 3a. For a $q\bar{q}g$-system the two string pieces will produce hadrons such that the momentum four vectors lie around two hyperbolae (fig. 3b). This corresponds to three jets along the parton momenta, and the jets are smoothly connected to each other.

For a multigluon system it is possible to generalize the hyperbolae in the $q\bar{q}$- and $q\bar{q}g$-cases and define a timelike curve in energy-momentum space [8]. This curve (which we call the x-curve) follows the (colour-ordered) parton momenta in such a way that the corners are smoothed out with a resolution power given by a parameter m_0 (see fig. 3c). (The x-curve was introduced in ref. [9] for a different purpose.)

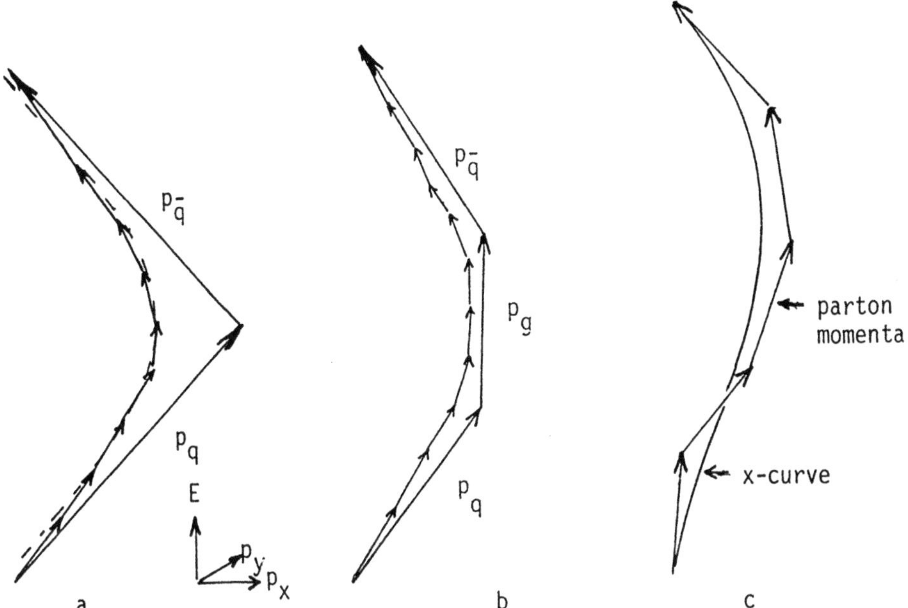

Fig. 3a. In a $q\bar{q}$ -system the hadron momenta are distributed around a hyperbola in energy-momentum space.

b. For a $q\bar{q}g$ - system the hadron momenta are distributed around two hyperbolae.

c. For a multigluon state the hadron momenta are distributed around a curve in energy-momentum space (called the x-curve) which smoothly follows the (colourordered) parton momenta.

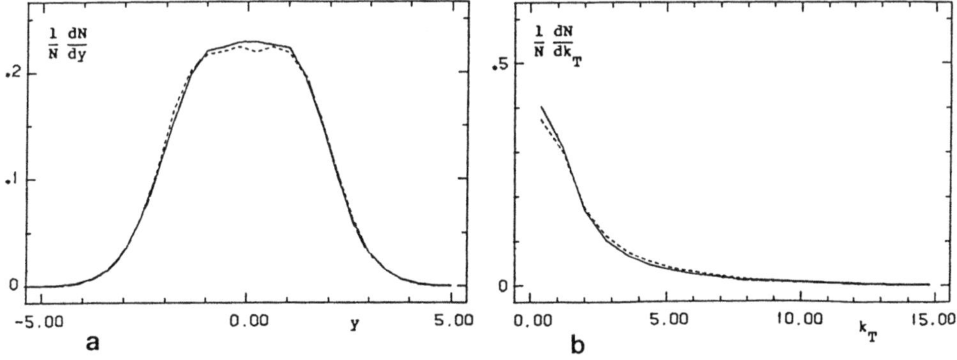

Fig. 4. Distributions in rapidity (a) and $p_\perp$ (b) for e^+e^--annihilation events at 200 GeV with sphericity >0.1. The dashed curve corresponds to a partitioning of the x-curve and the solid line to standard Lund fragmentation.

The infrared stable generalization of λ, mentioned above, is actually given by the invariant length of the x-curve. Thus the length of this curve corresponds to the total multiplicity. We also find [8], that if we just cut the x-curve into equal pieces, then we obtain an average momentum distribution of the hadrons. The soft hadronization just adds limited fluctuations around this average. These features give a quantitative meaning to the notion of local parton-hadron duality (LPHD) [10]. This is illustrated in fig. 4, which shows the distribution in rapidity and transverse momentum.

IV Gluon or jet multiplicity

As mentioned above, λ is the length of the curve obtained when we cut the surface in fig. 1c at the level $\ln k_\perp^2 = \ln(m_0^2 / \Lambda^2)$. We will also be interested in the length obtained when cutting at an arbitrary level $k_\perp^2 = k_{\perp c}^2$. In this way we can define $\lambda(s, k_{\perp c}^2)$. We can also study the number, $N(s, k_{\perp c}^2)$, of dipoles present at the level $k_\perp^2 = k_{\perp c}^2$. In this case $N+1$ is also equal to the number of jets resolved with a resolution defined by $k_{\perp c}^2$. We note that transverse momentum is the natural resolution measure in perturbative QCD, as was pointed out by Dokshitzer at the Durham Workshop [11].

It is possible to study the combined distribution in λ and N for fixed values of s and $k_{\perp c}^2$, $P(\lambda, N; s, k_{\perp c}^2)$. The Laplace transform

$$\mathcal{P}(\beta, \eta) = \int d\lambda dN \; e^{-\beta\lambda - \eta N} P(\lambda, N) \tag{14}$$

satisfies the differential equation

$$\frac{d^2(\ln \mathcal{P})}{dL^2} = \frac{\alpha_0}{L}(\mathcal{P} - 1) \quad ; L = \ln s$$

$$\left.\begin{array}{c} \mathcal{P} = e^{-\eta} \\ \dfrac{d\mathcal{P}}{dL} = -\beta \end{array}\right\} \; L = \ln k_{\perp c}^2 \tag{15}$$

From these relations it is possible to calculate different moments of N and λ, e.g. the average number of jets determined by

$$\bar{N} \sim (\ln s / \ln k_{\perp c})^{1/4} \; \exp(2\sqrt{\alpha_0 \ln s} - 2\sqrt{\alpha_0 \ln k_{\perp c}^2}) \tag{16}$$

We can also study the distribution in λ for fixed N, and derive e.g. the relations

$$\left.\begin{array}{c} \bar{\lambda}_N \approx N / \sqrt{\alpha_s(k_{\perp c}^2)} \\ \overline{\lambda_N^2} \approx N^2 / \alpha_s(k_{\perp c}^2) \end{array}\right\} \tag{17}$$

This result implies that $(\overline{\lambda_N^2} - \bar{\lambda}_N^2) / \bar{\lambda}_N^2$ is small and there is a very strong correlation between λ and N. This is seen in fig. 5 which shows results from MC simulations at the very high energy 200 TeV. The points in the scatter plot lie around a straight line with slope $\sim 1/\sqrt{\alpha_s(k_{\perp c}^2)}$. In the MLLA we obtain a correction and the slope is given by the expression

$$\frac{\lambda}{N} \approx \frac{1}{\sqrt{\alpha_s}} + \frac{11}{12} + \frac{1}{4\alpha_0} \approx \frac{1}{\sqrt{\alpha_s}} + 1.3 \tag{18}$$

In principle this gives an opportunity to study the running of α_s when $k_{\perp c}$ is varied.

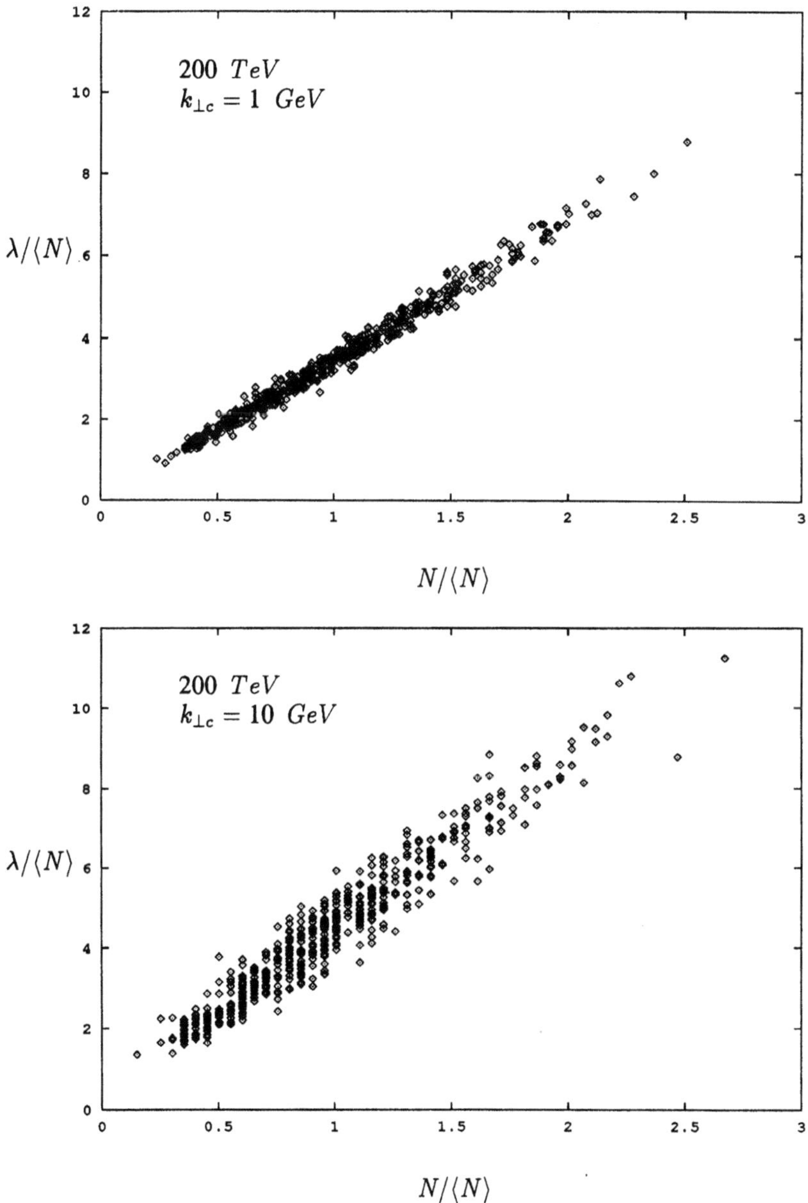

Fig 5. Scatter plot for the distribution in $\lambda/\langle N \rangle$ and $N/\langle N \rangle$ at $W = 200\ TeV$ and for two different values of $k_{\perp c}$, $1\ GeV$ and $10\ GeV$. The slope is related to $1/\sqrt{\alpha_s(k_{\perp c}^2)}$.

V. Fractal structures

Distributions in small angular regions [12]

The number of jets in an event depends upon the resolution. With increasing resolution one finds more but thinner jets. Thus jets have fractal properties similar to a Cantor dust. This property is related to what has been dubbed intermittency [13]. If the scaled multiplicity moments

$$C_q = <n^q> / <n>^q \tag{19}$$

or factorial moments

$$F_q = <n(n-1)\cdots(n-q+1)> / <n^q> \tag{20}$$

in small rapidity intervals δy have a scaling property

$$C_q \sim (1/\delta y)^\rho \tag{21}$$

then this can be described by a multifractal dimension [14, 15]

$$D_q = 1 - \rho/(q-1) \tag{22}$$

Here D_0 equals the Hausdorff dimension.

If we study normal moments rather than factorial moments it is possible to study also noninteger variables, e.g. the piece of the x-curve which has its tangent within an interval in y. This would represent the hadron distribution without the noise from the soft hadronization, and thus reveal the properties of the perturbative cascade.

The feature of QCD with jets within jets within jets, similar to a Cantor dust, gives fractal properties. If we look at $<\lambda^q>$ we see that for large q-values this is dominated by few events with large values of λ. These are events with a hard jet where the tip of the jet is inside the y-range δ. We can calculate the angle between an emitted gluon in a gluon jet and the main jet axis, and see that for the situation shown in fig. 6 the piece indicated is contained within a y-region $\delta \approx e^{-\Delta}$, where Δ is defined in the figure. If we look at fig. 1c, we realize that the λ-value for this piece is just the same as for a full e^+e^--annihilation event with an energy given by $\ln s' = \ln k_\perp^2 - 2\Delta = \ln(k_\perp^2 \delta^2)$, where $k_\perp$ is the transverse momentum of the jet. To get the distribution $P_\delta(\lambda)$ of λ within the y-region δ we note that the logarithm of the Laplace transform is obtained by adding the contribution from jets with different $k_\perp$. Thus in a symbolic way we can write

$$\text{distr. in } \lambda \sim \int (\cdots) \frac{dk_\perp^2}{k_\perp^2} \text{ (distr. for an } e^+e^- - \text{ann. event at } s' = k_\perp^2 \delta^2) \tag{23}$$

In this way we can show that the contribution to $<\lambda^q>$ is given by

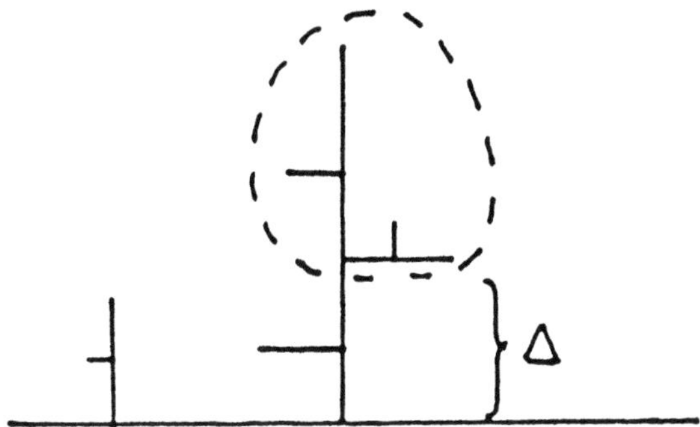

Fig. 6. The baseline of fig. 1c . The length of the curve corresponds to the number of hadrons. The branches and twigs show the QCD structure with jets within jets. The piece indicated is contained within a y-region $\delta \approx e^{-\Delta}$.

$$< \lambda^q >_{tip} \sim \delta(\overline{\lambda}_{e^+e^-}(s\delta^2))^q \sim \overline{\lambda}^q_{e^+e^-}(s)\delta^{1+2q\sqrt{\frac{3\alpha_s}{2\pi}}} \tag{24}$$

Here $\overline{\lambda}_{e^+e^-}$ is the average λ in an e^+e^--annihilation event.

If there are no jets we would obtain a background contribution $\lambda = \delta$ corresponding to a straight string. This would give

$$< \lambda^q >_{BG} = \delta^q \tag{25}$$

We see that when q and s are large and δ is small, then the jet tips dominate and from the result in eq. (24) we get the following dimension

$$D_q \approx 2\sqrt{\frac{3\alpha_s(s\delta^2)}{2\pi}} \quad q, \ s \text{ large}, \ \delta \text{ small} \tag{26}$$

In the square root we recognize the anomalous dimension of QCD. When, on the other hand, q and s are small and δ large, the background dominates. From eq. (25) we then get the dimension

$$D_q \sim 1 \quad q, \ s \text{ small}; \ \delta \text{ large} \tag{27}$$

If we include the rest of the jets and not only the tips, we obtain a sum of terms with dimensions between $2\sqrt{\frac{3\alpha_s}{2\pi}}$ and 1. Therefore the result will be as seen in fig. 7, which shows MC results for the scaled moments of the λ-distribution at 200 GeV. Here the slope of the curve is given by $(q-1)(1-D_q)$. We note that for δ large the curve is rather flat because $D \sim 1$. For smaller δ it becomes steeper and D is closer to $2\sqrt{\frac{3\alpha_s}{2\pi}}$. For very small δ it flattens out slightly because the running α_s becomes larger when $s\delta^2$ becomes small.

If we look at how the dimensions vary with q we see in fig. 8 how for small q the terms

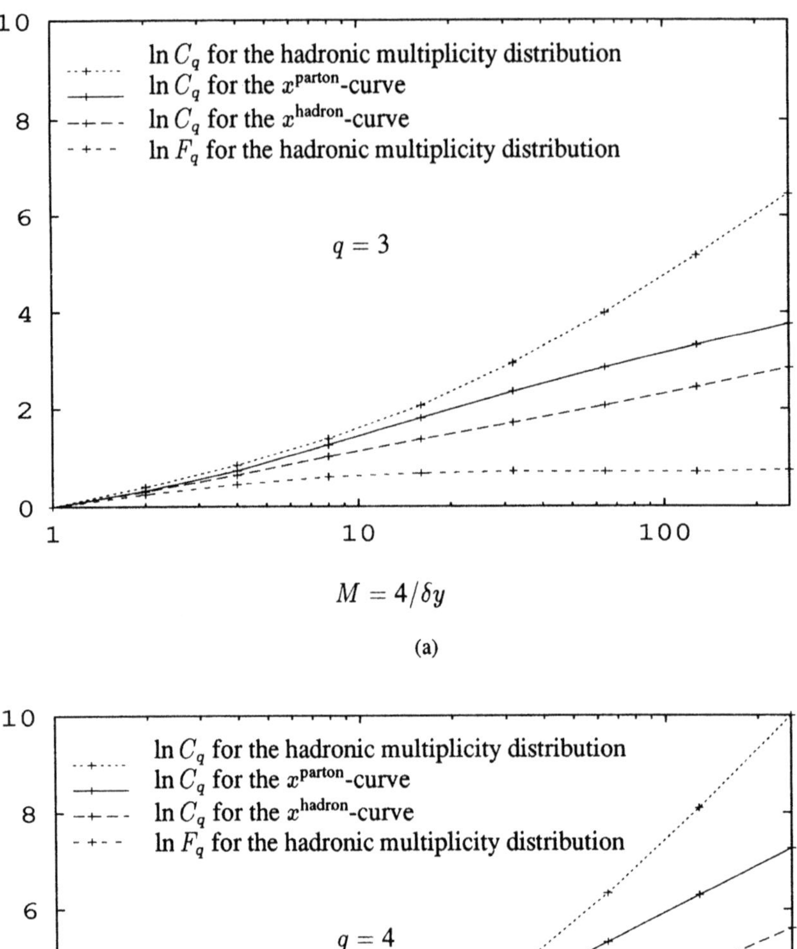

(a)

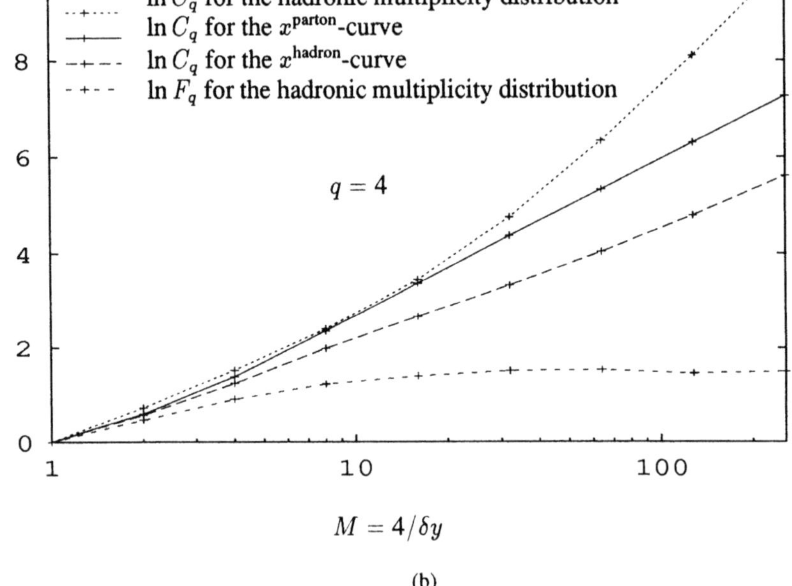

(b)

Fig 7. Normal multiplicity moments $C_q = \langle n^q \rangle / \langle n \rangle^q$ and factorial moments F_q for the hadron distribution compared with the moments $C_q = \langle \lambda^q \rangle / \langle \lambda \rangle^q$ for the x-curve, in (a) for $q = 3$ and in (b) for $q = 4$. The analysis is done for the energy 200 GeV and the rapidity region $|y| < 2$, $M = 4/\delta y$.

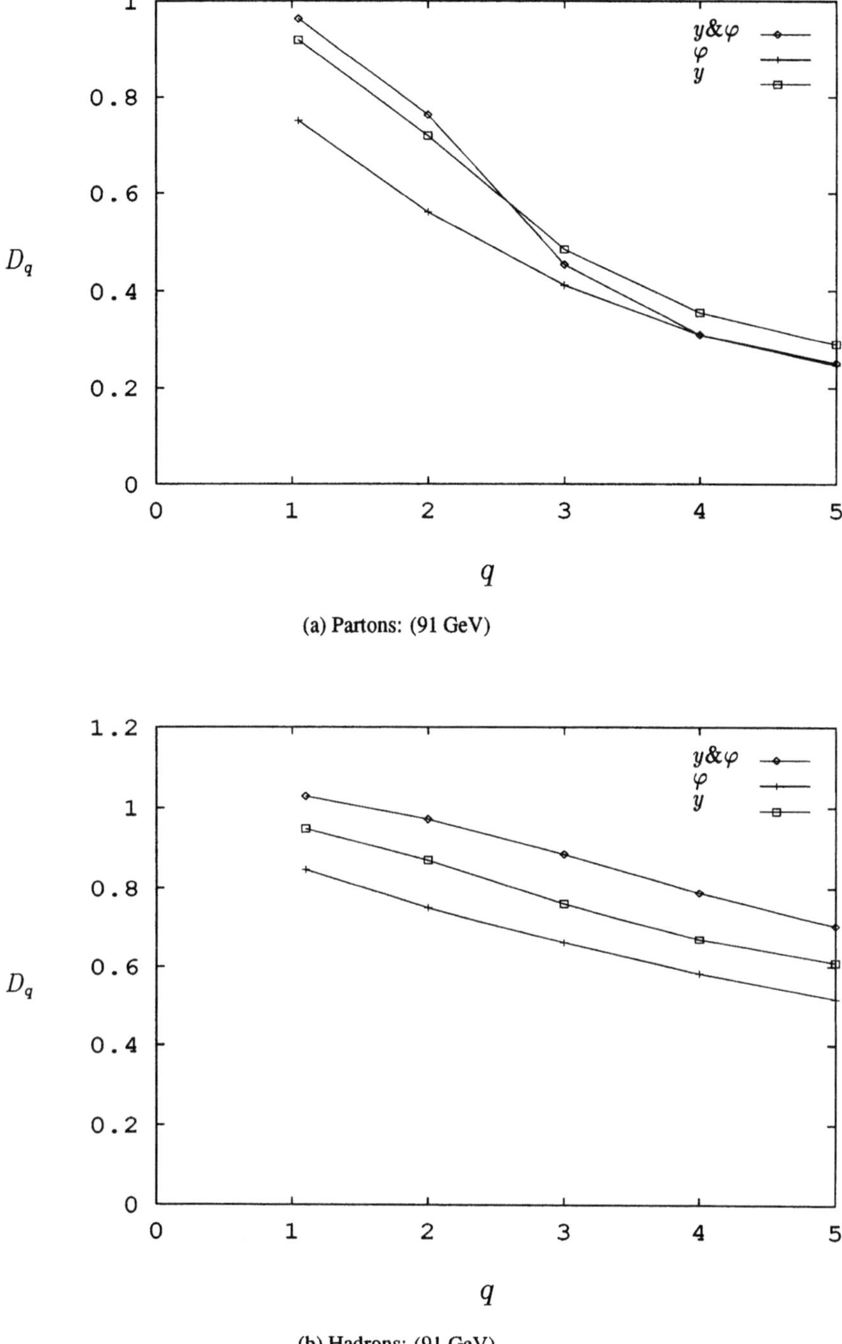

(a) Partons: (91 GeV)

(b) Hadrons: (91 GeV)

Fig 8. Average multifractal dimensions D_q from moments of the x-curve when binned in y, in φ or simultaneously in y and φ. In (a) using the x-curve as defined from the partons, in (b) as defined from the hadrons ordered in rapidity. The energy is 91 GeV.

with higher dimensions dominate while for larger q the jet tips dominate giving lower dimensions. We also note that if we divide the phase space in bins in y or φ or in twodimensional bins in y and φ, we obtain essentially the same dimensions [12]. This is confirmed in the MC calculations shown in fig. 8.

Observables on the hadronic state

Although related to the hadronic multiplicity, the λ-measure is defined for a partonic state and thus not directly observable. The hadronic multiplicity moments do depend on the soft hadronization process. If the hadronization was a purely Poissonian process with an average multiplicity given by $<n>=$ const $\cdot \lambda$, then the factorial moments $f_q = <n(n-1)...(n-q+1)>$ would be directly given by the moments of λ

$$f_q = (\text{const})^q \cdot <\lambda^q> \tag{28}$$

In this case the normal multiplicity moments $\langle n \rangle$ would blow up for small values of δy (small values of $\langle n \rangle$), while the underlying parton dynamics should be reflected in the factorial moments. However, in string fragmentation the fluctuations are smaller than in a Poisson distribution (the variance V is approximately equal to $\frac{2}{3} V_{Poisson}$), and the production of particles in neighbouring bins are correlated. Therefore the scaled factorial moments are much smaller than in the corresponding moments of λ, and they do not scale well when δy is varied. This is seen in fig. 7, which shows MC results for scaled moments $\langle \lambda^q \rangle / \langle \lambda \rangle^q$ together with scaled factorial moments $F_q = f_q / \langle n \rangle^q$.

The scaled normal multiplicity multiplicity moments $C_q = \langle n^q \rangle / \langle n \rangle^q$ are closer to the λ-moments, but as expected they become larger for small values of δy (cf fig. 7). This implies that also these moments do not scale so well and it is difficult to specify a dimension.

The x-curve is defined from the colour ordered parton momenta, and the final state hadron momenta closely follow this curve. It is possible to define a corresponding curve x^{hadron} if we replace the parton momenta by the hadron momenta. The curves x and x^{hadron} will then lie close to each other. A problem is that, for the observed hadrons we do not know their ordering along the string, i.e. their rank ordering. This ordering is, however, strongly correlated with the ordering in rapidity. In ref. [12] we showed that fractal properties of the x-curve are reflected in the corresponding x^{hadron}-curve, obtained if the hadrons are ordered in rapidity.

Fig. 7 also shows scaled moments $\langle \lambda^q \rangle / \langle \lambda \rangle^q$ obtained from the x^{hadron}-curve. We see that they indeed are more similar to the moments obtained from the original x-curve. They scale in a similar way, although the slopes in fig. 7 are somewhat smaller than those of the x^{parton}-curve.

A word of warning is appropriate here. String fragmentation produces hadrons in the same regions where soft gluons are emitted in perturbative QCD. A higher cutoff in the cascade can be compensated by a change in the fragmentation parameters. Second order matrix elements plus string fragmentation describes most observables rather well. In a situation without the triple gluon coupling it is possible to show, using the arguments described above, that the dimensions D_q are given by

$$D_q \sim \frac{2}{\ln(s\delta^2 / \Lambda^2)} \qquad s, q \text{ large} \tag{29}$$

In this case the gluon jets do not have the fractal structure. Nevertheless, at LEP energies the result from the second order MC for the moments is not so different from the real QCD case. At higher energies, however, the dimensions will in this case become lower than for the cascading case, as should be expected from eq. (29).

The x-curve has a fractal structure [16]

The baseline curve in fig. 1c, a piece of which is seen in fig. 6, has a fractal structure similar to that of Koch's snowflake curve. If studied with increasing resolution more twigs are resolved and the length appears larger. The length of a twig is given by $\ln k_{\perp i}^2 / \Lambda^2$. If we include all twigs with $k_{\perp i}^2 > k_{\perp res}^2$ where $k_{\perp res}$ defines our resolution, the length C of the curve will be

$$C = \lambda(L, \kappa_{res}) + N(L, \kappa_{res})\kappa_{res} \; ; \; L = \ln(s / \Lambda^2) \, , \; \kappa_{res} = \ln(k_{\perp res}^2 / \Lambda^2) \, . \tag{30}$$

For $1 << \kappa_{res} << L$ we obtain for the average length $\overline{C}$

$$\overline{C} = L^{\frac{1}{4}} \exp(2\sqrt{\alpha_0 L}) \; \kappa_{res}^{\frac{3}{4}} \; \exp(-2\sqrt{\alpha_0 \kappa_{res}}) \tag{31}$$

It is possible to define a fractal dimension D by the relation

$$D = 1 - \frac{d \ln \overline{C}}{d \ln k_{\perp res}^2} \approx 1 + \sqrt{\frac{\alpha_0}{\kappa_{res}}} = 1 + \sqrt{\frac{3\alpha_s(\kappa_{res})}{2\pi}} \tag{32}$$

In the square root we recognize the anomalous dimension of QCD.

The curve in fig. 6 is not directly observable, and thus this fractal structure is not so interesting. However, the x-curve discussed above (fig. 3c) is directly related to the momenta of the final state hadrons. Also this x-curve is "longer" when studied with higher resolution, so that more gluons are resolved.

If we measure the x-curve with a measuring stick of invariant length $\hat{s}$, this implies that we resolve gluons corresponding to a dipole mass $\sqrt{\hat{s}}$. If the length of the curve is determined by the number of pieces times the invariant length of each piece, then the total length will decrease with increasing resolution. This is therefore not a suitable way to measure the length of the more twisted curve. Instead we use as definition of the length of a piece with invariant length $\hat{s}$, the length of a hyperbola passing through the endpoints. This length is given by $\ln(\hat{s} / m_0^2)$ (if $\hat{s}$ is not too small). Thus we define the length S obtained with the resolution given by $\hat{s}$ in the following way

$$S \sim (\text{no of pieces}) \cdot \ln(\hat{s} / m_0^2) \tag{33}$$

For asymptotic energies ($s >> \hat{s} >> m_0^2$) we obtain the result

$$S \sim (\ln s)^{\frac{1}{4}} \exp(2\sqrt{\alpha_0 \ln s})(\ln \hat{s})^{\frac{3}{4}} \exp(-2\sqrt{\alpha_0 \ln \hat{s}}) \tag{34}$$

Thus we see that the fractal dimension D is given by the following expression

$$D = 1 - \frac{d(\ln S)}{d(\ln \hat{s})} \approx 1 + \sqrt{\frac{\alpha_0}{\ln \hat{s}}} = 1 + \sqrt{\frac{3\alpha_s(\hat{s})}{2\pi}} \qquad (35)$$

In the square root we see again the anomalous dimension of QCD. Thus the x-curve gives a geometrical interpretation of this anomalous dimension. We note that the dimension is not constant, because the theory is not perfectly scaling due to the running coupling constant.

It is possible to calculate $S(s,\hat{s})$ also with the Monte Carlo simulation program, which takes recoils and kinematical constraints better into account. The result is shown in fig. 9 together with the analytic solution. We note in particular that the derivative of $\ln S$ with respect to $\ln \hat{s}$ is decreasing with increasing $\hat{s}$, which reflects the running coupling constant in the expression for the dimension in eq. (35). Thus the fractal dimension is also varying and approaching 1 for very coarse resolutions.

As the partons are not directly observable, we have also calculated the corresponding results for the x^{hadron}-curve, defined from the observable hadrons as discussed above. The result using the MC is also shown in fig. 9, and we see that the fractal structure of the curves x and x^{hadron} are very similar, and thus accessible for experimental investigation.

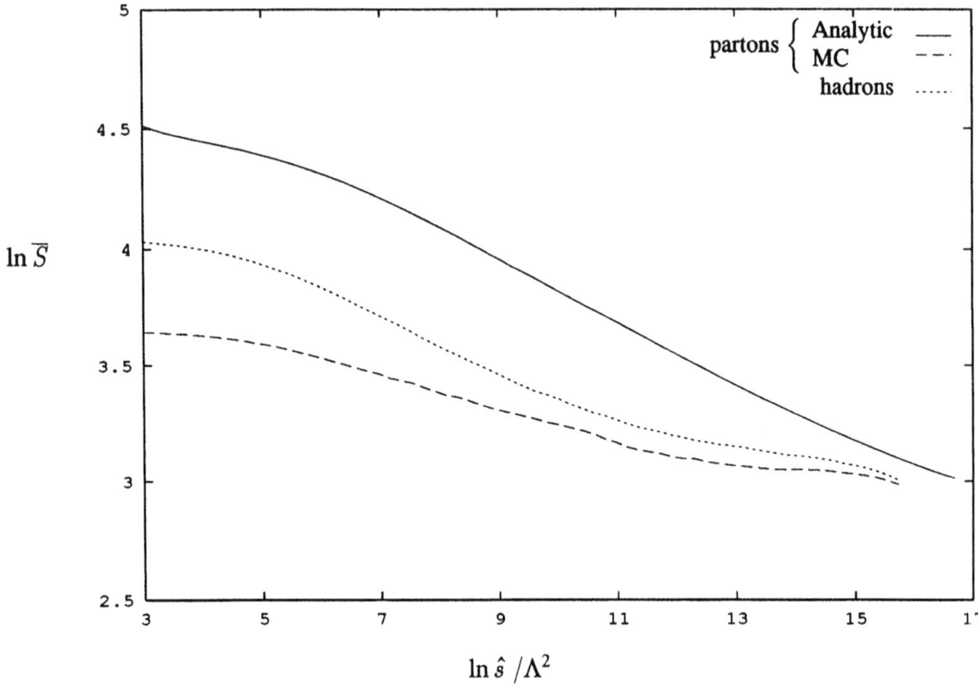

Fig 9. The average length $\overline{S}$ of the x-curve, as defined from the parton momenta, as a function of the resolution $\hat{s}$, in e^+e^--annihilation at 1000 GeV. The solid line is the analytic approximation. The dashed line is the MC result and the dotted line shows the MC results for the x^{hadron}-curve defined from the hadron momenta.

VI. Conclusions

i) The dipole formalism describes the QCD cascade as a branching process, where all the branchings are independent of each other. Soft gluon coherence is automatically taken into account. It is suitable both for analytic calculations and for MC simulations.

ii) Relation partons-hadrons: We have introduced infrared stable variables defined on the parton state: A measure, λ, related to the hadron multiplicity, and a curve, x, in energy-momentum space related to the hadron momenta.

We have also introduced a new observable, the x^{hadron}-curve, which reveals the underlying parton structure.

iii) Gluon or jet multiplicity: There is a strong correlation between the λ-measure and the jet multiplicty for fixed jet resolution. This makes it possible to check that α_s runs with $k_\perp^2$.

iv) Fractal structures: The distribution in narrow angular regions shows a fractal structure similar to a Cantor dust. The multifractal dimensions are given by $D_q \sim 2\sqrt{3\alpha_s / 2\pi}$. The x-curve has a fractal structure like Koch's snowflake curve, with a dimension $D = 1 + \sqrt{3\alpha_s / 2\pi}$. This offers a geometrical interpretation of the anomalous dimension of QCD.

References

[1] Ya.I. Azimov, Yu.L. Dokshitzer, V.A. Khoze, S.I. Troyan, Coherence effects in QCD jet, Leningrad preprint 1051 (1985); *Phys. Lett.* **B165** (1985) 147

[2] G. Gustafson, *Phys. Lett.* **B175** (1986) 453
G. Gustafson, U. Pettersson, *Nucl. Phys.* **B306** (1988) 746

[3] L. Lönnblad, U. Pettersson, ARIADNE 2, Lund preprint LU TP 88-15
L. Lönnblad, ARIADNE 3, Lund preprint LU TP 90-10

[4] B. Andersson, P. Dahlqvist, G. Gustafson, *Phys. Lett.* **B214** (1988) 604

[5] A.H. Mueller, *Nucl. Phys.* **B213** (1983) 85, **B228** (1983) 351
A. Bassetto, M. Ciafaloni, G. Marchesini, *Phys. Rep.* **100** (1983) 201

[6] E.C. Malaza, B. Webber, *Nucl. Phys.* **B267** (1986) 702

[7] B. Andersson, P. Dahlqvist, G. Gustafson, *Z. Physik* **C44** (1989) 455

[8] B. Andersson, P. Dahlqvist, G. Gustafson, *Z. Physik* **C44** (1989) 461

[9] B. Andersson, G. Gustafson, B. Söderberg, *Nucl. Phys.* **B264** (1986) 29

[10] Ya.I. Azimov, Yu.L. Dokshitzer, V.A. Khoze, S.I. Troyan, *Z. Physik* **C27** (1985) 65
L. van Hove, A. Giovannini, *Acta Physica Polonica* **B19** (1988) 917, 931

[11] See summary by J. Sterling, Workhop on Jet Studies at LEP and HERA, Durham, December 1990

[12] G. Gustafson, A. Nilsson, Multifractal dimensions in QCD cascades, Lund preprint LU TP 91-5, to be published in *Z. Physik* C

[13] A. Bialas, R. Peschanski, *Nucl. Phys.* **B273** (1986) 703, **B308** (1988) 857

[14] See e.g. H.G.E. Hentschel, I. Procaccia, *Physica* **8D** (1983) 435
T. Halsey et al., *Phys. Rev.* **A33** (1986) 1141

[15] P. Dahlqvist, G. Gustafson, B. Andersson, *Nucl. Phys.* **B328** (1989) 76
P. Lipa, B. Buschbeck, *Phys. Lett.* **B223** (1989) 465
R.C. Hwa, *Phys. Rev.* **D41** (1990) 1456

[16] G. Gustafson, A. Nilsson, *Nucl. Phys.* **B355** (1991) 106

HERETICAL STRUCTURE FUNCTIONS

Bo Andersson [1]

Department of Theoretical Physics
University of Lund
Sölvegatan 14A
S-223 62 Lund, Sweden

Work done together with Leif Lönnblad [2] [3]
Presented at Ettore Majorana Centre for Scientific Culture
17th Workshop: "QCD at 200 TeV"

Abstract

A model based upon string dynamics is presented for the structure functions of a
hadron. There are major differences in our approach compared to earlier work and I
comment upon them.

1 Introduction

We will use the following assumptions as basic input into the model:

A On the short time scale available in an inelastic leptoproduction event a hadron can
be treated (in a semi-classical way) as a Lund String state. In such a state the gluons
and the ocean $q\bar{q}$−pairs are internal excitations in the string and the states are taken
as superpositions of the available radiation.

---

[1] thepba@seldc52 (bitnet) bo@thep.lu.se (internet)
[2] thepll@seldc52 (bitnet) leif@thep.lu.se (internet)
[3] Address after Sept 1991: Theory Division, DESY, Hamburg

QCD at 200 TeV, Edited by L. Cifarelli
and Y. Dokshitzer, Plenum Press, New York, 1992

B The gluon emission and gluon splitting processes in connection with color separation is well described by The Dipole Cascade Model ([1, 2]). Similarly the particular color coherence properties of an inelastic leptoproduction event will be described by The Soft Radiation Model (SRM)([3].

C Inelastic leptoproduction corresponds to a measuring process in which a basically pointlike probe, the field pulse q, resolves the hadronic state up to a virtuality level given by q. If the initial hadron moves with a large negative lightcone energy momentum P_- and the field pulse is described by $q = (Q_+, -Q_-, \vec{0})$ then the typical interaction (lightcone) time is $1/Q_+$ and the corresponding virtual hadronic states are $(P_+, P_-, \vec{0})$ with $P_+ \leq Q_+$.

D The field pulse then interacts with an available parton in such a state with an energy momentum fraction x with $xP_- = Q_-$. We therefore *define the structurefunctions as the inclusive sum of these partons over all such states.*

The main point in our approach is the assumption that the structure functions of a hadron are given by an inclusive sum over all the *realisable* radiation in these states. We are consequently defining the virtuality as *the virtuality of the total state* and not in terms of the individual partons (cf below). Therefore only those virtual fluctuations are allowed, which are in agreement with the total coherence requirements of the stringlike force field.

In the ordinary approach (cf eg Ref [15], where a particularly lucid physics description is given) it is assumed that the partonic quanta of a fastmoving hadron will (independently) have long virtual lifetimes. Each one can give rise to a cascade chain, which later reassembles in accordance with the coherence properties of the field. In this way there are in general many chains available and the interaction probe picks out one parton with x, thereby breaking the coherence in that particular chain and realising the corresponding radiation state.

For the ocean parton region (small x, but still above the scale of the vacuum fluctuations) there may in the approach of Ref [15] sometimes occur such a large density of virtual quanta that they will start to interact, thereby losing their independence. In that way the number of available virtual chains will diminish. Our approach can in that language be described as a kind of "maximal interaction", so that only a single "chain", corresponding to a total possible string state in The Dipole and Soft Radiation Models, is realisable at a time.

Therefore there will be essentially fewer low $x-$partons in our approach ([17] as compared to the ordinary perturbative QCD approaches ([13, 18, 14, 15]). We obtain eg for the ocean parton distributions f_o´

$$f_o \propto \frac{g(Q^2)}{x} \quad \text{if } x < \left(\frac{\mu_0}{Q}\right)^4 \tag{1}$$

as compared to behaviours like $x^{-(1+\epsilon)}$ with $\epsilon \sim 0.5$ in other approaches.

This lower density, "the final state single chain density" will always occur in an approach like ours. The reason is that the radiation emitted in the Dipole Cascade Model is of a local character in angles and consequently in rapidity. For a given rapidity the amount of radiation is basically given by the largest possible transverse momentum that can occur for this rapidity ([10]). For sufficiently small $x-$values we are considering rapidity regions which are so far away from "the front" of the hadron that the radiation is no longer influenced by the (initial) hadron energy momentum P_-. (The particular numbers in the inequality for x $((\mu_0/Q)^4)$ will, however, be different depending upon the way The Soft Radiation Model is implemented).

The evolution equations of a $DGLAP$-type ([13]) corresponds to looking for the possible gluon emissions, which can produce a final state parton with x at a given virtuality Q^2. We will term this approach Initial State Bremsstrahlung (ISB) models and the natural kinematical variables are in that case the pair (x, Q^2). In our approach it is rather the values of the total state variables $\nu = 2Pq$ and $s = W^2 \simeq (1-x)\nu$ that play the main role in the investigation.

Thus in a Monte Carlo simulation (where we use ARIADNE ([4], which contains The Dipole and The Soft Radiation Models) states are produced with given values of (W^2, ν). The structure functions of the ocean partons are then sampled by summing over those partons which have the right values of (x, Q^2) (cf Ref [17]).

Our model does not, however, provide the distributions of the valence constituents, ie the large x distributions of the hadrons. They correspond to large fluctuations in the wave functions.

It is, however, evident that the virtual emissions in our model demand that some energy must be transferred to the radiation. We will consequently also in our model have evolution equations in case *we assume that this energy is taken from the valence constituents*.

There is actually inside the SRM a particular mechanism which is very similar to the ISB mechanism. A brief description of this mechanism will be included in the end of section 4, based upon a paper which is under publication ([7]). Sections 2 and 3 contains some details in connection with the Dipole Cascade Model and with the SRM. In section 5, finally, the results for the small x-region will be shown.

2   The Dipole Cascade Model

The Lund Dipole Model [1, 2] for gluon radiation is based upon the repeated application of the well-known dipole radiation formula

$$d\sigma \propto \alpha_S \frac{x_1^u + x_3^w}{(1-x_1)(1-x_3)} dx_1 dx_3 \tag{2}$$

with the $x_j, j = 1, 3$ the (final state) center of mass energy fractions of the emitters and $x_2 = 2 - x_1 - x_3$ the corresponding energy fraction for the emitted gluon. The exponents u and w are 2 or 3 depending upon the nature of the emitters [1] and α_S is the ordinary running coupling constant:

$$\frac{3\alpha_S}{2\pi} \equiv \frac{\alpha_0}{\ln(k_\perp^2/\Lambda^2)} \tag{3}$$

The transverse momentum $k_\perp$ and the (pseudo)rapidity of the gluon may be defined in terms of the dipole mass square s_{di} and the x_j as

$$\begin{aligned} k_\perp^2 &= s_{di}(1-x_1)(1-x_3) \\ y &= 1/2\ln\left(\frac{1-x_1}{1-x_3}\right) \end{aligned} \tag{4}$$

In terms of these variables it is easy to see from energy momentum conservation that the available emission region is approximately

$$|y| \le \ln\left(\frac{\sqrt{s_{di}}}{k_\perp}\right) \tag{5}$$

It corresponds to the inside of a triangular region in the $(\kappa \equiv \ln(k_\perp^2/\Lambda^2), y)$ plane (cf. figure 1, where the scale in κ is half of that in y to save space). The crossection above is in the same approximation

$$d\sigma \simeq \frac{\alpha_0}{\kappa} d\kappa dy \qquad (6)$$

In an e^+e^--annihilation event the original $q\bar{q}$-pair starts to emit a gluon in accordance with the emission formula above. The resulting state containing a color3, color$\bar{3}$ and a color8 can (for a second emission) to a good approximation be considered as *two independent dipoles, one between the q and the g (gluon) and one between the g and the $\bar{q}$*.

The process is repeated and one obtains an ever increasing number of independent dipoles with decreasing masses and consequently (in general) quickly decreasing $k_\perp$ of the "last"

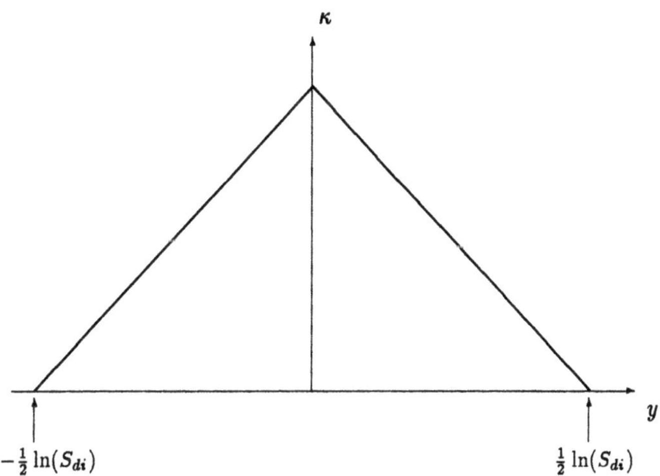

Figure 1. The available phase space for dipole emission.

emitted gluon. The phase space limitations in equation (5) is equivalent to the "strong angular ordering" condition in the QCD cascades [5].

It is worthwhile to notice that the variables $(k_\perp, y)$ to some extent are "theoretical" variables, ie they do not, in particular further down the emission chain, necessarily describe the "true" transverse momentum and rapidity with respect to eg the thrust axis. The reason for that is that they are at every step in the cascade defined iteratively from the earlier steps. Therefore they are only "good" variables with respect to the emitting dipole, which in general is moving with respect to the total cms.

In the Lund Gluon Model the gluons are local kink excitations on the string. Therefore in the Lund String picture the variables are good local descriptions of "the bending" of the string due to a gluonic excitation. We also note that the dipoles are stretched and moving in the same way as the segments of the corresponding Lund String. This is one reason why so many observables are predicted to be the same in the Lund Model as in the Webber-Marchesini Model ([18]), which are conceptually so different.

In ref. [2] we have introduced an extension to The Dipole Model, which also contains the gluon splitting process into $q\bar{q}$-pairs. This is in general a small correction, of the order of 10%. It is a bit model dependent and we tend to produce somewhat more $q\bar{q}$-pairs than the cascades implemented in eg JETSET 7.3[6].

In the e^+e^--annihilation reactions the emitters can be considered as essentially pointlike objects but this may no longer be the case in connection with inelastic lepto-production. The probe, ie. the field pulse q mentioned above is well-defined in size but the target hadron is extended.

3 The Soft Radiation Model

The Soft Radiation Model suggests one possible way to treat this extension. The starting point is to note that the color radiation can be described in two different (complementary) ways. One can consider the emission in terms of an initial state (space-like) cascade and a final state (time-like) cascade. But it can also be argued that, as the hadron is in a bound state, there is no radiation before the interaction and that all radiation stems from the color dipole formed between the struck parton and the hadron remnant.

In case eg. a q-parton is struck out then the whole of the remainder contains the corresponding energy and $\bar{3}$ color. Thus the initial dipole contains one pointlike object and one extended, in particular with respect to the carrying of energy momentum.

A gluon emitted in this case in the phase space element $(k_\perp, y)$ will need both a positive and a negative energy momentum light-cone component

$$k_\pm = k_\perp \exp(\pm y) \tag{7}$$

While it may easily pick up the positive component from the large energy concentration in the struck q, it will have to obtain the negative component from the extended object. It is a well-known property that a coherent wave, built in an emitter of size l, with a wave-length $\lambda << l$ only stems from a fraction of the emitter comparable to λ.

Thus a large $k_\perp$ gluon (corresponding to a short wave length $\lambda \sim 1/k_\perp$) will have to obtain its total negative component from a fraction of the emitter P_{-r} (the index r stands for the remainder of the hadron). This corresponds to some strong damping in parts of the phase space.

In the SRM [3] we have assumed that this changes the phase space limit into

$$\begin{aligned} k_- &< \left(\frac{\mu}{k_\perp}\right)^\alpha P_{-r} \\ k_+ &< Q_+ \end{aligned} \tag{8}$$

with a parameter μ corresponding to the inverse size of the source and α a number describing the dimensionality of the source.

We have found that the EMC data prefer a value of $\mu \sim 1GeV/c$ and $\alpha \sim 1$. This would mean that the mean transverse jextension of the hadron is of the order $\pi/\mu \sim 0.6$ fm and that the energy is basically continuously distributed in a one-dimensional region, which

we will identify with a string-like shape. Although the EMC data cover a rather small
(Q^2, ν)-region we will extend this result to arbitrary energies.

We will make one adjustment, however. In the case when there is a large x-interaction
the remnant only contains $P_{-r} = (1 - x)P_-$, ie essentially less than the incoming hadronic
state P_-. Assuming that it is *the energy density* in the rest system that is a constant we
are lead to expect that the effective l is correspondently smaller and thus that μ behaves
as

$$\mu \equiv \mu(x) = \frac{\mu_0}{(1 - x)} \tag{9}$$

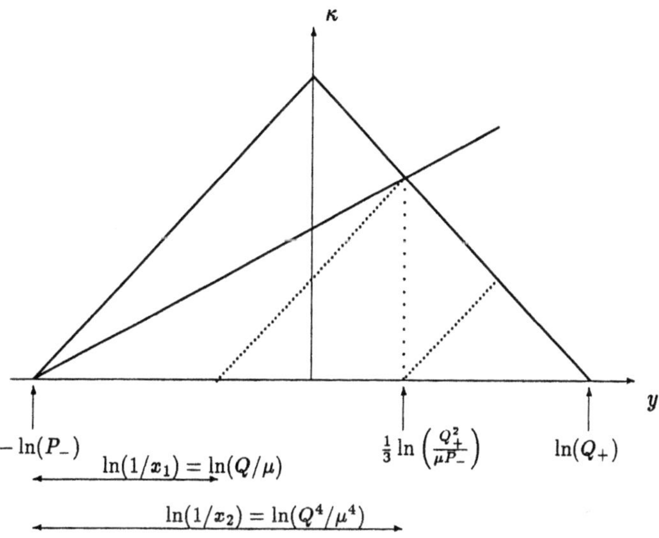

Figure 2. The available phase space for dipole emission in inelastic lepto-production.

This will mean that *the radiation in the target fragmentation region is the same, independent
of the interaction*, ie the damping is governed by $(\mu_0/k_t)P_-$, independent of x and Q^2. The
available energy will, however, only permit an emission in case the energy momentum
component of the emitted gluon, k_-, is smaller than $P_{-r} = (1 - x)P_-$.

The allowed emission region in an e^+e^--annihilation event is in this way changed (sometimes very much) in the target fragmentation region of an inelastic leptoproduction event.
We obtain the region, depicted in figure 2, as the allowed region in an inelastic lepto-
production event. (The notations $x_j, j = 1, 2$ are explained later).

The damping we have introduced above is of a step function character. In case there
would be a power suppression in $k_\perp$ (as in an ordinary form factor) then the situation
would correspond to an exponential suppression in the (κ, y) plane. Most of our results are

insensitive to the appearance of such an exponential tail into the depletion region.

One notes that the forward region, along the struck q-particle is essentially unaffected by the damping mechanism. Actually, there may be more radiation in this region than in some other models because it is the maximum allowed $k_\perp$ which governs the radiation. It is easily seen that the maximum is in The Soft Radiation model obtained for

$$k_{\perp,max} = (\mu^\alpha \nu)^{\frac{1}{2+\alpha}} \tag{10}$$

with $\nu = 2pq$. This may be larger in some kinematical situations than eg. Q^2.

4   The Recoil Structure and a Comparison to an ISB model

There are two kinds of recoil phenomena in connection with the emission of radiation, ie the energy loss of the emitters and the necessary momentum compensation. The energy loss corresponds to the obvious requirements that the variables $x_j < 1$. The x_j-variables also determines the relative angles in the emission, eg

$$\sin^2(\theta_{12}/2) = \frac{(1 - x_3)}{x_1 x_2} \tag{11}$$

The relative orientation between the original dipole direction and the final state partons is, however, not fixed by the x_j variables but instead according to the transition matrix elements by the directions of the currents "before" and "after" the emission.

If the beam direction is along the z-direction then the annihilation current of an (e^+e^-)-state is in the xy-plane. The overlap between this current and the produced $(q\bar{q})$-state current (which also lies in a plane perpendicular to the direction of motion of the $(q\bar{q})$-pair) then results in the well-known $(1 + \cos^2(\theta))$ factor in the cross section. The angle θ is the angle between the two "beam" directions.

If a (hard) gluon emission occurs then the resulting color current is more complex. But there is still an alignment between the original current direction and a particular direction related to the final state. It can most easily be described as the axis along which the emitters in the final state has a minimum for the transverse momentum combination $(k_{\perp 1}^2 + k_{\perp 3}^2)$ with respect to the dipole axis.

A prescription given by Kleiss ([19]) allows a simple implementation of this alignment in Monte Carlo simulation programs. There is, however, no prescription known at present for handling the emission of two or more gluons. In the ARIADNE default version the emission from a dipole composed of two gluons is handled by aligning the original dipole direction with this "minimal $k_\perp$-axis". In a dipole composed of a gluon and a q or $\bar{q}$ the total recoil is in the default version given to the q or $\bar{q}$, however.

For the particular case when the softening mechanism in The Soft Radiation Model is used, ie when the radiation occurs between the remnant and an earlier emitted gluon it is necessary to introduce further assumptions. It is natural to treat the remnant fraction

$$(\mu_0/k_\perp)P_- \equiv x_r P_- \tag{12}$$

as an initial state gluon for this $k_\perp$-emission. In case we allow it to recoil then this emission will actually lead to the occurrence of *two* gluons. The extra gluon, which I will term a "recoil gluon", is not emitted in the standard fashion but rather behaves as a collective effect.

Before considering the properties of this recoil gluon I will present a simple ISB scenario. Assume that x is large but not too close to 1. Then in an ISB model a parton will come in with a fraction $x_- \equiv x'$, emit a gluon with $x_r \equiv (1-\zeta)x'$, thereby turning into a virtual parton with a lightcone fraction $x = \zeta x'$. Its (virtual) mass square Q_I^2 is related to the transverse momentum of the emission $k_{\perp,I}$ by

$$Q_I^2 = \frac{k_{\perp I}^2}{\zeta(1-\zeta)} \tag{13}$$

The probability for this emission is (for fixed x)

$$dP(\zeta, Q_I^2) \propto \alpha_s \frac{dQ_I^2}{Q_I^2}\frac{d\zeta}{\zeta(1-\zeta)} \tag{14}$$

which is identical to the cross section given in Eq (6) (note that the ζ–dependent part is equal to $dx_r/x_r = dy$).

There is an essential prescription with regard to *the ordering* of the emissions in an ISB. Thus in the $DGLAP$–approach the Q^2 (or essentially equivalently the $k_\perp^2$) are increasing while (in practice) the x_r are decreasing along the emission chain. This feature is in general handled by means of an appropriate Sudakov factor.

It is shown in Ref [7] that

1. in case one requires that the fraction x_r is determined by

$$x_r \equiv \frac{\mu_0}{k_{\perp 3}} = \frac{k_{-2} + k_{-3}}{P_-} \tag{15}$$

2. and for a particular recoil strategy, ie a particular distribution of the momentum recoils between the "really" emitted gluon (with energy momentum vector k_3) in the SRM and the above-mentioned recoil gluon (with energy momentum vector k_3):

$$k_{\perp 2}^2 \exp(-y_2) = k_{\perp 3}^2 \exp(-y_3) \tag{16}$$

then

I the probability (according to Eq (6)) to produce the recoilgluon 2 and obtain the gluon indexed 3 is equal to the probability to produce 3 and obtain 2, ie it does not matter which one is "the primary" one,

II the probability to emit both of them "together" (as a (time-like) gluon with energy momentum vector $(k_2 + k_3)$) in an ISB model (according to Eq (14)) is equal to emit the gluon 3 (ie the "really" emitted one) in the SRM scenario (according to Eq (6)) and obtain the recoil gluon 2.

The significance of this result is that the SRM is equivalent to a (simple) ISB model. In such a model the Sudakov ordering would lead to the emission of 2 "before" the emission of 3, while in the SRM there will be the opposite ordering (I above).

The fact that according to II the emission of both of them in the ISB is equal to the direct probability in the SRM scenario means, coupled with the condition in Eq (16), that there is a particular density of gluons characteristic for the SRM in such an ISB scenario (cf ref [7]). Actually the combined emission means that a total dipole is emitted, which may continue to radiate

The condition in Eq (16) corresponds to a balancing requirement for the two bending kinks along the color force field which is a very reasonable one. It means that when the rapidity gap between the two gluons is large then there is less alignment of their transverse momenta. It is often close to but not identical to the recoil strategy used in ARIADNE for a gluon gluon dipole emission.

In ref ([7]) it is also shown that the mechanism, described above, leads to a $DGLAP-$like equation for the changes in the valence quark structure functions.

5   The Results for the Small x distributions

The density of partons occurring for small values of x, ie the ocean quark and small$x-$gluons have very different properties in our model built upon the Dipole cascade and the SRM than in a model built upon the ISB scenario.

In our approach a parton at small values of x (which in the kinetic scenario we have defined above in general corresponds to large rapidities) stems from a decay chain which starts for large values of $k_\perp$ and develops "downwards". The occurrence of a hard gluon means that the original dipole decays into two independent dipoles which later produces new gluons and further dipoles.

The rapidity region available for emission at a certain $k_\perp$ (corresponding to κ) is for a single dipole of mass square s_{di} given by Eq (5):

$$\Delta y \simeq \ln s_{di} - \kappa \tag{17}$$

In case there is an emission at eg $k_{\perp 1}$ (correponding to κ_1) the original dipole is divided into two with mass squares $s_{dij}, j = 1, 2$. Then the available phase space for gluon emission increases essentially into

$$\Delta y_{12} \simeq \sum_{1}^{2} (\ln s_{dij} - \kappa) = \Delta y + \kappa_1 - \kappa \tag{18}$$

The procedure can be generalised (cf Ref [10]) in connection with multigluon emission into a functional λ with

$$\lambda \sim \Delta y + \sum (\kappa_j - \kappa) \tag{19}$$

We have shown ([10]) that λ is a measure on a fractal curve (with the dimensions given by the anomalous dimensions of QCD) along which there is in general a constant density of gluonic kinks ([16]).

We have used the properties of the $\lambda-$measure in Ref [17] in order to obtain a simplified analytical description of the gluon density in $(\kappa, \ln(1/x))$ for different values of Q^2. The qualitative features of these results also evolve from the simulation results with ARIADNE although the quantitative results are different due to the recoil effects.

It is, however evident that the maximal $k_\perp$ in Eq (10) for the emission will play a major role (cf Fig 2). We may immediately deduce that there is a forward region

$$x \geq \frac{\mu_0}{Q} \equiv x_1 \tag{20}$$

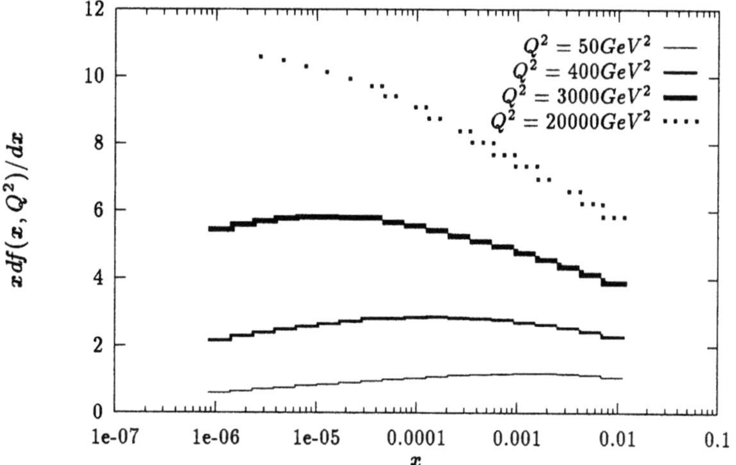

Figure 3. Gluon structure function as a function of x for different Q^2.

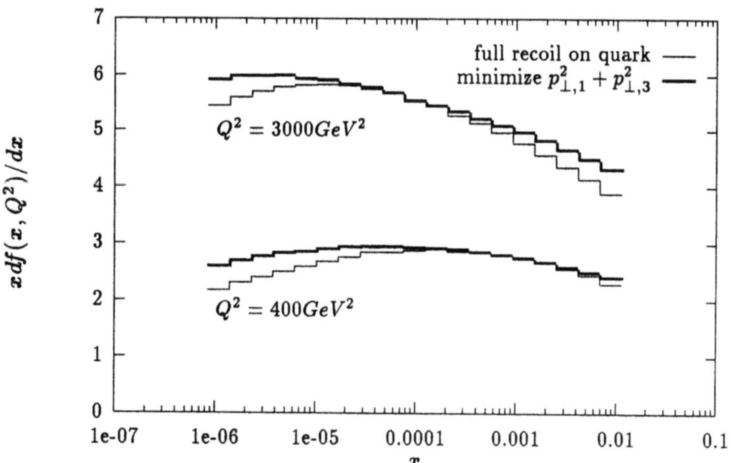

Figure 4. Gluon structure function for different recoil strategies.

inside which there is no influence from the largest possible $k_\perp$ (Eq (10)). This region is mostly populated by the large x-fluctuations of the wave function and we note that x_1 is a very natural scale in our one-dimensional string scenario. It is that part of the state that a probe with wavelength $\sim 1/Q$ will be able to resolve.

There is then an intermediate region, in Fig 2 shown between $\ln(1/x_1)$ and the value $\ln(1/x_2)$ with $x_2 = (\mu_0/Q)^4$ (cf Eq (1)). Inside this region the proliferation effect mentioned above is very noticable, ie the structure functions should increase. After that, ie for $x \leq x_2$ they will saturate in accordance with the results mentioned in Eq (1).

We will next consider the results of a simulation of the structure functions from ARIADNE.

We show in figure 3 the results (after integration over $k_\perp$) for the behavior of the densities for different $(l_Q \equiv \ln(Q^2/\Lambda^2), l_x \equiv \ln(1/x))$-values. The most noticeable features is the (slow) rise with decreasing l_x towards a l_Q-dependent plateau (both with respect to size and height) and then some decrease for even smaller x-values.

In practice, however, the recoils from the emissions affects the string state also in this region and we obtain from the ARIADNE events even a small decrease in this region (cf. figure 3). The effect depends upon "the recoil strategy" used in the cascade calculations. In figure 4 we compare the result of letting the q take the full (momentum) recoil in case emission occurs in a (q, g)- dipole to the case when the two emitters "share" the recoil so that the sum $k_{\perp,1}^2 + k_{\perp,3}^2$ is minimized. In the last case the decrease with respect to the $(1/x)$-behavior is essentially smaller.

This soft behavior for small x is a definite prediction from our model and although it corresponds to very tiny x-values it should be possible to test it at HERA.

The behaviour of the ocean $(q\bar{q})$-pairs is basically proportional to the gluon distributions both for the totally inclusive distribution and the "light-mass" ocean q distributions. There are some kinematical special properties of the heavy ocean q but I will not go into them here.

References

[1] Ya. I. Azimov, Yu. L. Dokshitzer, V. A. Khoze, S. I. Troyan, *Phys. Lett.* **B165** (1985) 147,
G. Gustafson, *Phys. Lett.* **B175** (1986) 453,
G. Gustafson, U. Pettersson, *Nucl. Phys.* **B306** (1988) 746.

[2] B. Andersson, G. Gustafson, L. Lönnblad, *Nucl. Phys.* **B339** (1990) 393.

[3] B. Andersson, G. Gustafson, L. Lönnblad, U. Pettersson, *Z. Phys.* **C43** (1989) 621.

[4] L. Lönnblad, ARIADNE 3 - a Monte Carlo for QCD cascades in the Colour Dipole Formulation, Lund Preprint LU TP 89-10 (1989).

[5] A. H. Mueller, *Phys. Lett* **B104** (1981) 161,
B. I. Ermolaev, V. S. Fadin, *JETP Lett.* **33** (1981) 269,
A. Bassetto, M. Ciafaloni, G. Marchesini, A. H. Mueller, *Nucl. Phys.* **B207** (1982) 189,
G. Marchesini, B. Webber, *Nucl. Phys* **B238** (1984) 1.

[6] T. Sjöstrand, JETSET 7.3 program and manual see e.g. B. Bambah et. al., QCD Generators for LEP, CERN-TH.5466/89,
T. Sjöstrand, *Computer Phys. Comm.* **39** (1986) 347;
T. Sjöstrand and M. Bengtsson, *Computer Phys. Comm.* **43** (1987) 367.

[7] B. Andersson, L. Lönnblad, (to be published).

[8] B. Andersson, G. Gustafson, G. Ingelman, T. Sjöstrand, *Phys. Rept.* **97** (1983) 31.

[9] B. Andersson, G. Gustafson, B. Söderberg, *Z. Phys.* C20 (1983) 317.

[10] B. Andersson, P. Dahlqvist, G. Gustafson, *Phys. Lett.* **B214** (1988) 604.

[11] B. Andersson, G. Gustafson, A. Nilsson, C. Sjögren, "Fluctuations and Anomalous Dimensions in QCD cascades", Lund Preprint LU TP 90-4 (1990).

[12] B. Andersson, "On the Dipole Structure of QCD", to be published in *Nucl. Phys.* **B**.

[13] V. N. Gribov, L. N. Lipatov, *Sov. J. Phys.* **15** (1972) 438 and 675,
L. N. Lipatov, *Sov. J. Phys.* **20** (1975) 94,
G. Altarelli, G. Parisi, *Nucl. Phys.* **B126** (1977) 298,
Yu. L. Dokshitser, *Sov. Phys. JETP* **46** (1977) 641.

[14] E. A. Kuraev, L. N. Lipatov, V. S. Fadin, *Sov. Phys. JETP* **45** (1977) 199.

[15] V. Gribov, E. Levin, M. Ruskin, *Physics Reports*.

[16] G.Gustafson, These Proceedings and Refs in there.

[17] B.Andersson, L.Lönnblad, Lund preprint LU-TP 91-16 (submitted to *Phys. Lett. B*).

[18] G.Marchesini, B.Webber, These Proceedings and Refs in there.

[19] R.Kleiss, *Phys. Lett.* **B180** (400) 1986.

PARTICIPANTS

291

J. ALBERTY	ECP Division, CERN, Geneva, Switzerland
F. ANSELMO	ECP Division, CERN, Geneva, Switzerland
B. ANDERSSON	University of Lund, Lund, Sweden
J. BARTELS	University of Hamburg, Hamburg, Germany
S. BETHKE	University of Heidelberg, Heidelberg, Germany
E. BLANCO	World Laboratory, Lausanne, Switzerland
L. BRANNSTRÖM	University of Lund, Lund, Sweden
G. BRUGNOLA	University of Bari, Bari, Italy
V. BUZULOIU	World Laboratory, Lausanne, Switzerland and Institutul Politehnic, Bucuresti, Romania
S. CATANI	INFN-Florence, Florence, Italy
M. CIAFALONI	University of Florence, Florence, Italy
L. CIFARELLI	University of Naples, Naples, Italy and PPE Division , CERN, Geneva, Switzerland
D. DENEGRI	CEN-SACLAY, Gif-sur-Yvette, France
O. DI ROSA	ECP Division, CERN, Geneva, Switzerland
Yu. DOKSHITZER	Leningrad Institute for Nuclear Physics, Gatchina, USSR and University of Lund, Lund, Sweden
M. GOURDIN	Université Pierre et Marie Curie, Paris, France
V. GRIBOV	Landau Inst. Theor. Physics, Moscow, USSR
J.F. GUNION	University of California, Davis, USA
G. GUSTAFSON	University of Lund, Lund, Sweden
D. HATZIFOTIADOU	PPE Division, CERN, Geneva, Switzerland
A.B. KAIDALOV	ITEP-Moscow, USSR and LPTHE-Orsay, Paris, France
V. A. KHOZE	Leningrad Institute for Nuclear Physics, Gatchina, USSR; INFN ELOISATRON Project and World Laboratory, Spb.
V.V. KHOZE	M.I.T., Cambridge, Massachusetts, USA
P. LECOMTE	ETH, Zurich, Switzerland
L. LÖNNBLAD	University of Lund, Lund, Sweden
M.L. MANGANO	INFN-Pisa, Pisa, Italy
G. MARCHESINI	University of Parma, Parma, Italy
B. MELE	INFN-Rome, Rome, Italy
J. NYIRI	Central Research Inst. Physics, Budapest, Hungary
C. PETERSON	University of Lund, Lund, Sweden
Yu. SHABELSKI	Leningrad Institute for Nuclear Physics, Gatchina, USSR

W.J. STIRLING University of Durham, Durham, UK
L. TERMINIELLO World Laboratory, Lausanne, Switzerland and
Universidad Nacional, La Plata, Argentina
B.R. WEBBER Cavendish Laboratory, Cambridge, UK
C. WILLIAMS PPE Division, CERN, Geneva, Switzerland
A. ZICHICHI PPE Division, CERN, Geneva, Switzerland